High Accuracy Detection of Mobile Malware Using Machine Learning

High Accuracy Detection of Mobile Malware Using Machine Learning

Editor

Suleiman Yerima

MDPI • Basel • Beijing • Wuhan • Barcelona • Belgrade • Manchester • Tokyo • Cluj • Tianjin

Editor
Suleiman Yerima
Faculty of Computing,
Engineering and Media
De Montfort University
Leicester
United Kingdom

Editorial Office
MDPI
St. Alban-Anlage 66
4052 Basel, Switzerland

This is a reprint of articles from the Special Issue published online in the open access journal *Electronics* (ISSN 2079-9292) (available at: www.mdpi.com/journal/electronics/special_issues/ accuracy_detection_mobile_malware).

For citation purposes, cite each article independently as indicated on the article page online and as indicated below:

LastName, A.A.; LastName, B.B.; LastName, C.C. Article Title. *Journal Name* **Year**, *Volume Number*, Page Range.

ISBN 978-3-0365-7175-1 (Hbk)
ISBN 978-3-0365-7174-4 (PDF)

© 2023 by the authors. Articles in this book are Open Access and distributed under the Creative Commons Attribution (CC BY) license, which allows users to download, copy and build upon published articles, as long as the author and publisher are properly credited, which ensures maximum dissemination and a wider impact of our publications.
The book as a whole is distributed by MDPI under the terms and conditions of the Creative Commons license CC BY-NC-ND.

Contents

About the Editor vii

Preface to "High Accuracy Detection of Mobile Malware Using Machine Learning" ix

Suleiman Y. Yerima
High Accuracy Detection of Mobile Malware Using Machine Learning
Reprinted from: *Electronics* 2023, 12, 1408, doi:10.3390/electronics12061408 1

Suleiman Y. Yerima and Abul Bashar
A Novel Android Botnet Detection System Using Image-Based and Manifest File Features
Reprinted from: *Electronics* 2022, 11, 486, doi:10.3390/electronics11030486 5

Yuxin Ding, Miaomiao Shao, Cai Nie and Kunyang Fu
An Efficient Method for Generating Adversarial Malware Samples
Reprinted from: *Electronics* 2022, 11, 154, doi:10.3390/electronics11010154 23

Robertas Damaševičius, Algimantas Venčkauskas, Jevgenijus Toldinas and Šarūnas Grigaliūnas
Ensemble-Based Classification Using Neural Networks and Machine Learning Models for Windows PE Malware Detection
Reprinted from: *Electronics* 2021, 10, 485, doi:10.3390/electronics10040485 41

Suleiman Y. Yerima, Mohammed K. Alzaylaee, Annette Shajan and Vinod P
Deep Learning Techniques for Android Botnet Detection
Reprinted from: *Electronics* 2021, 10, 519, doi:10.3390/electronics10040519 65

Mathew Ashik, A. Jyothish, S. Anandaram, P. Vinod, Francesco Mercaldo and Fabio Martinelli et al.
Detection of Malicious Software by Analyzing Distinct Artifacts Using Machine Learning and Deep Learning Algorithms
Reprinted from: *Electronics* 2021, 10, 1694, doi:10.3390/electronics10141694 83

Muath Alrammal, Munir Naveed and Georgios Tsaramirsis
A Novel Monte-Carlo Simulation-Based Model for Malware Detection (eRBCM)
Reprinted from: *Electronics* 2021, 10, 2881, doi:10.3390/electronics10222881 111

Patricia Iglesias, Miguel-Angel Sicilia and Elena García-Barriocanal
Detecting Browser Drive-By Exploits in Images Using Deep Learning
Reprinted from: *Electronics* 2023, 12, 473, doi:10.3390/electronics12030473 121

Moutaz Alazab, Ruba Abu Khurma, Albara Awajan and Mohammad Wedyan
Digital Forensics Classification Based on a Hybrid Neural Network and the Salp Swarm Algorithm
Reprinted from: *Electronics* 2022, 11, 1903, doi:10.3390/electronics11121903 135

Janaka Senanayake, Harsha Kalutarage and Mhd Omar Al-Kadri
Android Mobile Malware Detection Using Machine Learning: A Systematic Review
Reprinted from: *Electronics* 2021, 10, 1606, doi:10.3390/electronics10131606 153

Hany F. Atlam and Olayonu Oluwatimilehin
Business Email Compromise Phishing Detection Based on Machine Learning: A Systematic Literature Review
Reprinted from: *Electronics* 2022, 12, 42, doi:10.3390/electronics12010042 187

About the Editor

Suleiman Yerima

Dr. Suleiman Y. Yerima is currently a Senior Lecturer in Cybersecurity at the Faculty of Computing, Engineering and Media, De Montfort University, Leicester, United Kingdom. He is a Certified Ethical Hacker (CEH), a Certified Information Systems Security Professional (CISSP) and a Fellow of the British Advance HE (FHEA). He holds a Ph.D. in Computing from the University of South Wales (formerly University of Glamorgan), UK (2009); an MSc (with distinction) in Personal, Mobile and Satellite Communications from the University of Bradford, UK (2004); and a first-class honours degree in Electrical and Computer Engineering from the Federal University of Technology, Minna, Nigeria (2000). In 2017, he received the IET Information Security Journal Premium (Best Paper) Award, and in 2021 he won the Best Paper Award at the 18th ACS/IEEE International Conference on Computer Systems and Applications (AICCSA 2021). Dr. Yerima has led various research projects in Android malware detection, which were supported by McAfee (Intel Security). He is the principal investigator of various projects funded by UKRI, Innovate UK and GCRF including: AVAC—Enhancing agricultural value chains in Nigeria using data analytics and GIS; ADRELO—Advancing resilience in low-income housing using climate-change science and big data analytics; and STEGA—Secure and trustworthy e-governance in Africa. His current research interests include malware and cyber threat detection, mobile and IoT security, and applied machine learning.

Preface to "High Accuracy Detection of Mobile Malware Using Machine Learning"

The increase in the volume and sophistication of malicious software (malware) and malware-related attacks targeting modern networks and systems calls for new, more advanced and effective solutions. Recently, there has been a growing interest in developing and advancing machine learning-based approaches to detect malware, which are constantly evolving and becoming more evasive. This Special Issue aims to present recent advances and high-quality research on machine learning-based approaches to detect attacks and malicious software that pose severe threats to the mobile platforms of today.

Suleiman Yerima
Editor

Editorial

High Accuracy Detection of Mobile Malware Using Machine Learning

Suleiman Y. Yerima

Cyber Technology Institute, School of Computer Science and Informatics, De Montfort University, The Gateway, Leicester LE1 9BH, UK; syerima@dmu.ac.uk

Introduction

As smartphones and other mobile and IoT devices have become pervasive in everyday life, malicious software (malware) authors are increasingly targeting the operating systems that are at the core of these mobile systems. Malware targeting mobile platforms has witnessed an explosive growth in the last decade. As a result of this rapid increase in mobile malware, the limits of traditional signature-based antivirus scanning have been stretched. This has led to the emergence of machine learning-based detection as a complementary solution to traditional antivirus scanning. Although machine learning-based malware detection has continued to attract great research interest, many challenges remain as emerging malware families continue to evolve with more sophisticated capabilities and stealthy evasive techniques.

This Special Issue in Electronics presents some of the most recent research results and innovative machine learning-based approaches to detecting malicious software and attacks that can compromise mobile platforms.

The authors of [1] proposed a novel Android botnet detection system based on image features and manifest file features. The method aims to overcome the limitations of hand-crafted features for machine learning-based botnet detection. Their proposed approach employs Histogram of Oriented Gradients, together with byte histograms obtained from images representing the app executables, and these are subsequently combined with the features derived from manifest files. The proposed system was evaluated using the ISCX botnet dataset, and the experimental results demonstrate its effectiveness with F1 scores ranging from 0.923 to 0.96 using popular machine learning algorithms.

In [2], the authors present a study on generating malware adversarial samples using deep learning models. Gradient-based methods are usually employed in generating adversarial samples; however, they generate the samples on a case-by-case basis, which is very time-consuming for large scale sample generation. To address this issue, a novel method was proposed, which extracts feature byte sequences from benign samples using deep learning. Feature byte sequences represent the characteristics of benign samples and can affect classification decisions. The feature byte sequences are directly injected into malware samples to generate adversarial samples. The proposed method is compared with random injection and gradient-based methods, and the experimental results show that the new method is suitable for generating a large number of adversarial samples.

The authors of [3] propose an ensemble classification-based approach for malware detection. The first-stage classification is performed by a stacked ensemble of dense (fully connected) and convolutional neural networks (CNN), while the final stage classification is performed by a meta-learner. For the meta-learner, 14 classifiers are explored and compared. For baseline comparison, 13 machine learning methods are used: K-Nearest Neighbors, Linear Support Vector Machine (SVM), Radial basis function (RBF) SVM, Random Forest, AdaBoost, Decision Tree, ExtraTrees, Linear Discriminant Analysis, Logistic, Neural Net, Passive Classifier, Ridge Classifier and Stochastic Gradient Descent classifier. The results of experiments performed on the Classification of Malware with PE headers (ClaMP) dataset are presented. The best performance is achieved by an ensemble of five dense and CNN neural networks, and the ExtraTrees classifier as a meta-learner.

In [4], the authors presented a comparative study of deep learning techniques with the aim of investigating their efficacy for Android botnet detection, based on static features. To create the deep learning-based botnet detection system, a bespoke tool for automated reverse engineering of Android Package Files (APKs) was developed and used to extract 342 features, which are then used to represent the application as a vector of binary vectors. These vectors were used to train several deep learning models including: Convolutional Neural Networks (CNN), Dense Neural Networks (DNN), Gated Recurrent Units (GRU), Long Short-Term Memory (LSTM), as well as more complex networks like CNN-LSTM and CNN-GRU. Evaluation experiments were conducted using 6802 Android applications out of which 1920 were botnet samples from the ISCX botnet dataset. The results showed that the deep-learning models outperformed classical machine learning classifiers and achieved very high accuracy, as well as high precision, recall and F1 scores.

The authors of [5] investigated the relevance of the features of unpacked malicious and benign executables such as mnemonics, instruction opcodes, and API calls to identify a feature that classifies the executable. Prominent features were extracted using Minimum Redundancy and Maximum Relevance (mRMR) and Analysis of Variance (ANOVA). Experiments were conducted on four datasets using machine learning approaches such as Support Vector Machine (SVM), Naïve Bayes, J48, Random Forest (RF), and XGBoost. In addition, they evaluated the performance of deep neural networks such as Deep Dense Network (DDN), One-Dimensional Convolutional Neural Network (1D-CNN), and CNN-LSTM in classifying unknown samples, and observed promising results using APIs and system calls. On combining APIs/system calls with static features, a marginal performance improvement was attained compared to models trained only on dynamic features. Moreover, to improve accuracy, the solution was implemented using distinct deep learning methods and demonstrated a fine-tuned deep neural network that resulted in an F1-score of 99.1% and 98.48% on Dataset-2 and Dataset-3, respectively.

In [6], the authors presented an approach called eRBCM to detect malware. The eRBCM system was designed using the reinforcement learning approach, which utilizes the strength of Monte–Carlo simulations and builds a strong machine learning model to detect complex malware patterns. It combines the most beneficial elements of MOCART's reinforcement learning and RF's exploration capabilities. A large number of experiments were conducted using different malware benchmarks, including ARP attack, ICMP attack, and Microsoft Malware. eRBCM was consistently better than its competitors in terms of learning the new malware patterns and detecting unknown malware. This was mainly explained by eRBCM's self-adaptability to exploration and intelligent tuning of the balance for the trade-off between exploration and exploitation.

The authors of [7] present a study on detecting drive-by exploits in images using deep learning. With steganographic techniques being combined with polyglot attacks to deliver exploits in web browsers, machine learning approaches have been proposed for detecting steganography in images. However, exploit code hiding has not been systematically addressed; hence the paper proposes the use of deep learning methods for such detection, accounting for the specifics of the situation in which the images and the malicious content are delivered using Spatial and Frequency Domain Steganography algorithms. The methods were evaluated by using benchmark image databases with collections of JavaScript exploits, for different density levels and steganographic techniques in images. A convolutional neural network was built to classify the infected images with a validation accuracy around 98.61% and a validation AUC score of 99.75%.

In [8], the authors propose a Salp Swarm Algorithm (SSA) as a trainer for Multilayer perceptron (MLP) in the context of digital forensics. SSA is an effective meta-heuristic algorithm that belongs to the swarm-based family. It has a single parameter that decreases in an adaptive manner relative to increasing iteration. It also performs an extensive exploration in the initial iterations and then adaptively switches to exploit the most promising areas of the search space. Furthermore, SSA also preserves the best-found solution so that it never loses the optimal solution. Lastly, follower salps change their locations adaptively following other members of the population, so it has the power to alleviate the local minima problem. In this paper, seven metaheuristic algorithms are compared to the proposed approach. Particle Swarm Optimization (PSO), Ant Colony Optimization (ACO), Genetics Algorithm (GA), Differential Evolution algorithm (DE), and BackPropagation. In the majority of cases, the SSA-based MLP outperformed the other approaches when evaluated on the digital forensics dataset created from audit logs, registry, and file system.

The authors of [9] provide a systematic review of machine learning-based Android malware detection techniques. This paper aims to enable researchers to acquire in-depth knowledge in the field and to identify potential future research and development directions. The paper critically evaluates 106 carefully selected articles and highlights their strengths and weaknesses as well as potential improvements. Finally, the machine learning-based methods for detecting source code vulnerabilities are discussed, because it might be more difficult to add security after the app is deployed.

In [10], the authors present a systematic literature review and examination of the state of the art of Business Email Compromise (BEC) phishing detection techniques with the aim of providing a detailed understanding of the topic to allow researchers to identify the main principles of BEC phishing detection. Based on a selected search strategy, 38 articles (of 950 articles) were chosen for closer examination. The selected articles were discussed and summarized to highlight their contributions as well as their limitations. In addition, the features of BEC phishing used for detection were provided, and the ML algorithms and datasets that were used in BEC phishing detection models were discussed. In the end, open issues and future research directions of ML-based BEC phishing detection were also discussed.

Acknowledgments: I would like to thank all the authors for the papers they submitted to this Special Issue. I would also like to acknowledge all the reviewers for their careful and timely reviews to help improve the quality of this Special Issue.

Conflicts of Interest: The author declares no conflict of interest.

References

1. Yerima, S.; Bashar, A. A Novel Android Botnet Detection System Using Image-Based and Manifest File Features. *Electronics* **2022**, *11*, 486. [CrossRef]
2. Ding, Y.; Shao, M.; Nie, C.; Fu, K. An Efficient Method for Generating Adversarial Malware Samples. *Electronics* **2022**, *11*, 154. [CrossRef]
3. Damaševičius, R.; Venčkauskas, A.; Toldinas, J.; Grigaliūnas, Š. Ensemble-Based Classification Using Neural Networks and Machine Learning Models for Windows PE Malware Detection. *Electronics* **2021**, *10*, 485. [CrossRef]
4. Yerima, S.; Alzaylaee, M.; Shajan, A.; Vinod, P. Deep Learning Techniques for Android Botnet Detection. *Electronics* **2021**, *10*, 519. [CrossRef]
5. Ashik, M.; Jyothish, A.; Anandaram, S.; Vinod, P.; Mercaldo, F.; Martinelli, F.; Santone, A. Detection of Malicious Software by Analyzing Distinct Artifacts Using Machine Learning and Deep Learning Algorithms. *Electronics* **2021**, *10*, 1694. [CrossRef]
6. Alrammal, M.; Naveed, M.; Tsaramirsis, G. A Novel Monte-Carlo Simulation-Based Model for Malware Detection (eRBCM). *Electronics* **2021**, *10*, 2881. [CrossRef]
7. Iglesias, P.; Sicilia, M.; García-Barriocanal, E. Detecting Browser Drive-By Exploits in Images Using Deep Learning. *Electronics* **2023**, *12*, 473. [CrossRef]
8. Alazab, M.; Khurma, R.A.; Awajan, A.; Wedyan, M. Digital Forensics Classification Based on a Hybrid Neural Network and the Salp Swarm Algorithm. *Electronics* **2022**, *11*, 1903. [CrossRef]
9. Senanayake, J.; Kalutarage, H.; Al-Kadri, M. Android Mobile Malware Detection Using Machine Learning: A Systematic Review. *Electronics* **2021**, *10*, 1606. [CrossRef]
10. Atlam, H.; Oluwatimilehin, O. Business Email Compromise Phishing Detection Based on Machine Learning: A Systematic Literature Review. *Electronics* **2023**, *12*, 42. [CrossRef]

Disclaimer/Publisher's Note: The statements, opinions and data contained in all publications are solely those of the individual author(s) and contributor(s) and not of MDPI and/or the editor(s). MDPI and/or the editor(s) disclaim responsibility for any injury to people or property resulting from any ideas, methods, instructions or products referred to in the content.

Article

A Novel Android Botnet Detection System Using Image-Based and Manifest File Features

Suleiman Y. Yerima [1,*] **and Abul Bashar** [2]

[1] Cyber Technology Institute, Faculty of Computing, Engineering and Media, De Montfort University, Leicester LE1 9BH, UK

[2] Department of Computer Engineering, Prince Mohammad Bin Fahd University, Khobar 31952, Saudi Arabia; abashar@pmu.edu.sa

* Correspondence: syerima@dmu.ac.uk

Abstract: Malicious botnet applications have become a serious threat and are increasingly incorporating sophisticated detection avoidance techniques. Hence, there is a need for more effective mitigation approaches to combat the rise of Android botnets. Although the use of Machine Learning to detect botnets has been a focus of recent research efforts, several challenges remain. To overcome the limitations of using hand-crafted features for Machine-Learning-based detection, in this paper, we propose a novel mobile botnet detection system based on features extracted from images and a manifest file. The scheme employs a Histogram of Oriented Gradients and byte histograms obtained from images representing the app executable and combines these with features derived from the manifest files. Feature selection is then applied to utilize the best features for classification with Machine-Learning algorithms. The proposed system was evaluated using the ISCX botnet dataset, and the experimental results demonstrate its effectiveness with F1 scores ranging from 0.923 to 0.96 using popular Machine-Learning algorithms. Furthermore, with the Extra Trees model, up to 97.5% overall accuracy was obtained using an 80:20 train–test split, and 96% overall accuracy was obtained using 10-fold cross validation.

Keywords: botnet detection; Histogram of Oriented Gradients; image processing; android botnets; machine learning

1. Introduction

The prevalence of mobile malware globally is a well-known phenomenon as increasing malware of different types continue to target mobile platforms and particularly Android. The McAfee Threat report of June 2021 stated that around 7.73 million new mobile malware samples were seen in 2020 alone [1]. The report further revealed that 2.34 million new mobile malwares had already been discovered in the wild during the first quarter of 2021.

Android, being an open source mobile and IoT platform that also permits users to install apps from diverse sources is the prime target for mobile malware. Unverified and/or re-packaged apps can be downloaded and installed on an Android device from virtually any online third-party source other than the official Google play store. Even though the Google play store benefits from screening services to prevent the distribution of malicious apps, cleverly crafted malware, such as the Chamois botnet [2–4], were still able to bypass protection mechanisms and infect millions of users worldwide.

Chamois was distributed through Google play and third-party app stores and infected over 20.8 million Android devices between November 2017 and March 2018. The first generation of the Chamois botnet was primarily distributed through fake apps, and initial eradication efforts by anti-malware professionals almost completely eliminated the threat. The creators of the botnet responded by adopting a more sophisticated distribution model

that bundled Chamois into a fake payment solution for device manufacturers and a fake advertising SDK for developers.

As mobile devices—especially smartphones—tend to be online for long periods, they provide a suitable platform for operating botnets when they have been compromised. Mobile botnets are controlled using SMS or web-based commands and control channels and are used for various attacks, such as Distributed Denial of Service (DDoS), phishing attacks, spam distribution, click fraud, credential stuffing etc. A study by Imperva on mobile botnet activity revealed that 5.8 million bot-infected mobile devices were used to launch credential stuffing attacks on websites and apps over a 45-day period on six major cellular networks [5].

DDoS attacks are high volume and high frequency and are thus easily detected by traditional network intrusion detection systems. By contrast, credential stuffing attacks from botnets are characterized by low frequency and low volume network traffic and are therefore more challenging to detect. Thus, complementary approaches to network-based detection are needed to strengthen defense against mobile botnet infection and attacks.

As mobile malware continues to increase and become more sophisticated, research efforts directed at detecting and mitigating Android malware has intensified in recent years. Several Machine-Learning-based detection systems have been proposed in the current literature to combat the rising incident of Android malware, including botnets [6–8]. Such systems rely on statically or dynamically extracted features for training the Machine-Learning models. In many cases, these features are either hand-crafted and/or depend heavily on domain expertise to effectively extract them. As the Android OS evolves, many of these hand-crafted features may become deprecated or obsolete and the entire feature extraction process will need to be re-engineered.

The utilization of image processing techniques to extract features from image-based representation of the application has the distinct advantage of eliminating the need to rely on hand-crafted features to build Machine-Learning models. Moreover, with image-based approach, little or no modification will be required to adapt to platform/OS evolution, and this leads to long-term efficiency compared to systems based on hand-crafted features.

Hence, in this paper, we propose a system that utilizes an image processing technique called Histogram of Oriented Gradients, to extract features for training Machine-Learning models to detect Android botnets. Our proposed system is a novel scheme that detects Android botnets based on Histogram of Oriented Gradients (HOG). In the scheme, the HOG features are combined with byte histograms and features from the app manifest file to improve prediction accuracy. Furthermore, we demonstrate the feasibility of our approach using a dataset of Android botnets and benign samples.

The rest of the paper is organized as follows: In Section 2, we provide an overview of related work. Section 3 describes our proposed system, while in Section 4, we outline the study undertaken to evaluate the system. Section 5 presents and discusses the results of the evaluation. Finally, in Section 6, we conclude the paper and give an outline of future work.

2. Related Work

There is extensive literature regarding the Machine-Learning-based detection of mobile malware, and [9,10], provide recent surveys on the topic. Here, we provide an overview of related works in Android botnet detection as well as image-based detection of malicious applications.

2.1. Image-Based Analysis of Malicious Applications

In [11], a method for image-based malware classification using an ensemble of CNN architectures was proposed. This was based on the malimg dataset where the raw images were used as input to the CNN-based classification system. Additionally, a malware dataset of 96 packed executables was also used and converted into images to evaluate the proposed system. The images were divided into training and validation sets based on a 70:30 split.

The method consisted of using transfer learning with fine-tuned ResNet-50 and VGG16 models that were pre-trained on ImageNet data. The output of these models obtained through SoftMax classifiers were fused with a version of the output that had been reduced using PCA and fed to a one-vs-all multiclass SVM classifier. In their experiments, they obtained a classification accuracy of up to 99.50% with unpacked samples and 98.11% and 97.59% for packed and salted samples, respectively.

In [12], the authors presented a method to recognize malware by capturing the memory dump of suspicious processes and representing them as RGB images. The study was based on 4294 malware samples consisting of 10 families and benign executables and several Machine-Learning classifiers, including J48, SMO with RBF kernel, Random Forest, Linear SVM and XGBoost. Dimensionality reduction was achieved using UMAP based manifold learning strategy. A combination of GIST and HOG features were used to extract features from the RGB images. The method yielded the highest prediction accuracy of up to 96.39% using the SMO classifier.

In [13], Bozkir et al. evaluated several CNN architectures for PE malware classification using coloured images. They used the Malevis Dataset containing 12,394 malware files and split this into 8750 training and 3644 testing samples. From their experiments, they obtained an accuracy of 97.48% using the DenseNet architecture.

Nataraj et al. [14] used grayscale images to visualize malware binaries to distinguish between different families. GIST was used to extract features from the images, and using KNN as a classifier they achieved 97.18% classification accuracy on experiments with a dataset consisting of 9458 malware samples from 25 families. In another paper [15] by the same authors, similar results were obtained when they applied image processing with dynamic analysis, in order to address both packed and unpacked malware.

Kumar et al. [16] proposed a method that uses an autoencoder enhanced deep convolutional neural network (AE-DCNN) to classify malware images into their respective families. A novel training mechanism is proposed where a DCNN classifier is trained with the help of an encoder. The encoder is used to provide extra information to the CNN classifier that may be lost during the forward propagation thus resulting in better performance. On the standard malimg dataset, 99.38% accuracy and F1-score of 99.38% were reported.

In [17], Fine Tuning and Transfer Learning approaches were used for multi-class classification of malware images. Eight different fine-tuned CNN-based transfer learning models were developed for vision-based malware multi-classification applications. These included VGG16, AlexNet, DarkNet-53, DenseNet-201, Inception-V3, Places365-GoogleNet, REsNet-50 and MobileNet-V2. Experiments based on the malimg dataset showed high performance with 99.97% accuracy.

Similarly, in [18], the IMCFN system i.e., image-based malware classification using fine-tuned convolutional neural network architecture, was presented. IMCFN converts raw malware binary into color images that are used by fine-tuned CNN architecture to classify malware. It fine-tunes a previously trained model based on ImageNet dataset and uses data augmentation to address class imbalance. The method was evaluated using malimg and an IoT-android mobile dataset containing 14,733 malware and 2486 benign samples. With the malimg dataset, an accuracy of 98.82% was obtained, while 97.35% accuracy was obtained for the IoT-android mobile dataset.

Xiao et al. [19] proposed a malware classification framework, MalFCS based on malware visualization and automated feature extraction. Malware binaries are visualized as entropy graphs based on structural entropy, while a deep convolutional neural network is used as a feature extractor to automatically extract patterns shared by a family from entropy graphs. An SVM classifier was used to classify malware based on the extracted features. The method achieved 99.7% accuracy when evaluated on the malimg dataset and 100% accuracy when evaluated on the Microsoft dataset.

Awan et al. also proposed an image-based malware classification system, which was investigated using malimg data [20]. The VGG19 model was used with transfer learning as a feature extractor, while a CNN model enhanced by a spatial attention mechanism was used to enhance the system. The attention-based model achieved an accuracy of 97.68% in the classification of the 25 families using a 70:30 training and testing split.

In [21], DenseNet was used with the final classification layer adopting a reweighted class-balanced loss function in the final classification layer to address data imbalance issues and improve performance. Experiments performed on malimg dataset yielded 98.23% accuracy while 98.46%, 98.21% and 89.48% accuracies were obtained with BIG 2015, MaleVis and Malicia datasets, respectively.

In [22,23], local binary patterns (LBP) were used while [24] used Intensity, Wavelet and Gabor to extract grayscale image features. Han et al. [25], used entropy graphs and similarity measures between entropy images for malware family classification. They obtained an accuracy of 97.9% by experimenting with 1000 malware samples from 50 families. In [26], the authors first disassembled binary executables and then converted the opcode sequences into RGB images. They evaluated their approach on 9168 malware and 8640 benign binaries achieving 94.8% to 96.5% accuracy.

Dai et al. [27] proposed a method for identifying malware families aimed at addressing the deficiencies of dynamic analysis approaches, by extracting a memory dump file and converting it to a grayscale image. They used the Cuckoo sandbox and built the procdump program into the sandbox, while using the ma command to extract the dump of the monitored process. Histogram of Gradient (HOG) was used to extract features from the image file and train KNN, Random Forest and MLP classifiers. Experiments were performed on 1984 malware samples from the Open Malware Benchmark dataset, and MLP performed best with an accuracy of 95.2% and F1-score of 94.1%.

Although these works highlight the success of employing image-based techniques in malware related work, their focus has largely been on Windows (PE) malware and family classification. By contrast, this paper uses image-based techniques for detection of botnets on the Android platform based on a novel approach to utilize HOG with manifest file features.

In a recent paper that focused on Android, Singh et al. [28] proposed a system called SARVOTAM that converts malware non-intuitive features into fingerprint images to extract quality information. Automatic extraction of rich features from visualized malware is then enabled using CNN, ultimately eliminating feature engineering and domain expert cost. They used 15 different combinations of Android malware image sections to identify and classify malware and replaced the softmax layer of CNN with ML algorithms like KNN, SVM and Random Forest to analyze grayscale malware images. It was observed that CNN-SVM outperformed the original CNN as well as CNN-KNN and CNN-RF. The experiments performed on the DREBIN dataset achieved 92.59% accuracy using Android certificates and manifest malware images.

2.2. Botnet Detection on Android

In [29], the authors proposed a signature-based, real-time SMS botnet detection system that applies pattern-matching for incoming and outgoing SMS messages. This is followed by a second step that uses rule-based techniques to label SMS messages as suspicious or normal. They performed experiments to evaluate their system with more than 12,000 messages. The system detected all 747 malicious SMS messages but also had a high false positive rate with 349 normal SMS messages misclassified as malicious.

Jadhav et al. presented a cloud-based Android botnet detection system in [30], based on strace, netflow, logcat, sysdump and tcpdump. Although this is a real-time dynamic analysis system, one major drawback is the ability of sophisticated botnets to detect and evade the cloud environment. Moreover, detecting Android botnets using a cloud-based dynamic analysis system based on several types of traces is more resource intensive compared to an image-based static analysis system.

Moodi et al. [31], presented an approach to detect Android botnets based on traffic features. Their method was based on SVM where a new approach called smart adaptive particle swarm optimization support vector machine (SAPSO-SVM) is developed to adapt the parameters of the optimization algorithm. The proposed approach identified the top 20 traffic features of Android botnets from the 28-SABD Android botnet dataset.

Bernardeschia et al. [32], used model checking to identify Android botnets. Static analysis is used to derive a set of finite state automata from the Java byte code that represents approximate information about the run-time behaviour of an app. However, the authors only evaluated their approach using 96 samples from the Rootsmart botnet family and 28 samples from the Tigerbot family in addition to 1000 clean samples.

Anwar et al. [33], proposed a static technique that consists of four layers of botnet security filters. The four layers consist of MD5 signatures, permissions, broadcast receiver and background services modules. Based on these, classification models were built using SVM, KNN, J48, Bagging, Naive Bayes and Random Forest. Experiments were performed on 1400 mobile botnet applications from the ISCX Android botnet dataset and 1400 benign applications. They observed the best result of 95.1% accuracy from the results of their experiments. In [34], the Android Botnet Identification System (ABIS) was proposed based on static and dynamic features using API calls, network traffic and permissions. These features were used to train several Machine-Learning classifiers, where Random Forest showed the best performance by obtaining a precision score of 0.972 and a recall score of 0.960.

Yusof et al. proposed a botnet classification system based on permission and API calls in [35]. They used feature selection to select 16 permissions and 31 API calls that were subsequently used to train Machine-Learning algorithms using the WEKA tool. The experiments were performed on 6282 benign and malicious samples using Naive Bayes, KNN, J48, Random Forest and SVM. Using both permission and API call features, Random Forest obtained the best results with 99.4% TP rate, 16.1% FP rate, 93.2% precision and 99.4% recall. This work was extended in [36] to include system calls and this resulted in improved performance with Random Forest achieving 99.4% TP rate, 12.5% FP rate, 93.2% precision, 99.4% recall and 97.9% accuracy.

In [37], a system for Android botnet detection using permissions and their protection levels were proposed. Random Forest, MLP, Naive Bayes and Decision Trees were used as Machine-Learning classifiers, with the experiments conducted using 1635 benign and 1635 botnet applications from the ISCX botnet datasets. Random Forest achieved 97.3% accuracy, 98.7% recall and 98.5% precision as the best result.

In [38], Android botnet classification (ABC) was proposed as a Machine-Learning-based system using requested permissions as features with Information Gain feature selection applied to select the most significant requested permissions. Naive Bayes, Random Forest and J48 were used as classifiers and experiments showed that Random Forest had the highest detection accuracy of 94.6%, lowest FP rate of 9.9%, with precision of 93.1% and recall of 94.6%. The experiments were performed on 2355 Android applications (1505 samples from the ISCX botnet dataset and 850 benign applications).

Karim et al. proposed DeDroid in [39], as a static analysis approach to extract critical features specific to botnets that can be used in the detection of mobile botnets. They achieved this by observing the code behaviour of known malware binaries that possess command and control features. In [40], an Android botnet detection system based on deep learning was proposed. The system is based on 342 static features including permissions, API calls, extra files, commands and intents. The model was evaluated using 6802 samples including 1929 ISCX botnet dataset samples and 4873 clean applications.

The performance of CNN was compared to Naive Bayes, Bayes Net, Random Forest, Random Tree, Simple Logistic, ANN and SVM. The CNN-based model achieved the best performance with 98.9% accuracy, 98.3% precision, 97.8% recall and 98.1% F1-score. In [8], a comprehensive study of deep learning techniques for Android botnet detection was presented using the same dataset and static features utilized in [40]. CNN, DNN, LSTM, GRU, CNN-LSTM and CNN-GRU models were studied, and the overall best result from DNN was 99.1% accuracy, 99% precision, 97.9% recall and 98.1% F1-score.

This cross-section of Android botnet detection systems summarized above indicates that in the current literature, most proposed solutions are based on hand-crafted (static or dynamic features) or rely on in-depth (Android) domain knowledge, unlike the system proposed in this paper. Furthermore, compared to image-based approaches, hand-crafted features may not be sustainable in the long run because as the Android OS evolves, new features are added while some old ones may become deprecated. This will require significant re-engineering of hand-crafted based systems to cope with the OS/platform evolution.

Some recent papers have begun exploring image-based techniques for Android botnet detection. In [41], the Bot-IMG framework was used to extract HOG descriptors and train Machine-Learning-based classifiers to distinguish botnets from benign applications. An enhanced HOG scheme was proposed, which enabled improved accuracy performance with the use of autoencoders. The system was evaluated with experiments performed using 1929 ISCX botnet applications and 2500 benign applications.

KNN, SVM, Random Forest, XGBoost and Extra Trees learning algorithms were trained using the HOG-based schemes. With Extra Trees, the best result from 10-fold cross validation was obtained using autoencoder and gave 93.1% accuracy with 93.1% F1-score. In [42], the authors used permissions to generate images based on a co-occurrence matrix. The images were used to train a CNN model to classify applications into benign or botnet. The experiments were performed on 3650 benign applications and 1800 botnet applications from the ISCX dataset. Their best result was 97.2% accuracy, 96% recall, 95.5% precision and 95.7% F1-score.

Different from [41,42], the system presented and evaluated in this paper is a novel botnet detection system based on image features (i.e., HOG, byte histograms) and manifest features (i.e., permissions, intents). All of these features come from a single pre-processed composite image derived from automated reverse engineering of the Android applications. In this paper, we demonstrate the feasibility and performance of the proposed scheme by using it to train and evaluate several popular Machine-Learning classifiers on a dataset of 1929 ISCX botnet applications and 2500 benign applications.

3. Proposed HOG-Based Android Botnet Detection System

Our proposed system is based on the Bot-IMG framework [41], which enables automated reverse engineering of the Android applications, image generation and subsequent extraction of image-based and manifest features. Figure 1 shows an overview of the system for HOG-based Android botnet detection. As shown in the figure, the first step involves reverse engineering the apks to extract the various files contained in the application.

Out of all the files present in an apk, only the manifest file and the Dalvik executable (dex) file are utilized in the proposed system. The manifest file is processed using AXMLPrinter2 tool, which converts it into a readable text file that is scanned to generate a set of 187 features consisting of permissions and intents. These features extracted from the manifest file are encoded for gray-scale representation.

Thus, the presence of a feature is denoted by 255 (or white), while a 0 (i.e., black) is recorded if the feature is absent and these are stored in an array of manifest features. The dex file is converted to a byte array consisting of integer encoded bytes ranging from 0 to 255. This byte array from the executable is combined with the array of manifest features. The combined array is then used to generate a composite gray-scale image representing the application.

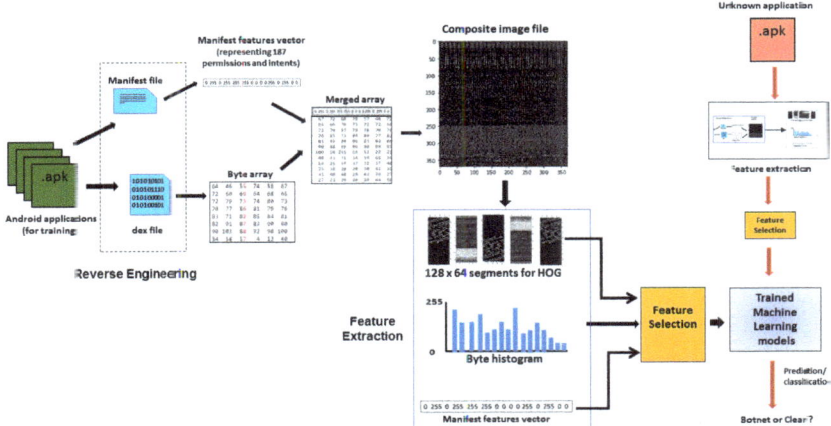

Figure 1. Overview of the different steps involved in building the image-based Android botnet detection system.

The image files are processed by the feature extraction engine using the algorithm described in Section 3.2 to generate feature vectors for each application used in the training of the Machine-Learning models. During the training of a model, a feature selection algorithm is applied to select the best features. The trained model is then used to detect botnet apps by classifying an unknown application into 'botnet' or 'benign'. The proposed system is based on HOG, byte histograms and manifest features. We provide a brief description of HOG in the following section.

3.1. Histogram of Oriented Gradients

HOG, first proposed for human detection by Dalal and Triggs [43] is a popular image descriptor that has found wide application in computer vision and pattern recognition. For example, it has been applied to handwriting recognition [44], recognition of facial expressions [45], pedestrian detection system for autonomous vehicles [46]. HOG is considered to be an appearance descriptor because it counts occurrences of gradient orientation in localized portions of an image. Due to the simple computations involved, HOG is generally a fast descriptor compared to Local Binary Patterns (LBP) or Scale Invariant Feature Transforms (SIFT)

HOG descriptors are computed on a dense grid of uniformly spaced cells and overlapping local contrast normalizations are used for improved performance. For each pixel, magnitude and orientation can be computed using the following formulae:

$$g = \sqrt{g_x^2 + g_y^2} \qquad (1)$$

$$\theta = tan^{-1}\left(\frac{g_y}{g_x}\right) \qquad (2)$$

where g_x and g_y are calculated from the neighboring pixels in the horizontal and vertical directions respectively. Figure 2 illustrates how the histograms are generated for a cell, using the highlighted pixel as an example. For the pixel represented by number 65, the change in x direction g_x is $69 - 54 = 15$, and the change in y direction g_y is $78 - 30 = 48$. Using the Equations (1) and (2), the total magnitude $g = 50.3$ while the orientation $\theta = 72.65°$. To generate the histogram for the cell, using nine bins representing the orientations separated 20 degrees apart, each pixel's contribution will be added to the bin according to orientation.

For example, in Figure 2, the orientation is 72.65°, which is between 60 degrees and 80 degrees. Thus, the magnitude is split between these two bins by using the following weighting approach where the distances from the bin orientations are used. Hence, we have $(72.65 - 60)/20$ and $(80 - 72.65)/20$ as the weights that will be used to split the 50.3 magnitude between the bins. This means that the split will result in 31.7 and 18.6 being placed in the 4th and 5th bins respectively. The process is repeated for all the pixels in the cell.

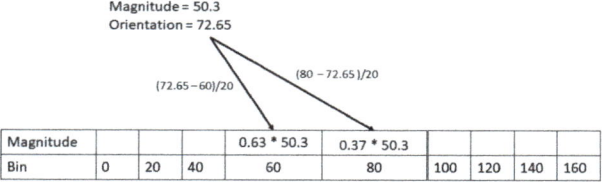

Figure 2. Building a Histogram of Oriented Gradients using nine bins representing positive orientations spaced 20 degrees apart.

The binning of the magnitudes by taking the orientations into consideration, produces a histogram of gradient directions for the pixel magnitudes in a cell. If the number of bins is taken as 9, then, each cell will be represented by a 9×1 array. In order to make the gradients less sensitive to scaling, the histogram array is normalized in blocks, where each block is made up of $b \times b$ cells. Hence, taking $b = 2$ will result in 4 cells per block. This means that each block will be represented by a 36×1 vector or array (i.e., 4 cells × nine bins). Block normalization is based on the L2-norm computed as in Equation (3), where ϵ is a small constant:

$$v \leftarrow \sqrt{||v||_2^2 + \epsilon^2} \qquad (3)$$

In the default situation, the HOG algorithm takes an input image whose size is 128×64. Therefore, in the 128×64 pixel image, it turns out that if we take 8×8 pixels in each cell and 2×2 cells in each block, then this will result in 7 horizontal block positions and 15 vertical block positions. Hence, we get a total HOG vector length of 3780 (which is $36 \times 7 \times 15$). As such, to get a HOG descriptor vector of length 3780 for an image, we are required to choose the following parameters: $n = 9$ (number of orientations); ppc = 8×8 (number of pixels per cell) and cpb = 2×2 (number of cells per block).

3.2. Characterizing Apps with Image and Manifest Features

The methodology for extracting the HOG descriptors and using them together with byte histogram and manifest features to characterize the apps, is discussed in this section. The steps involved in deriving the composite features for Machine-Learning-based detection approach are shown in Algorithm 1.

Algorithm 1: Extracting image-based and manifest-based features.

Input: $D = \{(I_1, y_1), (I_2, y_2) \ldots (I_n, y_n)\}$ set of images with their class labels
Output: $V = \{(V_1, y_1), (V_2, y_2) \ldots (V_n, y_n)\}$ class labelled set of output vectors

01:    Initialize $X1, X2, X3, X4$ and $X5$, the arrays of size $K = 8192$ with zeros
02:    Initialize BH the byte histogram array of size 256 with zeros
03:    Initialize HOG parameters: $n = 9$, dim $= 128 \times 64$; ppc $= 8 \times 8$; cpb $= 2 \times 2$
04:    **for** each image I **do**
05:       Slice the image to separate the first 187 pixels
06:       Copy first slice having 187 pixels into a manifest vector $\overline{M}$
07:       // Obtain the HOG vector for the image
08:       Convert second slice into an array P of pixel decimal values
09:       Copy the first $K * 5$ bytes into arrays $X1, X2, X3, X4, X5$
10:       Reshape $X1, X2, X3, X4, X5$ to 128×64 size arrays
11:       Convert $X1, X2, X3, X4, X5$ into images
12:       **for** each sub-image j **do**
13:          $H_j \leftarrow getHogVector(n, dim, pbc, cpb)$
14:          $\overline{H_j} \leftarrow subsample(H_j)$
15:       **end**
16:       $HV \leftarrow concatenate(\overline{H_1}, \overline{H_2}, \overline{H_3}, \overline{H_4}, \overline{H_5})$
17:       // Obtain the byte histogram vector for the image
18:       $X \leftarrow concatenate(X1, X2, X3, X4, X5)$
19:       **for** index $= 0, 1, 2 \ldots 255$ **do**
20:          count $= 0$
21:          **for** $t = 0, 1, 2 \ldots Length(X) - 1$ **do**
22:             count = count +1
23:             **if** count > max
24:             count = max
25:             **end if**
26:          **end**
27:          $BH(index) \leftarrow 255 * \left(\frac{\log\left(\frac{count+1}{max}\right) - \log\left(\frac{1}{max}\right)}{-\log\left(\frac{1}{max}\right)} \right)$
28:       **end**
29:       // Obtain the overall output vector for the image
30:       $V \leftarrow concatenate(\overline{M}, HV, BH)$
31:    **end**

The required input is the set of images from benign and botnet apps, while the output will be a high dimensional vector V. The images generated from the apps are of different sizes. Thus, in order to utilize the original HOG descriptors generation approach proposed by Dalal and Triges, the images must be reshaped to 128×64 pixels. However, it has been found that resizing the images diminishes the performance of the trained Machine-Learning models [41]. We therefore adopt a methodology that uses five patches or segments from the image with each segment being 128×64 pixels in size. In line 1 of Algorithm 1, the five arrays $X1, X2, X3, X4$ and $X5$ that will hold the pixels of the five segments are initialized with zeros.

This approach utilizes only the first 40,960 pixels of the images (after extracting the manifest features) making for a fast and efficient system. Once the five segments from the image are copied into the arrays, they are reshaped into 128×64 arrays and converted into 5 separate images. This is because the HOG descriptor function only takes images as input and not arrays. From line 12 to line 15 of Algorithm 1, five different HOG vectors are generated for each segment image, and in line 14, we sub-sample each of them to retain 500 descriptors. The 500 descriptors from each batch are then concatenated into a 2500 descriptor vector HV in line 16.

From line 17 to line 28 of Algorithm 1, a byte histogram is generated, but only for the same combined area where the HOG descriptors were extracted, i.e., the first 40,960 pixels of the application's image. The byte histogram will consist of a vector of dimension 256 that will hold the occurrences of bytes (pixels) within that region. The occurrences are clipped and log-scaled as depicted in lines 23, 24 and 27 respectively, to keep the values between zero and 255. Finally, the overall feature vector V of dimension 2943 is generated by concatenating the extracted manifest vector $\overline{M}$ (MV) with the final HOG vector (HV) and the log-scaled byte histogram (BH).

3.3. Feature Selection Using CHI Square Algorithm

Since the image processing resulted in a high dimensional vector, we apply feature selection for dimensionality reduction and to improve the performance of the Machine-Learning classifiers. As we know, if there are more features resulting a high dimensional vector as an input to the classifier during the training phase, they will contribute to algorithmic complexity in terms of data storage and processing. As not all features contribute to the model's performance in the classification phase, it is suitable that they be removed from the training phase as well. This process is termed as 'Dimensionality Reduction'.

Dimensionality reduction can be achieved by "measuring" the contribution of each of the features to the model's prediction performance. Those features that have insignificant contributions can be safely removed to enhance the training speed of the Machine-Learning model. Dimensionality reduction chooses those features, which are good contributors to the model performance, and hence this process is also called Feature Selection.

Various approaches for Feature Selection have been presented in the literature, such as Information Gain, Mutual Information, Principal Component Analysis and the Chi-Square test [47]. In our research, we chose to use the Chi-Square test, which results in a better prediction performance for our ML classifiers. The Chi-Square test is represented by the formula given in Equation (4):

$$\chi^2 = \sum_i (O_i - E_i)^2 / E_i \qquad (4)$$

where
χ^2 = Chi-Squared value
O_i = Observed value
E_i = Expected value

In our case, the observed value could take one of the values of the input features variable and the expected value would be another feature variable. If there is a strong correlation between them (that is χ^2 is too low) then it is enough to consider only one out of them and hence reduce one feature. Similarly, all possible combinations of the feature variables can be compared and sorted according to their Chi-Square values. Then, we can choose those feature variables that have high Chi-Square values from the list.

4. Experiments and Evaluation of the System

4.1. Dataset Description

The ISCX botnet dataset obtained from [48] has been used to evaluate the proposed system. The dataset consists of 1929 botnet apps of 14 different families. We complemented this dataset with 2500 clean apps from different categories on the Google play store and used VirusTotal for verification. Thus, our experiments were based on a total of 4429 applications from, which the images were generated and processed using the Bot-IMG framework. The clean applications can be made available to third parties on request.

4.2. Evaluation Metrics

The following metrics were used in measuring the performance of the models: accuracy, precision, recall and F1-score. All the results of the experiments are from 10-fold cross validation where the dataset is divided into 10 equal parts with 10% of the dataset held out for testing, while the models are trained from the remaining 90%. This is repeated until all of the 10 parts have been used for testing. The average of all 10 results is then taken to produce the final result. We also employed the 80:20 split approach where 80% of the samples were used for training and 20% for testing.

4.3. Machine-Learning Classifiers

In this section, a brief overview of the Machine-Learning classifiers is presented, which were used to distinguish between botnet and clean apps. In general these are algorithms, which are trained on the labelled data (input) and then the learned model is used for estimating the target variable (output, in this case, malicious botnet or clean app).

1. **K-Nearest Neighbor (KNN):** KNN is a supervised classifier that classifies an input data into a specific set of classes based on the distance metric among its nearest neighbors [49]. Various distance metrics are possible candidates for the K-NN algorithm, such as the Euclidean distance, Manhattan distance, City block distance and Hamming distance. Due to its simplicity, Euclidean distance is the preferred choice among these distance measures. The K-NN algorithm uses vectors in a multidimensional feature space as training examples, each having a class label. During the training phase the algorithm stores the feature vectors and their class labels for the purpose of learning the model. During the classification phase, an unlabeled vector is classified by assigning the label, which is most frequent among the k training samples. Here k is a user defined constant whose choice depends on the type of data to be classified.

2. **Support Vector Machines (SVM):** SVM classifies the input data into different classes by finding a hyperplane in a higher dimension space of the feature set to distinguish among various classes [50]. This technique transforms the input data, which is divided into separate classes non-linearly, by applying various types of kernel functions, such as linear, polynomial, Gaussian and radial basis functions. SVM follows the concept of minimizing the classification risk as opposed to optimizing the classification accuracy. As a result, SVMs have a better generalization capability and hence can be used in situations where the number of training samples are less and the data has large number of features. SVMs have been popularly used in text and image classification problems and also in voice recognition and anomaly detection (e.g., security, fraud detection and healthcare).

3. **Decision Trees (DT):** A Decision Tree uses a tree-like structure that models a labelled data [51]. Its structure consists of leaves and branches, which actually represent the classifications and the combinations of features that lead to those classifications, respectively. During the classification, an unlabeled input is classified by testing its feature values against the nodes of the decision tree. Two popular algorithmic implementations of Decision Trees are the ID3 and C4.5, which use the information entropy measurements to learn the tree from the set of the training data. The procedure followed when building the decision tree, is to choose the data attributes that most efficiently splits its set of inputs into smaller subsets. Normalised information gain is used as the criteria for performing the splitting process. Those attributes that have the highest normalized information gain are used in making the splitting decision.

4. **Random Forest (RF):** Random Forest belong to the class of classifiers that are known as the Ensemble Learning classifiers [52]. As the name suggests, RF is a collection of several decision trees that are created first and are then combined in a random manner to build a "forest of trees". A random sample of data from the training set is utilised for training the constituent trees of the RF. It is observed that due the presence of multiple DTs in the RF, it circumvents the over-fitting problem encountered in DTs. This is due to the fact that RF performs a "bagging" step that uses bootstrap aggregation to deal with the over-fitting problem. During the classification phase, the RF takes the test features as an input and each DT within the RF is used to predict the desired target variable. The final outcome of the algorithm is achieved by taking the prediction with maximum votes among the constituent DTs.

5. **Extra Trees (ET)**: Extra Trees is also an ensemble Machine-Learning algorithm that combines the predictions from many decision trees [53]. The concept is similar to the Random Forests, however there are certain key differences between them. One of the difference lies in how they take the input data to learn the models. RF uses bootstrap replicas (sub-sampling of the data), where as the Extra Trees use the whole input data as it is. Another difference lies in how the the cut points are selected in order to split the nodes of the tree. RF chooses the split in an optimal manner, however the Extra Trees do it randomly. That is why another name for Extra Trees is Extremely Randomised Trees. As such, Extra Trees add randomisation to the training process, but at the same time maintains the optimization. In other words, Extra Trees reduce both the bias and variance and are a good choice for classification tasks as compared to Random Forests.

6. **XGBoost (XGB)**: XGBoost also belongs to the category of Ensemble Learning classifiers similar to RF and ETs, mentioned above [54]. However, they are based on the concept of Boosting, rather than Bagging (which is implemented in RF). Boosting is a process of increasing the prediction capabilities of an ensemble of weak classifiers. It is actually an iterative process where the weights of the each of the constituent weak classifiers are adjusted based on their performance in making the predictions of the target variable. Boosting is an iterative method that uses random sampling of the data without replacement (as opposed to replacement used during the bagging process in the RF). In boosting, errors that occur in the prediction of earlier models are reduced by the predictions of future models. This step is very much different from the bagging process used in Random Forest classifiers that use an ensemble of "independently" trained classifiers.

5. Results and Discussions

In this section we present the results of the experiments performed to evaluate the performance of the proposed scheme described in Section 3. The proposed scheme was implemented with Python and the following libraries were utilized: OpenCV, PIL, Scikit-learn, Scikit-image, Pandas, Keras, Numpy, Seaborn and Matplotlib. The experiments were performed on an Ubuntu Linux 16.04 64-bit machine with 8 GB RAM.

Six popular Machine-Learning classifiers were used to evaluate the proposed scheme. These include: K-Nearest Neighbor (KNN), Random Forest (RF), Support Vector Machines (SVM), Decision Trees (DT), Extra Trees (ET) and XGBoost (XGB). We implemented two other schemes for baseline comparison of the Machine-Learning classifier performance. The first baseline scheme was the original HOG scheme where all the images in the training and test sets were resized to the standard 128×64 pixels and resulting in vectors of size 3780 used in training the models. The second baseline scheme used five segments to extract HOG descriptors in an identical way to our proposed scheme described by Algorithm 1 and used them to train the models without adding byte histograms or manifest features. We call the second baseline scheme the 'enhanced HOG' method.

In Table 1, the results of our proposed scheme using 10-fold cross validation are shown for the six Machine-Learning classifiers. We present the precision and recall for both malicious botnet class (M) and the benign or clean class (C). Note that the F1-scores presented in the table are weighted values, due to the difference in the numbers of samples in each class. The table shows that all of the classifiers obtained an overall accuracy performance of 92.3% or above, indicating that our proposed approach enables the training of high performing machine learning classifiers. The Extra Tree classifier had the highest weighted F1-score of 0.96, followed by Random Forest with 0.958 and XGBoost with 0.952. The lowest weighted F1-score was for KNN with 0.923 while SVM obtained a weighted F1-score of 0.926.

The Extra Trees classifier had the best precision and recall values except in the case of malware recall, which was 94.2% compared to that of Random Forest, which had 94.4%.

SVM had the lowest malware class recall of 92.1% while KNN had the lowest benign class recall of 90.7%.

Table 1. Classifier performance with permissions, byte histograms and HOG descriptors (10-fold cross validation results).

	Precision (M)	Recall (M)	Accuracy	Precision (C)	Recall (C)	F1-Score
Extra Trees	0.965	0.942	0.960	0.955	0.974	0.960
SVM	0.911	0.921	0.926	0.938	0.930	0.926
KNN	0.889	0.942	0.923	0.953	0.907	0.923
XGBoost	0.950	0.935	0.952	0.950	0.962	0.952
RF	0.962	0.944	0.958	0.955	0.970	0.958
DT	0.913	0.940	0.936	0.954	0.933	0.936

In Table 2, the results of the proposed scheme using a train–test split of 80:20 are presented. The table shows that all of the classifiers resulted in an overall accuracy of 93.7% or higher, with the Extra Tree classifier yielding an accuracy of 97.5%. The Extra Tree classifier had the highest weighted F1-score of 0.980 followed by XGBoost and Random Forest with 0.970 and Decision Trees with 0.950. SVM and KNN had the lowest weighted F1-score of 0.940. Extra Trees had the highest malware class recall of 97% while SVM had the lowest recall of 92% for malware. Extra Tree, XGBoost and RF had the highest recall for benign class with 98%, while SVM had the lowest one, with 92%. Based on these results Extra Trees model will be the classifier of choice for our proposed Android botnet detection system.

Table 2. Classifier performance with permissions, byte histograms and HOG descriptors (train–test split results).

	Precision (M)	Recall (M)	Accuracy	Precision (C)	Recall (C)	F1-Score
Extra Trees	0.970	0.970	0.975	0.980	0.980	0.980
SVM	0.890	0.920	0.937	0.940	0.920	0.940
KNN	0.860	0.940	0.944	0.960	0.900	0.940
XGBoost	0.970	0.940	0.966	0.960	0.980	0.970
RF	0.970	0.960	0.973	0.970	0.980	0.970
DT	0.920	0.950	0.953	0.960	0.950	0.950

The results presented in Tables 1 and 2 demonstrates the effectiveness of our proposed scheme. This is evident in the performance of the strongest and the weakest classifiers in the group. SVM and KNN were the weakest classifiers but still managed to yield quite high accuracies and F1-scores in both 10-fold cross validation and the split based evaluation. On the other hand, the strongest classifiers Extra Trees, RF and XGBoost produced results that were comparable to the state-of-the art in the literature.

It is possible that the few malicious botnets that were not detected by the system had characteristics that made them resemble benign apps. For example, botnets with relatively few permissions and intents, or those with HOG representation were very close to those of benign training examples. This could be addressed in future work by extracting additional types of features or complementing our proposed method with alternative methods, for example through an ensemble or voting approach.

In Table 3 and Figure 3, we compare the performance of the proposed scheme with the two baseline schemes (HOG original and HOG enhanced) using the overall classification accuracy as the metric. The accuracies of each of the Machine-Learning classifiers for the compared schemes can be seen side by side in Figure 3. The proposed scheme outperforms the baseline schemes in all of the Machine-Learning classifiers.

From Table 3, HOG original obtained the highest classification accuracy of 89.2% with the XGBoost classifier. This suggests that the resizing of the images during pre-processing has adverse effects on the performance of the models. The enhanced HOG scheme reached a highest accuracy of 92.7% also with XGBoost classifier. The scheme proposed in this paper, which additionally leverages byte histograms and encoded manifest features led to significantly improved performance.

Table 3. Comparison of the baseline HOG schemes with the proposed method using the overall accuracy (10-fold cross validation results).

	HOG (Original)	HOG (Enhanced)	HOG + BH + MF
XGBoost	**0.892**	**0.927**	0.952
Extra Trees	0.863	0.925	**0.960**
RF	0.871	0.919	0.958
KNN	0.877	0.877	0.920
SVM	0.811	0.866	0.926
DT	0.773	0.835	0.935

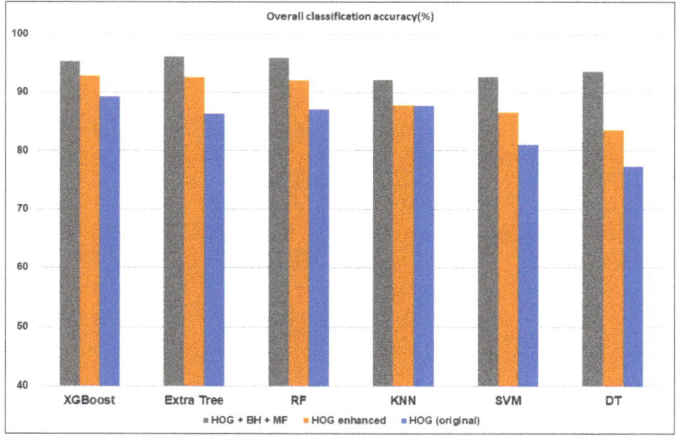

Figure 3. Overall classification accuracy for the various classifiers using the three compared schemes.

In Figure 4, the average training times for the samples trained during the 10-fold cross validation experiments are shown. Note that the training includes the feature selection step. The XGBoost classifier needed about 14.49 s to train 3987 samples in the training set, equivalent to an average of 3.6 milliseconds per sample. The rest of the classifiers were much faster and required significantly lower average training times for the training sets as shown in Figure 4.

The highest accuracy classifier, Extra Trees, needed an average of 1.72 s for the training sets—equivalent to 0.43 milliseconds per sample. In the pre-processing stage, the average amount of time taken to extract the features per application was approximately 1.37 s. These relatively low pre-processing and training times required per application indicates that the proposed approach is feasible in practice. The fact that we successfully utilized off-the-shelf Python libraries to build and evaluate the proposed system also indicates that commercial implementation is viable.

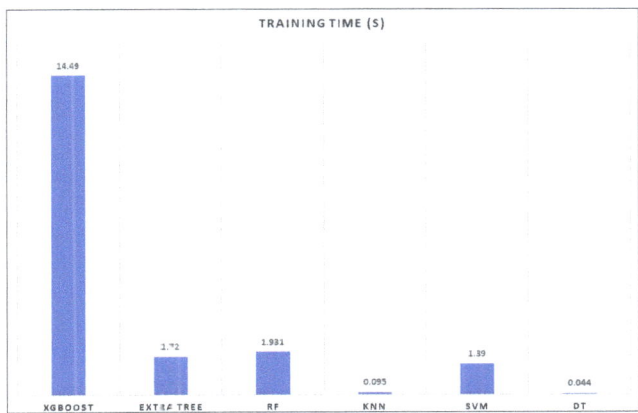

Figure 4. Average training times for the training set samples in seconds for each of the ML classifiers using our proposed scheme.

Although we demonstrated the effectiveness of our proposed method by experimental results showing high performance with several classifiers, our observed results also compare favorably with existing works. Due to variations in environments, datasets, the numbers of samples, reported metrics etc., direct comparison is not always possible. However, the set of results reported in this paper either exceeds or is similar to what has been reported in recent related works.

For example, the work in [41] was based on the same dataset used in this paper but had a lower performance with 93.1% as the highest accuracy. The papers [33,38] also reported lower accuracies than our results. However, these works were based on a different dataset and used hand-crafted features. As mentioned before, such features on Android have the disadvantage of maintenance overhead in the long run. Moreover, it was shown in [55] that the performance of hand-crafted features used to build machine learning models declined over time.

6. Conclusions and Future Work

In this paper, we proposed a novel approach for the detection of Android botnets based on image and manifest features. The proposed approach removes the need for detection solutions to rely on extracting hand-crafted features from the executable file, which ultimately requires domain expertise. The system is based on a Histogram of Oriented Gradients (HOG) and additionally leverages a byte histogram and the manifest features. We implemented the system in Python and evaluated its performance using six popular Machine-Learning classifiers.

All of them exhibited good performance with Extra Trees, XGBoost and Random Forest obtaining better performance as compared to the state-of-the-art results. These results demonstrate the effectiveness of the proposed approach. An overall accuracy of 97.5% and F1-score of 0.980 were observed with Extra Trees when evaluated with the 80:20 split approach; while a 96% accuracy and 0.960 F1-score were observed when evaluated using a 10-fold cross validation approach. In future work, we plan to explore other types of image descriptors and investigate whether they could be leveraged to improve the performance of the HOG-based scheme.

Author Contributions: Conceptualization, S.Y.Y.; methodology, S.Y.Y.; software, S.Y.Y. and A.B.; validation, A.B. and S.Y.Y.; formal analysis, S.Y.Y.; investigation, S.Y.Y.; resources, A.B. and S.Y.Y.; data curation, S.Y.Y.; writing—original draft preparation, S.Y.Y.; writing—review and editing, A.B.; visualization, A.B.; supervision, A.B. and S.Y.Y.; project administration, A.B.; funding acquisition, A.B. and S.Y.Y. All authors have read and agreed to the published version of the manuscript.

Funding: There is no external funding.

Institutional Review Board Statement: Ethical review and approval were waived for this study, due to the usage of the dataset available from the public domain (University of New Brunswick) governed by the ethics and privacy laws at: https://www.unb.ca/cic/datasets/android-botnet.html, accessed on 28 December 2021.

Informed Consent Statement: Since the dataset was taken from University of New Brunswick (public domain) the informed consent was not applicable in our case.

Data Availability Statement: The botnet dataset used in this research work was taken from the public domain (University of New Brunswick) at: https://www.unb.ca/cic/datasets/android-botnet.html, accessed on 28 December 2021.

Acknowledgments: This work is supported in part by the 2021 Cybersecurity research grant from the Cybersecurity Center at Prince Mohammad Bin Fahd University, Al-Khobar, Saudi Arabia.

Conflicts of Interest: The authors declare no conflict of interest.

References

1. McAfee. McAfee Labs Threat Report 06.21. Available online: https://www.mcafee.com/enterprise/en-us/assets/reports/rp-threats-jun-2021.pdf (accessed on 28 December 2021).
2. Rashid, F.Y. Chamois: The Big Botnet You Didn't Hear About. Available online: https://duo.com/decipher/chamois-the-big-botnet-you-didnt-hear-about (accessed on 28 December 2021).
3. Brook, C. Google Eliminates Android Adfraud Botnet Chamois. Available online: https://threatpost.com/google-eliminates-android-adfraud-botnet-chamois/124311/ (accessed on 28 December 2021).
4. Grill, B.; Ruthven, M.; Zhao, X. Detecting and Eliminating Chamois, a Fraud Botnet on Android. Available online: https://android-developers.googleblog.com/2017/03/detecting-and-eliminating-chamois-fraud.html (accessed on 28 December 2021).
5. Imperva. Mobile Bots: The Next Evolution of Bad Bots. In *Imperva*; Report, 2019. Available online: https://www.imperva.com/resources/resource-library/reports/mobile-bots-the-next-evolution-of-bad-bots/ (accessed on 28 December 2021).
6. Feng, P.; Ma, J.; Sun, C.; Xu, X.; Ma, Y. A Novel Dynamic Android Malware Detection System With Ensemble Learning. *IEEE Access* **2018**, *6*, 30996–31011. [CrossRef]
7. Wang, W.; Li, Y.; Wang, X.; Liu, J.; Zhang, X. Detecting Android malicious apps and categorizing benign apps with ensemble of classifiers. *Future Gener. Comput. Syst.* **2018**, *78*, 987–994. [CrossRef]
8. Yerima, S.Y.; Alzaylaee, M.K.; Shajan, A.; Vinod, P. Deep Learning Techniques for Android Botnet Detection. *Electronics* **2021**, *10*, 519. [CrossRef]
9. Senanayake, J.; Kalutarage, H.; Al-Kadri, M.O. Android Mobile Malware Detection Using Machine Learning: A Systematic Review. *Electronics* **2021**, *10*, 1606. [CrossRef]
10. Liu, K.; Xu, S.; Xu, G.; Zhang, M.; Sun, D.; Liu, H. A Review of Android Malware Detection Approaches Based on Machine Learning. *IEEE Access* **2020**, *8*, 124579–124607. [CrossRef]
11. Vasan, D.; Alazab, M.; Wassan, S.; Safaei, B.; Zheng, Q. Image-Based malware classification using ensemble of CNN architectures (IMCEC). *Comput. Secur.* **2020**, *92*, 101748. [CrossRef]
12. Bozkir, A.S.; Tahillioglu, E.; Aydos, M.; Kara, I. Catch them alive: A malware detection approach through memory forensics, manifold learning and computer vision. *Comput. Secur.* **2021**, *103*, 102166. [CrossRef]
13. Bozkir, A.S.; Cankaya, A.O.; Aydos, M. Utilization and Comparison of Convolutional Neural Networks in Malware Recognition. In Proceedings of the 27th Signal Processing and Communications Applications Conference (SIU), Sivas, Turkey, 24–26 April 2019.
14. Nataraj, L.; Karthikeyan, S.; Jacob, G.; Manjunath, B.S. Malware Images: Visualization and Automatic Classification. In Proceedings of the 8th International Symposium on Visualization for Cyber Security, Pittsburgh, PA, USA, 20 July 2011.
15. Nataraj, L.; Yegneswaran, V.; Porras, P.; Zhang, J. A Comparative Assessment of Malware Classification Using Binary Texture Analysis and Dynamic Analysis. In Proceedings of the 4th ACM Workshop on Security and Artificial Intelligence, Chicago, IL, USA, 21 October 2011.
16. Kumar, S.; Meena, S.; Khosla, S.; Parihar, A.S. AE-DCNN: Autoencoder Enhanced Deep Convolutional Neural Network For Malware Classification. In Proceedings of the 2021 International Conference on Intelligent Technologies (CONIT), Hubli, India, 25–27 June 2021; pp. 1–5.
17. El-Shafai, W.; Almomani, I.; AlKhayer, A. Visualized Malware Multi-Classification Framework Using Fine-Tuned CNN-Based Transfer Learning Models. *Appl. Sci.* **2021**, *11*, 6446. [CrossRef]
18. Vasan, D.; Alazab, M.; Wassan, S.; Naeem, H.; Safaei, B.; Zheng, Q. IMCFN: Image-based malware classification using fine-tuned convolutional neural network architecture. *Comput. Netw.* **2020**, *171*, 107138. [CrossRef]

19. Xiao, G.; Li, J.; Chen, Y.; Li, K. MalFCS: An effective malware classification framework with automated feature extraction based on deep convolutional neural networks. *J. Parallel. Distrib. Comput.* **2020**, *141*, 49–58. [CrossRef]
20. Awan, M.J.; Masood, O.A.; Mohammed, M.A.; Yasin, A.; Zain, A.M.; Damaševičius, R.; Abdulkareem, K.H. Image-Based Malware Classification Using VGG19 Network and Spatial Convolutional Attention. *Electronics* **2021**, *10*, 2444. [CrossRef]
21. Hemalatha, J.; Roseline, S.A.; Geetha, S.; Kadry, S.; Damaševičius, R. An Efficient DenseNet-Based Deep Learning Model for Malware Detection. *Entropy* **2021**, *23*, 344. [CrossRef] [PubMed]
22. Yan, H.; Zhou, H.; Zhang, H. Automatic Malware Classification via PRICoLBP. *Chinese J. Chem.* **2018**, *27*, 852–859. [CrossRef]
23. Luo, J.S.; Lo, D.C.T. Binary malware image classification using machine learning with local binary pattern. In Proceedings of the 2017 IEEE International Conference on Big Data (Big Data), Boston, MA, USA, 11–14 December 2017; pp. 4664–4667.
24. Kancherla, K.; Mukkamala, S. Image visualization based malware detection. In Proceedings of the 2013 IEEE Symposium on Computational Intelligence in Cyber Security (CICS), Singapore, 16–19 April 2013; pp. 40–44.
25. Han, K.S.; Lim, J.H.; Kang, B.; Im, E.G. Malware analysis using visualized images and entropy graphs. *Int. J. Inf. Secur.* **2015**, *14*, 1–14. [CrossRef]
26. Wang, T.; Xu, N. Malware variants detection based on opcode image recognition in small training set. In Proceedings of the IEEE 2nd International Conference on Cloud Computing and Big Data Analysis (ICCCBDA), Chengdu, China, 28–30 April 2017; pp. 328–332.
27. Dai, Y.; Li, H.; Qian, Y.; Lu, X. A malware classification method based on memory dump grayscale image. *Digit. Investig.* **2018**, *27*, 30–37. [CrossRef]
28. Singh, J.; Thakur, D.; Ali, F.; Gera, T.; Kwak, K.S. Deep Feature Extraction and Classification of Android Malware Images. *Sensors* **2020**, *20*, 7013. [CrossRef] [PubMed]
29. Alzahrani, A.J.; Ghorbani, A.A. Real-time signature-based detection approach for SMS botnet. In Proceedings of the 13th Annual Conference on Privacy, Security and Trust (PST), Izmir, Turkey, 21–23 July 2015; pp. 157–164.
30. Jadhav, S.; Dutia, S.; Calangutkar, K.; Oh, T.; Kim, Y.H.; Kim, J.N. Cloud-based Android botnet malware detection system. In Proceedings of the 17th International Conference on Advanced Communication Technology (ICACT), PyeongChang, Korea, 1–3 July 2015; pp. 347–352.
31. Moodi, M.; Ghazvini, M.; Moodi, H.; Ghawami, B. A smart adaptive particle swarm optimization–support vector machine: Android botnet detection application. *J. Supercomput.* **2020**, *76*, 9854–9881. [CrossRef]
32. Bernardeschia, C.; Mercaldo, F.; Nardonec, V.; Santonod, A. Exploiting Model Checking for Mobile Botnet Detection. *Procedia Comput. Sci.* **2019**, *159*, 963–972. [CrossRef]
33. Anwar, S.; Zain, J.M.; Inayat, Z.; Haq, R.U.; Karim, A.; Jabir, A.N. A static approach towards mobile botnet detection. In Proceedings of the 3rd International Conference on Electronic Design (ICED), Phuket, Thailand, 11–12 August 2016; pp. 563–567.
34. Tansettanakorn, C.; Thongprasit, S.; Thamkongka, S.; Visoottiviseth, V. ABIS: A prototype of Android Botnet Identification System. In Proceedings of the Fifth ICT International Student Project Conference (ICT-ISPC), Nakhonpathom, Thailand, 27–28 May 2016; pp. 1–5.
35. Yusof, M.; Saudi, M.M.; Ridzuan, F. A new mobile botnet classification based on permission and API calls. In Proceedings of the Seventh International Conference on Emerging Security Technologies (EST), Canterbury, UK, 6–8 September 2017; pp. 122–127.
36. Yusof, M.; Saudi, M.M.; Ridzuan, F. Mobile Botnet Classification by using Hybrid Analysis. *Int. J. Eng. Technol.* **2018**, *7*, 103–108. [CrossRef]
37. Hijawi, W.; Alqatawna, J.; Faris, H. Toward a Detection Framework for Android Botnet. In Proceedings of the International Conference on New Trends in Computing Sciences (ICTCS), Amman, Jordan, 11–13 October 2017; pp. 197–202.
38. Abdullah, Z.; Saudi, M.M.; Anuar, N.B. ABC: Android Botnet Classification Using Feature Selection and Classification Algorithms. *Adv. Sci. Lett.* **2017**, *23*, 4717–4720. [CrossRef]
39. Karim, A.; Salleh, R.; Shah, S.A.A. DeDroid: A Mobile Botnet Detection Approach Based on Static Analysis. In Proceedings of the 7th International Symposium on UbiCom Frontiers—Innovative Research, Systems and Technologies, Beijing, China, 10–14 August 2015; pp. 1327–1332.
40. Yerima, S.Y.; Alzaylaee, M.K. Mobile Botnet Detection: A Deep Learning Approach Using Convolutional Neural Networks. In Proceedings of the 2020 International Conference on Cyber Situational Awareness (Cyber SA 2020), Dublin, Ireland, 15–19 June 2020.
41. Yerima, S.Y.; Bashar, A. Bot-IMG: A framework for image-based detection of Android botnets using machine learning. In Proceedings of the 18th ACS/IEEE International Conference on Computer systems and Applications (AICCSA 2021), Tangier, Morocco, 3–30 November 2021; pp. 1–7.
42. Hojjatinia, S.; Hamzenejadi, S.; Mohseni, H. Android Botnet Detection using Convolutional Neural Networks. In Proceedings of the 28th Iranian Conference on Electrical Engineering (ICEE), Tabriz, Iran, 4–6 August 2020; pp. 1–6.
43. Dalal, N.; Triggs, B. Histograms of oriented gradients for human detection. In Proceedings of the IEEE Computer Society Conference on Computer Vision and Pattern Recognition (CVPR'05), San Diego, CA, USA, 20–26 June 2005; pp. 886–893.
44. Ebrahimzadeh, R.; Jampour, M. Article: Efficient Handwritten Digit Recognition based on Histogram of Oriented Gradients and SVM. *Int. J. Comput. Appl.* **2014**, *104*, 10–13.
45. Anu, K.A.; Akbar, N.A. Recognition of Facial Expressions Based on Detection of Facial Components and HOG Characteristics. In *Intelligent Manufacturing and Energy Sustainability*; Springer: Berlin/Heidelberg, Germany, 2021.

46. Arief, S.S.; Samratul, F.; Arumjeni, M.; Sari, Y.W. HOG Based Pedestrian Detection System for Autonomous Vehicle Operated in Limited Area. In Proceedings of the International Conference on Radar, Antenna, Microwave, Electronics, and Telecommunications (ICRAMET), Bandung, Indonesia, 23–24 November 2021.
47. Bahassine, S.; Madani, A.; Al-Sarem, M.; Kissi, M. Feature selection using an improved Chi-square for Arabic text classification. *J. King Saud Univ.-Comput.* **2020**, *32*, 225–231. [CrossRef]
48. ISCX. ISCX Android Botnet Dataset. Available online: https://www.unb.ca/cic/datasets/android-botnet.html (accessed on 28 December 2021).
49. Weiss, S. Small sample error rate estimation for k-NN classifiers. *IEEE T. Pattern. Anal.* **1991**, *13*, 285–289. [CrossRef]
50. Pontil, M.; Verri, A. Support vector machines for 3D object recognition. *IEEE Trans. Pattern. Anal.* **1998**, *20*, 637–646. [CrossRef]
51. Kruegel, C.; Toth, T. Using Decision Trees to Improve Signature-Based Intrusion Detection. In *Recent Advances in Intrusion Detection*; Springer: Berlin/Heidelberg, Germany, 2003; pp. 173–191.
52. Zhang, J.; Zulkernine, M.; Haque, A. Random-Forests-Based Network Intrusion Detection Systems. *IEEE Trans. Syst. Man. Cybern. Part C* **2008**, *38*, 649–659. [CrossRef]
53. Alsariera, Y.A.; Adeyemo, V.E.; Balogun, A.O.; Alazzawi, A.K. AI Meta-Learners and Extra-Trees Algorithm for the Detection of Phishing Websites. *IEEE Access* **2020**, *8*, 142532–142542. [CrossRef]
54. Podlodowski, L.; Kozłowski, M. Application of XGBoost to the cyber-security problem of detecting suspicious network traffic events. In Proceedings of the IEEE International Conference on Big Data (Big Data), Los Angeles, CA, USA, 9–12 December 2019; pp. 5902–5907.
55. Yerima, S.Y.; Khan, S. Longitudinal Performance Analysis of Machine Learning based Android Malware Detectors. In Proceedings of the 2019 International Conference on Cyber Security and Protection of Digital Services (Cyber Security), Oxford, UK, 3–4 June 2019.

Article

An Efficient Method for Generating Adversarial Malware Samples

Yuxin Ding *, Miaomiao Shao, Cai Nie and Kunyang Fu

Department of Computer Science and Technology, Harbin Institute of Technology (Shenzhen), Shenzhen 518000, China; 21B951007@stu.hit.edu.cn (M.S.); 19S051001@stu.hit.edu.cn (C.N.); 19S151080@stu.hit.edu.cn (K.F.)
* Correspondence: yxding@hit.edu.cn; Tel.: +86-755-2603-2193

Abstract: Deep learning methods have been applied to malware detection. However, deep learning algorithms are not safe, which can easily be fooled by adversarial samples. In this paper, we study how to generate malware adversarial samples using deep learning models. Gradient-based methods are usually used to generate adversarial samples. These methods generate adversarial samples case-by-case, which is very time-consuming to generate a large number of adversarial samples. To address this issue, we propose a novel method to generate adversarial malware samples. Different from gradient-based methods, we extract feature byte sequences from benign samples. Feature byte sequences represent the characteristics of benign samples and can affect classification decision. We directly inject feature byte sequences into malware samples to generate adversarial samples. Feature byte sequences can be shared to produce different adversarial samples, which can efficiently generate a large number of adversarial samples. We compare the proposed method with the randomly injecting and gradient-based methods. The experimental results show that the adversarial samples generated using our proposed method have a high successful rate.

Keywords: adversarial sample; malware detection; deep learning; convolutional neural network

Citation: Ding, Y.; Shao, M.; Nie, C.; Fu, K. An Efficient Method for Generating Adversarial Malware Samples. *Electronics* **2022**, *11*, 154. https://doi.org/10.3390/electronics11010154

Academic Editor: Suleiman Yerima

Received: 13 December 2021
Accepted: 1 January 2022
Published: 4 January 2022

Publisher's Note: MDPI stays neutral with regard to jurisdictional claims in published maps and institutional affiliations.

Copyright: © 2022 by the authors. Licensee MDPI, Basel, Switzerland. This article is an open access article distributed under the terms and conditions of the Creative Commons Attribution (CC BY) license (https://creativecommons.org/licenses/by/4.0/).

1. Introduction

Deep neural networks have been successfully applied in different fields, such as computer vision and natural language processing. Recently, deep neural networks have gained attention to improve the performance of malware detection [1–4]. Deep learning algorithms can automatically learn features from training data, so malware detectors can implement end-to-end training based on it. Most of the approaches directly use binary Windows portable executable (PE) files as input data for the malware detection model to distinguish malicious and benign samples. The experimental results show that deep learning-based malware detectors can achieve high detection accuracy.

Despite their successful application in different fields, deep learning methods are sensitive to small perturbations in input samples. Szegedy et al. [5] found that small changes on input samples can cause classification errors. These perturbed samples are called adversarial samples. In the field of malware, similar methods have been proposed to evade malware detectors [6–8]. These methods are usually optimized by computing the gradient of the objective function, with respect to each byte of a source malware binary. Gradient-based methods generate adversarial samples case-by-case. Each time they only translate a source malware sample into a corresponding adversarial malware sample. If the number of padding bytes needed to inject into a malware is large, the time cost for generating an adversarial sample is very high. Therefore, these methods are not suitable for generating a large number of adversarial samples.

In this paper, we propose an efficient deep learning-based method for generating malware adversarial examples. We firstly extracted the feature byte sequences from benign samples, according to their importance. The importance of a sequence for classification is evaluated by a feature weight calculation method. Feature byte sequences were then injected into malware samples to generate adversarial samples. Since benign sequences can be stored into a database and shared by different malware samples, our proposed method can generate adversarial samples more efficiently. We tried to use two different strategies, the end-of-file and the mid-file, to inject binary sequences into a PE file. The experimental results show that the adversarial samples generated using our proposed method have a high successful rate for attacking CNN-based malware detectors.

The rest of this paper is organized as follows. In Section 2, we introduce the related work. In Section 3, we propose the research motivation and the method for generating malware adversarial examples. The experiments and discussions are described in Sections 4 and 5, respectively. Finally, we give our conclusions in Section 6.

2. Related Work

Deep learning methods have been widely applied in many fields and achieved excellent experimental results. However, recent studies show that deep learning models are sensitive to small perturbations in the input data [6,9]. The data samples after adding perturbations are called adversarial samples. Adversarial samples may cause deep learning algorithms to make wrong decision. The methods for generating adversarial samples can be divided into two categories: black- and white-box algorithms. The white-box algorithms assume that attackers have detailed information about the structure and parameters of the deep learning model [5,10]. Such information can be exploited to calculate perturbations. For black-box algorithms [11,12], any information about deep learning models is unknown. The perturbations of adversarial samples are usually computed based on the gradients of the loss function, with respect to the input data and a target label.

Goodfellow et al. [9] made a point that adversarial samples are the result of the learning models being too linear, rather than too nonlinear, and proposed the fast gradient sign algorithm to generate adversarial examples (FGSM). They found that networks with hidden units which have unbounded active functions simply respond by making their hidden unit activations very large, so it is better to only change the original input. Papernot et al. [12] proposed the Jacobian matrix-based method (JSMA) to generate adversarial samples. JSMA constructs adversarial samples by computing forward derivatives of deep neural network. This model uses knowledge of the network architecture to create adversarial saliency maps. The saliency maps indicate which input features an adversary should perturb, in order to impact output result of classification. Xiao et al. [13] proposed an optimization framework for the attacker to find the near-optimal label flips that maximally reduces the classification performance. The framework simultaneously models the adversary's attempt and the defender's reaction in a loss minimization problem. Based on this framework, they developed an algorithm of attacking support vector machines (SVMs). Moosavi-Dezfooli et al. [11] proposed Deepfool, which is based on an iterative linearization of the classifier to generate minimal perturbations that are sufficient to change class labels. The experimental results show that Deepfool can generate smaller perturbations than that generated by FGSM.

Sometimes attackers cannot obtain the detail knowledge about the deep learning model. For example, only the network outputs on certain inputs can be observed. Under these cases, black-box algorithms are applied to adversarial samples generation. Black-box attack was firstly proposed by Papernot et al. [14]. They trained a substitute network to fit the unknown neural network, and then generated adversarial examples using the substitute neural network [12]. The substitute network is a simulator of the target network. Therefore, the success of the black-box attack depends on the transferability property to hold between the target and substitute network. Liu et al. [15] conducted an extensive study of the transferability over large models and a large-scale dataset. Their results prove that the transferability for non-targeted adversarial samples is prominent, even for large models and a large-scale dataset. They also presented novel, ensemble-based approaches to generate transferable adversarial samples.

In the malware detection field, different black- and white-box algorithms are also presented. Different from images, there are semantic dependencies between bytes in an executable, any modification to a byte value may cause the executable cannot be executed or loss its intrusive functionality. To avoid this problem, some methods [7,16] generate adversarial malware samples by appending specific bytes at the end of executables The input size of deep learning-based detector is fixed. If the size of an executable is bigger than the fixed size, it cannot be used to generate an adversarial sample. To solve this issue, padding bytes can be injected into the gaps between sections in a PE file [17].

Hu and Tan [18] used the generative adversarial network to generate adversarial samples. They constructed a substitute detector to fit the black-box malware detector. Then, the generative adversarial network is trained to minimize the probability that the generated adversarial samples are predicted as malware by the substitute detector. Al-Dujaili et al. [19] investigated the methods that reduce adversarial blind spots for DNN based detectors. They considered it a saddle-point optimization problem and used the inner maximize methods to improve the robustness of DNN. Hu and Tan [20] proposed a black-box algorithm to evade a RNN-based detector. They trained a substitute RNN to approximate the victim RNN, then used the generative RNN to output sequential adversarial samples. Chen et al. [21] proposed the adversarial crafting algorithm based on the Jacobian matrix to generate adversarial samples.

Bo an et al. [16] proposed a white-box algorithm for evading the deep learning-based detector MalConv [3]. The algorithm is a gradient-based method which aims to minimize the confidence associated to the malicious class. To preserve the intrusive functionality of an executable, they appended padding bytes at the end of each malware sample. Suciu et al. [7] also proposed a white-box algorithm to evade Malconv model. Based on FGSM, they proposed the one-shot FGSM append attack. The algorithm uses the gradient value of the classification loss, with respect to the target label to update the appended byte values.

Apart from the above-mentioned malware adversarial sample generation methods, there are some other methods. Kreuk et al. [22] proposed to generate adversarial examples by appending to the malware binary file a small section. Peng et al. [23] used a generative adversarial network to generate semantics aware adversarial malware samples, which can fool the detection algorithms. They trained a recurrent neural network BiLSTM based a substitute detector to fit the black-box malware detector. In [24], the authors proposed two white-box methods and one black-box method to attack the CNN-based malware detector MalConv [3]. Recently, Chen et al. [25] used the deep reinforcement learning to generate malware adversarial examples, which has high success rate. A comparison of typical methods for generating adversarial samples is given in Table A1 (see Appendix A).

3. Methodology for Generating Adversarial Malware Examples

3.1. Motivations

Different deep learning-based detectors have been proposed [3,20,26]. As one of the most popular algorithms in deep learning, convolutional neural network (CNN) is widely applied in these detectors. Since CNN can automatically learn features from training samples, these detectors directly use a binary executable file as input and classify it. In our work we focus on how to generate adversarial samples which can evade CNN-based malware detectors. The problem of generating adversarial malware samples can be formalized as follows.

An executable x is represented as a sequence of L binary bytes $x = (x_1, x_2, \cdots, x_L)$, where x_i is between 0 and 255 and L is the length of an executable. In our work we set $L = 2 \times 10^6$. If the length of an executable is less than 2×10^6, zeros are padded at the end of the file. The malware detector is denoted as $f_\theta(x) : x \to [0,1]$, where θ is the parameters of a detector, and $f_\theta(x)$ outputs the probability that x is malware. If $f_\theta(x) > 0.5$, x is classified as malware, otherwise x is classified as benign.

Given a malicious file which is correctly classified as malware, an adversarial sample generation method can inject carefully-selected bytes into an executable (while preserving its runtime functionality), so that the executable can be classified as benign.

Conventional methods use gradient-based algorithm to generate adversarial samples [7,16]. These approaches use the input gradient value to update the injected byte values. Gradient value is calculated by minimizing the classification loss function of a detector, with respect to the target label. The gradient-based algorithm is an iterative algorithm and only one byte value is computed per iteration. Therefore, the computation cost for generating an adversarial sample is high, which is not suitable for generating a large number of adversarial examples. The motivation of our research is to design a method which can generate adversarial samples efficiently.

3.2. Finding Data Area Important for Classification

To evade the detection of malware detectors, we need to inject padding bytes into a source malware binary to change its category. To avoid using gradient-based algorithms to calculate the values of injected padding bytes, the padding bytes we use are the byte sequences extracted from benign executables. If these byte sequences can represent the characteristics of benign executables, the probability that an adversarial malware sample can fool a detector will increase. Therefore, our main task is to extract byte sequences which can represent the characteristics of benign executables.

To evade the detection of malware detectors, we need to inject padding bytes into a source malware binary to change its category. To avoid using gradient-based algorithms to calculate the values of injected padding bytes, the padding bytes we use are the byte sequences extracted from benign executables. If these byte sequences can represent the characteristics of benign executables, the probability that an adversarial malware example can fool a detector will increase. Therefore, our main task is to extract byte sequences which can represent the characteristics of benign executables.

CNN-based detectors generate explicit feature maps for input samples. Figure 1 gives an example for CNN convolution operation. The input data is a sequence. When we apply convolution to the input data, we mix two buckets of information. The first bucket is the input data. The second bucket is the convolution kernel, a single matrix of floating-point numbers. The output of the kernel is the altered sequence which is often called a feature map. Usually there are multiple convolution kernels and each kernel outputs a feature map. Feature maps represent features of an input data at different level. Through analyzing feature maps, we can discover which features are more important for decision making, and the data corresponding to important features can be used to construct adversarial samples.

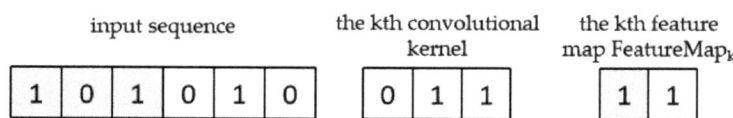

Figure 1. Convolution of a sequence with a convolution kernel.

Grad-CAM [27] algorithm provides explanations for decisions from a large class of CNN-based models. We use the Grad-CAM algorithm to evaluate the important values of each feature map for a target class c. The important value of a feature map, with respect to a specific class is computed as Equation (1). α_k^c indicates the importance of $FeatureMap_k$, with respect to class c.

$$\alpha_k^c = \frac{1}{Len\_FeatureMap_k} \sum_i \frac{\partial S^c}{\partial FeatureMap_k[i]} \qquad (1)$$

where $FeatureMap_k$ is the kth feature map, $FeatureMap_k[i]$ is the ith element of $FeatureMap_k$, $Len\_FeatureMap_k$ is the number of elements of $FeatureMap_k$, c is a class label, S^c is the input for class c in the softmax layer (classification layer in a CNN).

To discover the importance area of the input data for class c, the contributions of all feature maps need to be considered. The weighted sum of all feature maps is computed, which is defined as Equation (2). L^c is called the class-discriminative localization map, which has the same size as a feature map.

$$L^c = \text{ReLU}(\sum_k \alpha_k^c FeatureMap_k) \qquad (2)$$

In (2) the ReLU function ($\text{ReLU}(x) = \text{Max}(0, x)$) is applied to the linear combination of feature maps because only the features that have a positive impact on class c are considered. Without the ReLU function, the localization map sometimes highlights more than just the class of interest and performs worse at localization. Each element $L^c[i]$ can be seen as a feature extracted from the input data. The element $L^c[i]$, with a greater value, will also have more positive impact on class c. We can find the data area that is important for class c by mapping $L^c[i]$ back to the corresponding data area in the input.

3.3. Generating Adversarial Examples

In reality the structure and parameters of a malware detector are unknown. In order to obtain the feature maps, we have to create a pseudo detector, which can simulate the true detector. MalConv [3] is a typical CNN-based detector. In our work, we select MalConv network as the pseudo detector. The network structure of MalConv is shown in Figure 2.

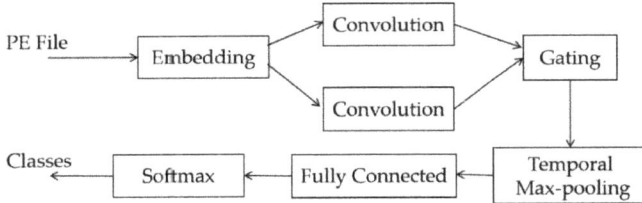

Figure 2. Structure of MalConv.

We regard an executable (PE file) as a byte stream. The input of MalConv is a fixed-length sequence from a PE file. If the length of an executable is shorter than the fixed-length, a number of zeros are inserted at the end of an executable. In MalConv, the first layer is an embedding layer, where each byte of an input sequence is converted into an 8-dimensional embedding vector. MalConv has two parallel convolutional layers. These embedding vectors are then transferred to two one-dimensional convolutional layers to generate feature maps, respectively. The next layer is a temporal max pooling layer, which combines the outputs of the two convolutional layers and passes them to a fully connected layer and a softmax layer for classification.

In our paper, we use Equation (1) to calculate the important value of each feature map, with respect to class c, denoted as $\alpha_{l,k}^c$, which is the important value of the kth feature map generated from the lth convolutional layer $FeatureMap_{l,k}$. MalConv has two parallel convolutional layers. We normalize $\alpha_{l,k}^c$ for each independent convolutional layer, respectively, which is shown as Equation (3).

$$w_{l,k}^c = \frac{\alpha_{l,k}^c}{\sum_k \alpha_{l,k}^c} \qquad (3)$$

The class-discriminative localization map is calculated as the weighted sum of the feature maps generated by the two parallel convolutional layers, which is shown as

Equation (4). Here, we set all convolution kernels to have the same size; thus, all feature maps, as well as the class-discriminative localization map, have the same size, which are one-dimensional vectors. Different CNN-based networks have different structures. Another key problem we should resolve is how to locate the byte sequences in a source binary file, according to the class-discriminative localization map.

$$L^c = \text{ReLU}(\sum_l \sum_k w^c_{l,k} FeatureMap_{l,k}) \qquad (4)$$

A MalConv model has two independent convolutional layers, and each convolution layer has multiple convolution kernels. To simplify data mapping, we set the kernel length equal to the kernel's moving stride, all kernels have the same length, and the length of the input data is 2×10^6 bytes. The mapping relationship between a feature map and an input data can be constructed as follows.

In [3], the authors tried different parameter settings to test the performance of MalConv. We followed [3] and set the length and the moving stride of a kernel as 500, and the kernel number of each convolutional layer as 128. Figure 3 shows the relationships between an input data and a features map. In Figure 3, each square in the first row represents an input byte, and each square in the second row represents the embedding vector of an input byte. Kernel1 is a one-dimensional convolution kernel of a convolutional layer, whose length is 500. Kernel1 is convolved across the embedding data, computing the dot product between the entries of the kernel and the embedding data and producing a one-dimensional feature map $FeatureMap_1$. If each convolutional layer has 128 kernels, we can obtain 128 one-dimensional feature maps from one convolutional layer. The embedding data has the same length as the input data. Therefore, each feature map has 4000 elements. In Figure 3, the fourth row shows the mapping relationship between an element of a feature map and a byte sequence in the input data. For example, the first element of $FeatureMap_1$, $FeatureMap_1[1]$, is calculated by convoluting Kernel1 with the first five hundred elements of the embedding vector, and each input byte corresponds to an element of the embedding vector. Therefore, $FeatureMap_1[1]$ is related with the first five hundred bytes of the input data. The class-discriminative localization map is the weighted sum of all feature maps, so it has the same mapping relationship as that of a features map.

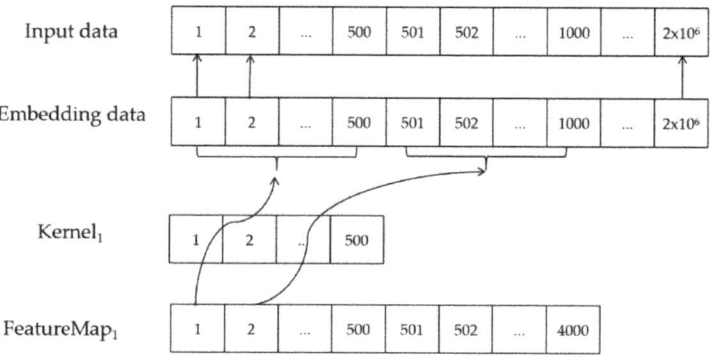

Figure 3. Mapping feature map back to raw data.

To generate adversarial examples, we firstly train a MalConv model as the pseudo detector. Then, we create a dataset for feature extraction. All samples in the dataset are benign samples and can be correctly classified as benign by a detector. We input a sample to the pseudo detector and obtain the class-discriminative localization map L^c of the sample. According to the mapping relationship between input data and the class-discriminative localization map, we can extract the byte sequences from the input data, which can represent the features of a sample. We usually extract the byte sequences corresponding to the elements having the greatest value in L^c. We call these byte sequences as feature byte sequences, which can be stored and shared by different adversarial samples. When generating an adversarial example, we randomly select one or multiple sequences and inject them into a malware sample.

Different from adversarial samples of image, feature byte sequences injected into a malware sample should have concrete program semantics. Sometimes the head and tail of a feature byte sequences are separated from other bytes of a program and cannot represent complete program semantics. In this case, we should extend a feature byte sequence to include the separate parts. For example, a feature byte sequence (bytes in the box), extracted according to the mapping relationship, is shown in Figure 4. The decompiling codes of the binary bytes are shown in Figure 5. We can see the head byte FF and the tail byte 45 cannot represent correct program semantics. To generate a feature byte sequence having correct program semantics, we should extend the feature byte sequences to include 8B and 08. From this point we can see the injected byte sequences, generated using our method, are explainable.

8B|FF 55 8B EC ... 83 EC 20 8B 45|08

Figure 4. A sample of a feature byte sequence.

8B FF	mov edi, edi
55	push ebp
8BEC	mov ebp, esp
......	
83EC20	sub esp, 00000020h
8B4508	mov eax, [ebp+08h]

Figure 5. Decompiling codes of a binary byte sequence.

To more accurately locate the important area in the input data, we train several MalConv models with different parameter settings and combine the class-discriminative localization map from all MalConv models to locate the important area of the input data.

Algorithm 1 gives the algorithm for extracting feature byte sequences from input data using multiple detection models. The length of convolution kernels in different MalConv models can be different. For the convenience of extracting feature byte sequences, we define a new data structure $byteWeightMap$. It is a vector having the same length as the input data. Each element in $byteWeightMap$ records the important value of the corresponding byte of the input data. The important values of input bytes are assigned according to L^c. According to the mapping relationship, we can find the byte sequence corresponding to L_i^c (the ith element of L^c); then, the values of the elements of $byteWeightMap$ corresponding to the byte sequence are set as L_i^c. The function $SetByteWeight()$ implements this objective. Due to multiple models used to locate feature byte sequences, we use L_i^c and $byteWeightMap_i$ represent the class-discriminative localization map and $byteWeightMap$, generated from model M_i (the ith detector). The vector $fByteWeightMap$ is the sum of all $byteWeightMap_i$, which stores the final important value of each byte of the input data.

In Algorithm 1, $ModelNum$ is the number of models, and $thresh$ gives the threshold of important value for selecting feature sequences. x_{benign} is the input data. The function $GetFeatureMap()$ returns all the feature maps generated by model M_i. $FeatureMap_{j,k}[n]$ is the nth element of the kth feature map generated by the jth convolutional layer of a MalConv model. The function $ExtFeaSeq()$ extracts all bytes whose important values are bigger than $thresh$ from x_{benign}, according to vector $fByteWeightMap$. The continuous bytes, having the same important value, consist of a feature byte sequence. Figure 6 shows a sample how to extract feature byte sequences from input data. We set *thresh* as 50; therefore, only two feature byte sequences (sequences in black box) are extracted from input data.

...... 8B FF 55 8B EC 83 EC 20 8B 45 08 56 57 65 08 59 ......

input data

...... 51 51 51 51 51 48 48 48 48 72 72 72 72 72 72 72 ......

fByteWeightMap

Figure 6. Extracting feature byte sequences.

Algorithm 1: Extracting feature byte sequences of a benign sample.

Input: x_{benign}, $c = benign$, $M_1, \cdots, M_{ModelNum}$, $ModelNum$, $thresh$
Output: $featureByteSequenceArray[]$

$fByteWeightMap = \vec{0}$;
for $i = 1$ to $ModelNum$ **do**
$\quad FeatureMapArray = GetFeatureMap(M_i, x_{benign})$;
$WeightVector = \vec{0}$;
for each $FeatureMap_{j,k}$ in $FeatureMapArray$ **do**
$\quad \alpha_{j,k}^c = \frac{1}{Len\_FeatureMap_{j,k}} \sum_n \frac{\partial S^c}{\partial FeatureMap_{j,k}[n]}$;
end
$\quad$ **for** each $\alpha_{j,k}^c$ **do**
$\quad\quad w_{j,k}^c = \text{Normalize}(\alpha_{j,k}^c)$;
$\quad\quad WeightVector = WeightVector + w_{j,k}^c FeatureMap_{j,k}$;
$\quad$ **end**
$\quad L_i^c = \text{ReLU}(WeightVector)$;
$\quad byteWeightMap_i = SetByteWeight(L_i^c)$;
$\quad fByteWeightMap = fByteWeightMap + byteWeightMap_i$;
end
$featureByteSequence[] = ExtFeaSeq(fByteWeightMap, thresh, x_{benign})$

3.4. Strategies for Injecting Feature Sequences

A malware adversarial sample should preserve the same semantics as that of a source file. It requires that any byte in the source executable cannot be changed. Therefore, feature sequences should be injected into the spare space of an executable, which cannot be executed by a computer. Two strategies can be adopted to locate spare space in an executable: mid-file and end-of-file injection. We apply both strategies to generate adversarial samples in our work.

Mid-file injection: we locate the gaps between neighboring PE sections by parsing a PE file header. The gaps are placed by the compiler, since the physical size allocated to a PE section is greater than its virtual size. The length of a gap is calculated as RawSize-VirtualSize. The index of the start address of a gap is computed as PointerToRawData (offset address of a section) + VirtualSize. We collect the start address and length of each gap in an executable, then inject the feature byte sequences with appropriate length into these gaps.

End-of-file injection: another strategy we use is adding new sections at the end of a PE file and injecting feature byte sequences into the newly added sections. Since the new sections are not accessed by program code, the semantics of the original PE file are preserved. The process of adding a new section block includes three steps. First, we modify the value of bytes, which store the number and size of sections in the PE file header and update the values of file alignment and section alignment. Then, we use the offset address of the last section block plus the offset address of the new block as the final offset address. Next, we set the attribute values of the new section, such as the section name, execution attributes, size of the hard disk, and size of the memory. Finally, we modify the offset address of the aligned section and the offset address of the file in the section table and modify the size of image in the PE header.

Similar to [17], our method adopting the mid-file injection generates adversarial samples by injecting perturbed bytes in the gaps between neighboring PE sections. The method adopting end-of-file injection generates malware adversarial examples by adding new sections at the end of PE file, which is similar to previous methods [7,16,22,24]. However, all these methods [7,16,17,22,24] are belong to gradient-based method, which is optimized by computing the gradient of the objective function, with respect to each byte of a source malware binary. The gradient-based algorithm is an iterative algorithm and only one byte value is computed per iteration. Generating an adversarial malware sample by gradient-based method spends much time, so it is not applicable for generating a large number of adversarial samples. To avoid using gradient-based algorithms to calculate the values of injected padding bytes, our methods use the byte sequences extracted from benign executables to generate adversarial samples. In addition, our methods aim to evade CNN-based malware detectors, which is similar to [23]. We make a more detailed comparison between our method and the gradient-based method [16] in Sections 4 and 5.

4. Experiments

4.1. Dataset Description

The malware samples we used came from the VirusShare project at http://virusshare.com/ (accessed on 1 December 2021). We downloaded 20,000 malicious samples, whose sizes were between 1 KB and 5 MB. The benign samples were collected from Windows platforms. We collected 20,000 benign Windows PE files in total. Two criteria were used to assess the quality of adversarial samples. The successful rate (SR) of the adversarial attack is defined as the percentage of the adversarial samples that can evade a detector. Another is the time cost for generating adversarial samples, which is used to evaluate the efficiency of the proposed algorithm. The experimental environment was 64-bit Ubuntu14 operating system, CPU Intel® Xeon Silver 4116 with 256 G memory.

4.2. Experimental Results

In the experiments, we trained four MalConv detectors. The description of parameter setting, training data, and detection accuracy is shown in Table 1. In Table 1, the column "Kernel Number" gives the kernel number for each convolutional layer. The training samples included fifty percent benign files and fifty percent malicious files, i.e., 5000 benign files and 5000 malicious files. The accuracy is defined as the percentage of the testing samples that can be correctly classified.

Table 1. Parameter setting for detectors.

Detector	Kernel Length	Moving Stride	Kernel Number	Training Samples	Accuracy
MalConv1	200	200	200	20,000	92.5%
MalConv2	400	400	150	20,000	92.6%
MalConv3	500	500	128	20,000	94.1%
MalConv4	800	800	100	20,000	91.8%

To objectively evaluate the successful rate that adversarial examples evade detection, in each experiment, we selected one MalConv model as the detector and used the remaining models to generate feature byte sequences. We repeated the experiments four times and used the average successful rate of four experiments to evaluate the performance of the proposed method. For each experiment, we randomly chose 100 benign samples from the testing set and use Algorithm 1 to extract the features sequences from benign samples.

Only the sequences with the highest important value in each sample were selected. We got about two thousand feature sequences per experiment. We randomly selected 1000 samples that were correctly classified as malware from the testing set and injected feature sequences into them, in order to generate adversarial samples.

To observe how the number of injected bytes affects the performance of the proposed method, we injected different numbers of bytes into a sample. The number of the injected bytes was set to 1000, 2000, 5000, 10,000, and 20,000, respectively. In our work, two injection strategies were applied to inject feature byte sequences.

The experimental results, adopting the mid-file and the end-of-file strategies, are shown in Tables 2 and 3, respectively. In two tables, "Avg Time Cost Per Sample" means the time cost for generating an adversarial sample.

Table 2. SR of the proposed method adopting the mid-file strategy.

No. of Injected Bytes	1000	2000	5000	10,000
SR of Experiment 1	0.42	0.55	0.78	0.88
SR of Experiment 2	0.46	0.57	0.77	0.86
SR of Experiment 3	0.45	0.59	0.78	0.90
SR of Experiment 4	0.41	0.61	0.76	0.89
Average SR	0.44	0.58	0.77	0.88
Avg Time Cost Per Sample(min)	0.2	0.5	1.1	2.1

Table 3. SR of the proposed method adopting the end-of-file strategy.

No. of Injected Bytes	1000	2000	5000	10,000	20,000
SR of Experiment 1	0.34	0.41	0.60	0.77	0.88
SR of Experiment 2	0.37	0.44	0.61	0.73	0.90
SR of Experiment 3	0.32	0.40	0.64	0.74	0.91
SR of Experiment 4	0.33	0.43	0.63	0.76	0.87
Average SR	0.34	0.42	0.62	0.75	0.89
Avg Time Cost Per Sample(min)	0.2	0.2	0.4	0.9	1.9

To verify whether the feature sequences can represent the characteristics of benign executables, we compared the proposed method with the randomly injecting method. The randomly injecting method randomly extracts byte sequences from benign executables and injects them into malware to generate adversarial samples. For the randomly injecting methods, we also used two different strategies to inject randomly extracted sequences. The experimental results are shown in Tables 4 and 5, respectively.

Table 4. SR of the randomly injecting method adopting mid-file strategy.

No. of Injected Bytes	1000	2000	5000	10,000
SR of Experiment 1	0.09	0.11	0.18	0.21
SR of Experiment 2	0.09	0.11	0.17	0.20
SR of Experiment 3	0.06	0.13	0.15	0.24
SR of Experiment 4	0.08	0.13	0.14	0.23
Average SR	0.08	0.12	0.16	0.22

From Tables 2–5, we can see that the successful rate of the proposed method was significantly higher than that of the randomly injecting method, which was about 30–60% higher than that of the corresponding randomly injecting method. It proves that the feature sequences injected into adversarial samples can reflect the characteristics of benign executables, which can influence the decision of the detectors. The injected sequences were extracted from benign executables. If more benign sequences were injected in a malware sample, a malware sample will be more similar as a benign sample. Therefore, we can see, for both methods, that the success rate increased with the length of the injected bytes increasing.

Table 5. SR of the randomly injecting method adopting end-of-file strategy.

No. of Injected Bytes	1000	2000	5000	10,000	20,000
SR of Experiment 1	0.03	0.07	0.07	0.10	0.13
SR of Experiment 2	0.05	0.05	0.09	0.12	0.15
SR of Experiment 3	0.03	0.06	0.09	0.11	0.15
SR of Experiment 4	0.05	0.06	0.07	0.11	0.19
Average SR	0.04	0.06	0.08	0.11	0.17

For the end-of-file strategy, all malicious features in malware samples are preserved and not modified. Compared with the end-of-file strategy, the mid-file strategy injects feature sequences into the gaps between sections, which destroys some malicious features of malware samples. To mislead the detector, the end-file strategy needs to inject more feature byte sequences to counteract the effects of the original malicious features. Therefore, from Tables 2–5 we can see when injecting the same number of benign bytes into malware samples, the successful rate of the method adopting the mid-file strategy is higher than that adopting the end-of-file strategy. For the proposed method, the successful rate adopting the mid-file strategy is about 3–23% higher than that adopting the end-of-file strategy. For the randomly injecting method, the successful rate adopting the mid-file strategy is about 4–11% higher than that adopting the end-of-file strategy.

We also compare the proposed method with the gradient-based method [16]. The end-of-file strategy is adopted to inject feature sequences. For the gradient-based method, the gradient is calculated by minimizing the classification loss of the detector, with respect to the target label. In the experiment we select two different classification loss functions to calculate the gradient. One is the softmax classification loss (see Equation (5)), which is used to train MalConv. The other is the mean-square error (see Equation (6)), which is often used to train conventional back propagation (BP) networks.

$$L_{softmax}(\theta) = -\frac{1}{m}\left[\sum_{i=1}^{m}\sum_{j=1}^{k} 1\{y^{(i)} = j\} log \frac{e^{\theta_j^T x^{(i)}}}{\sum_{l=1}^{k} e^{\theta_l^T x^{(i)}}}\right] \quad (5)$$

$$L_{ms}(\theta) = \frac{1}{2m}\sum_{i=1}^{m}(\hat{y}_i - y_i)^2 \quad (6)$$

Due to the limitation of computing cost, only 200 adversarial samples are generated for each experiment and the maximum number of the injected bytes is less than 10,000. The experimental results adopting two different classification loss functions are shown in Table 6. We can see the successful rate of the proposed method adopting the end-of-file strategy is about 6–10 percent higher than that of the gradient-based method adopting softmax classification loss. The successful rate of the gradient-based method adopting softmax classification loss is about 5–17 percent higher than the method adopting mean squared error loss.

Table 6. SR of the gradient-based method [16].

Byte seq. len.	Softmax Classification Loss				Mean Squared Error			
	1000	2000	5000	10,000	1000	2000	5000	10,000
Experiment 1	0.23	0.33	0.52	0.70	0.15	0.25	0.36	0.51
Experiment 2	0.26	0.39	0.55	0.66	0.21	0.27	0.40	0.49
Experiment 3	0.25	0.30	0.56	0.69	0.19	0.29	0.42	0.52
Experiment 4	0.22	0.32	0.53	0.71	0.22	0.30	0.41	0.54
Average SR	0.24	0.31	0.54	0.69	0.19	0.28	0.40	0.52
Avg Time Cost Per Sample(min)	25	51	99	239	23	47	100	240

5. Discussion

From the experiments we can see the gradient-based algorithm takes a relatively long time to generate an adversarial sample. In our work, 200 adversarial samples are generated for each experiment. The gradient-based method takes an average about 100 min to generate an adversarial sample (See Table 6), because it only generates one appended byte per iteration. In addition, it is hard to determine the iteration number when appended bytes converge to their optimal values. If we use the gradient-based algorithm to generate a large amount of adversarial samples, the time cost is very high. For the proposed method, most time is spent on training a CNN-based detector. In the experiments, we spent about 10 h training a MalConv model. The time for extracting feature sequences is about one hour. Injecting feature sequences into a PE file can be done in a very short time (the average time in our experiment is about one minute, see Tables 2 and 3). Because the feature sequences can be shared by all adversarial samples, the proposed method is suitable for generating a large number of adversarial samples.

Interpretability is another challenge faced by adversarial sample generation algorithms. The gradient-based methods calculate the value of injected bytes by minimizing the classification loss of a detector, with respect to the target label. These injected bytes have no explainable semantics and are only treated as binary values. Different from the gradient-based methods, the proposed method injects feature byte sequences into malware. A feature sequence is a byte sequence extracted from a benign executable. By decompiling the executable, the semantics of a feature byte sequences can be clearly defined. Therefore, using the proposed method we can explain the meaning of the injected bytes.

In our study, the proposed method is only designed to generate the adversarial samples for CNN-based detectors. The feature byte sequences are selected based on the convolution operation of CNN. This means that we need to know in advance which algorithms a detector uses. Compared with our proposed method, the gradient-based methods are more commonly used methods, which do not assume the classification methods a detector uses. So, they can be more widely used to generated adversarial samples for different neural networks, such as BP network, CNN [16], and RNN [20].

Generating malware adversarial samples is different from generating image adversarial samples. For image adversarial samples, we can directly update each pixel. For malware adversarial samples, we cannot modify any byte of a source executable, otherwise we cannot guarantee that it can be executed correctly. Therefore, we have to inject padding bytes into the gaps or the end of a PE file. The number of gaps and the length of each gap in a PE file are limited. Using the mid-file strategy, sometimes we cannot find enough gaps to store feature byte sequences in an executable, which may reduce the successful rate. For the end-of-file strategy, we can append any number of section blocks at the end of a PE file by modifying the PE file structure. Therefore, it is relatively easy for the end-file strategy to inject enough bytes to generate an adversarial sample. However, adversarial samples generated using the end-of-file strategy are prone to be detected by simply analyzing the PE section table or examining if such sections are accessed by program instructions. In addition, if the length of a malware sample is greater than the input length of a detector, and the end-of file strategy cannot be applied.

6. Conclusions

In this paper we study how to generate malware adversarial samples. Different from previous gradient-based methods, we generate malware adversarial examples by injecting byte sequences into a source executable. The injected byte sequences can be shared by different adversarial samples. Our proposed method is efficient and suitable for generating a large number of adversarial samples. We proposed the algorithm to extract feature byte sequences for CNN-based deep learning models. Feature byte sequences can represent the characteristics of benign samples. Compared with the padding bytes generated using gradient-based methods, the feature byte sequences are explainable. The experimental results show that the adversarial samples, generated using the proposed method, have a high successful rate, and the proposed method is suitable for generating a large number of adversarial samples. It is possible that a more robust malware detector can be trained using the generated adversarial samples and the original samples. In this work, we have not yet provided definitive evidence for the benefits of the generated adversarial samples in improving performance of malware detection, due to the complexity of adversarial training malware detectors. In our future work, we plan to investigate how to use the generated adversarial malware samples to improve the performance of malware detection models.

Author Contributions: Conceptualization, Y.D.; formal analysis, Y.D., M.S., C.N. and K.F.; methodology, Y.D and M.S.; software, C.N. and K.F.; validation, Y.D., M.S. and C.N. All authors have read and agreed to the published version of the manuscript.

Funding: This research was partially supported by the National Natural Science Foundation of China (Grant No. 61872107), Scientific Research Foundation of Shenzhen (Grant No. JCYJ20180507183608979).

Data Availability Statement: The data is available from http://virusshare.com/ (accessed on 1 December 2021).

Conflicts of Interest: The authors declare no conflict of interest.

Appendix A

Table A1. Typical adversarial samples generation methods.

Approach	Prior Knowledge	Descriptions & Advantages	Disadvantages
		Generating image adversarial samples	
Szegedy et al. [5]	White-box	Distortion rate of the generated adversarial sample is low.	Calculation process is complex and time-consuming.
Goodfellow et al. [9]	White-box	It can generate a large number of adversarial samples effectively and can be used in deep learning models.	Ease of optimization has come at the cost of models that are easily misled.
Moosavi-Dezfooli et al. [11]	Black-box	The modification to the original input is small, and the generated adversarial sample has good attack effect.	Calculation process is complex and time-consuming, and it is difficult to apply to large datasets.
Papernot et al. [12]	White-box	The original input is less modified and the process of generating adversarial samples is simple.	The method needs to be trained with large, labeled datasets.
Xiao et al. [13]	White-box	An optimization framework for the adversary to find the near-optimal label flips that maximally degrades the classifier's performance.	It can only be suitable for Support Vector Machines. Adversarial label noise is inevitable due to the limitation of quality control mechanisms.
Papernot et al. [14]	Black-box	An approach based on a novel substitute training algorithm using synthetic data generation to craft adversarial examples misclassified by black-box DNNs.	Construction process of the approach is complex and time-consuming. So, it is difficult to apply to large datasets.
Liu et al. [15]	Black-box	An ensemble-based approach can generate transferable adversarial examples which can successfully attack Clarifai.com.	Performance of generating targeted transferable adversarial examples of the model is poor, compared to other previous models.
		Generating malware adversarial samples	
Suciu et al. [7]	White-box	The one-shot FGSM append attack uses the gradient value of the classification loss, with respect to the target label to update the appended byte values.	The success rate of append attacks is relatively low.
Kolosnjaji et al. [16]	White-box	Adversarial malware samples are generated by injecting padding bytes at the end of file, which can preserve the intrusive functionality of an executable.	Applicable for the deep learning-based detector MalConv.
Kreuk et al. [17]	White-box	The same payload can be injected into different locations and can be effective when applied to different malware files.	Applicable for CNN-based malware detector.

Table A1. Cont.

Approach	Prior Knowledge	Descriptions & Advantages	Disadvantages
Hu et al. [18]	Black-box	An approach can decrease the detection rate to nearly zero and make the retraining based defensive method against adversarial examples hard to work.	Suitable for machine learning-based malware detector.
Hu et al. [20]	Black-box	The generated adversarial examples can attack a RNN-based malware detector.	Not applicable for attacking other systems except RNN-based malware detectors.
Chen et al. [21]	White-box	A method based on Jacobian matrix to generate adversarial samples.	It is not applicable for generating a large number of samples.
Kreuk et al. [22]	White-box	The method generates adversarial examples by appending to the binary file a small section and has high attack success rates.	The method heavily relies on the learned embeddings of the model, which can hinder the transferability of adversarial examples with different byte embeddings.
Peng et al. [23]	Black-box	It outruns other GAN based schemes in performance and has a lower overhead of API call inserting.	The generation process is complex and time-consuming, and it is applicable for CNN-based detectors.
Chen et al. [24]	White-box, Black-box	Attack success rate of the method is high, and it can be readily extended to other similar adversarial machine learning tasks.	Not applicable for generating a large number of samples.
Chen et al. [25]	Black-box	It uses reinforcement learning to generate malware adversarial samples which has high success rate of attack.	Not applicable for generating a large number of samples.

References

1. Alzaylaee, M.K.; Yerima, S.Y.; Sezer, S. DL-Droid: Deep learning based android malware detection using real devices. *Comput. Secur.* **2020**, *89*, 101663. [CrossRef]
2. Gibert, D.; Mateu, C.; Planes, J. HYDRA: A multimodal deep learning framework for malware classification. *Comput. Secur.* **2020**, *95*, 101873. [CrossRef]
3. Raff, E.; Barker, J.; Sylvester, J.; Brandon, R.; Catanzaro, B.; Nicholas, C.K. Malware detection by eating a whole exe. In Proceedings of the Workshops at the Thirty-Second AAAI Conference on Artificial Intelligence, New Orleans, LA, USA, 2–7 February 2018; pp. 268–276.
4. Wang, W.; Zhao, M.; Wang, J. Effective android malware detection with a hybrid model based on deep autoencoder and convolutional neural network. *J. Ambient Intell. Humaniz. Comput.* **2019**, *10*, 3035–3043. [CrossRef]
5. Szegedy, C.; Zaremba, W.; Sutskever, I.; Bruna, J.; Erhan, D.; Goodfellow, I.; Fergus, R. Intriguing properties of neural networks. *arXiv* **2013**, arXiv:1312.6199.
6. Biggio, B.; Nelson, B.; Laskov, P. Support Vector Machines Under Adversarial Label Noise. *J. Mach. Learn. Res.* **2011**, *20*, 97–112.
7. Suciu, O.; Coull, S.E.; Johns, J. Exploring adversarial examples in malware detection. In Proceedings of the 2019 IEEE Security and Privacy Workshops (SPW), San Francisco, CA, USA, 20–22 May 2019; pp. 8–14.
8. Maiorca, D.; Demontis, A.; Biggio, B.; Roli, F.; Giacinto, G. Adversarial detection of flash malware: Limitations and open issues. *Comput. Secur.* **2020**, *96*, 101901. [CrossRef]
9. Goodfellow, I.J.; Shlens, J.; Szegedy, C. Explaining and Harnessing Adversarial Examples. *arXiv* **2014**, arXiv:1412.6572.
10. Goodfellow, I. New CleverHans Feature: Better Adversarial Robustness Evaluations with Attack Bundling. *arXiv* **2018**, arXiv:1811.03685.
11. Moosavi-Dezfooli, S.M.; Fawzi, A.; Frossard, P. DeepFool: A simple and accurate method to fool deep neural networks. In Proceedings of the 2016 IEEE Conference on Computer Vision and Pattern Recognition (CVPR), Las Vegas, NV, USA, 27–30 June 2016; pp. 2574–2582.
12. Papernot, N.; McDaniel, P.; Jha, S.; Fredrikson, M.; Celik, Z.B.; Swami, A. The limitations of deep learning in adversarial settings. In Proceedings of the 2016 IEEE European Symposium on Security and Privacy (EuroS&P), Saarbrücken, Germany, 21–24 March 2016; pp. 372–387.
13. Xiao, H.; Xiao, H.; Eckert, C. Adversarial label flips attack on support vector machines. In Proceedings of the 20th European Conference on Artificial Intelligence (ECAI 2012), Montpellier, France, 27–31 August 2012; pp. 870–875.
14. Papernot, N.; McDaniel, P.; Goodfellow, I.; Jha, S.; Celik, Z.B.; Swami, A. Practical black-box attacks against machine learning. In Proceedings of the 2017 ACM on Asia Conference on Computer and Communications Security, Abu Dhabi, United Arab Emirates, 2–6 April 2017; pp. 506–519.
15. Liu, Y.; Chen, X.; Liu, C.; Song, D. Delving into transferable adversarial examples and black-box attacks. *arXiv* **2016**, arXiv:1611.02770.
16. Kolosnjaji, B.; Demontis, A.; Biggio, B.; Maiorca, D.; Giacinto, G.; Eckert, C.; Roli, F. Adversarial malware binaries: Evading deep learning for malware detection in executables. In Proceedings of the 2018 26th European Signal Processing Conference (EUSIPCO), Rome, Italy, 3–7 September 2018; pp. 533–537.
17. Kreuk, F.; Barak, A.; Aviv-Reuven, S.; Baruch, M.; Pinkas, B.; Keshet, J. Deceiving end-to-end deep learning malware detectors using adversarial examples. *arXiv* **2018**, arXiv:1802.04528v3. Available online: http://arxiv.org/abs/1802.04528v3 (accessed on 1 December 2021).
18. Hu, W.; Tan, Y. Generating adversarial malware examples for black-box attacks based on GAN. *arXiv* **2017**, arXiv:1702.05983.
19. Al-Dujaili, A.; Huang, A.; Hemberg, E.; O'Reilly, U.M. Adversarial deep learning for robust detection of binary encoded malware. In Proceedings of the 2018 IEEE Security and Privacy Workshops (SPW), San Francisco, CA, USA, 21–23 May 2018; pp. 76–82.
20. Hu, W.; Tan, Y. Black-box attacks against RNN based malware detection algorithms. In Proceedings of the Workshops at the Thirty-Second AAAI Conference on Artificial Intelligence, New Orleans, LA, USA, 2–7 February 2018; pp. 245–251.
21. Chen, S.; Xue, M.; Fan, L.; Hao, S.; Xu, L.; Zhu, H.; Li, B. Automated poisoning attacks and defenses in malware detection systems: An adversarial machine learning approach. *Comput. Secur.* **2018**, *73*, 326–344. [CrossRef]
22. Kreuk, F.; Barak, A.; Aviv-Reuven, S.; Baruch, M.; Keshet, J. Adversarial examples on discrete sequences for beating whole-binary malware detection. *arXiv* **2018**, arXiv:1802.04528v1. Available online: http://arxiv.org/abs/1802.04528v1 (accessed on 1 December 2021).
23. Peng, X.; Xian, H.; Lu, Q.; Lu, X. Semantics aware adversarial malware examples generation for black-box attacks. *Appl. Soft. Comput.* **2021**, *109*, 107506. [CrossRef]
24. Chen, B.; Ren, Z.; Yu, C.; Hussain, I. Adversarial examples for cnn-based malware detectors. *IEEE Access* **2019**, *7*, 54360–54371. [CrossRef]
25. Chen, J.; Jiang, J.; Li, R.; Dou, Y. Generating adversarial examples for static PE malware detector based on deep reinforcement learning. In Proceedings of the 5th Annual International Conference on Information System and Artificial Intelligence (ISAI2020), Hangzhou, China, 22–23 May 2020.

26. Krčál, M.; Švec, O.; Bálek, M.; Jašek, O. Deep convolutional malware classifiers can learn from raw executables and labels only. In Proceedings of the 6th International Conference on Learning Representation (ICLR 2018), Vancouver, BC, Canada, 30 April–3 May 2018.
27. Selvaraju, R.R.; Das, A.; Vedantam, R.; Cogswell, M.; Parikh, D.; Batra, D. Grad-CAM: Why did you say that? Visual Explanations from Deep Networks via Gradient-based Localization. In Proceedings of the IEEE International Conference on Computer Vision, Venice, Italy, 22–29 October 2017; pp. 618–626.

Article

Ensemble-Based Classification Using Neural Networks and Machine Learning Models for Windows PE Malware Detection

Robertas Damaševičius [1,*], Algimantas Venčkauskas [2], Jevgenijus Toldinas [2] and Šarūnas Grigaliūnas [2]

[1] Department of Software Engineering, Kaunas University of Technology, 44249 Kaunas, Lithuania
[2] Department of Computer Science, Kaunas University of Technology, 44249 Kaunas, Lithuania; algimantas.venckauskas@ktu.lt (A.V.); eugenijus.toldinas@ktu.lt (J.T.); sarunas.grigaliunas@ktu.edu (Š.G)
* Correspondence: robertas.damasevicius@ktu.lt

Abstract: The security of information is among the greatest challenges facing organizations and institutions. Cybercrime has risen in frequency and magnitude in recent years, with new ways to steal, change and destroy information or disable information systems appearing every day. Among the types of penetration into the information systems where confidential information is processed is malware. An attacker injects malware into a computer system, after which he has full or partial access to critical information in the information system. This paper proposes an ensemble classification-based methodology for malware detection. The first-stage classification is performed by a stacked ensemble of dense (fully connected) and convolutional neural networks (CNN), while the final stage classification is performed by a meta-learner. For a meta-learner, we explore and compare 14 classifiers. For a baseline comparison, 13 machine learning methods are used: K-Nearest Neighbors, Linear Support Vector Machine (SVM), Radial basis function (RBF) SVM, Random Forest, AdaBoost, Decision Tree, ExtraTrees, Linear Discriminant Analysis, Logistic, Neural Net, Passive Classifier, Ridge Classifier and Stochastic Gradient Descent classifier. We present the results of experiments performed on the Classification of Malware with PE headers (ClaMP) dataset. The best performance is achieved by an ensemble of five dense and CNN neural networks, and the ExtraTrees classifier as a meta-learner.

Keywords: malware analysis and detection; applied machine learning; mobile security; neural network; ensemble classification

1. Introduction

Many aspects of society have shifted online with the broad adoption of digital technology, from entertainment and social interactions to business, entertainment, industry and, unfortunately, crime as well. Cybercrime is rising in frequency and magnitude in recent years, with a projection of reaching USD 6 trillion by 2021 (up from USD 3 trillion in 2015) [1] and also taking on conventional crime both in number and revenues [2]. Additionally, these new cyber-attacks have become more complex [3], generating elaborate multi-stage attacks. By the end of 2018, about 9599 malicious packages appeared per day [4]. Such attacks also resulted in significant damage and major financial losses. Up to USD 1 billion was stolen from financial institutions around the world in two years due to malware [5]. In addition, Kingsoft estimated that between 2 and 5 million computers were attacked each day [6]. With cybercrime revenues reaching USD 1.5 trillion in 2018 [7] and cybercrime's global cost predicted to reach USD 6 trillion by 2021 [8], addressing cyber threats has become an urgent issue.

Moreover, the COVID-19 pandemic has delivered an extraordinary array of cybersecurity challenges, as most services have moved to online and remote mode, raising the danger of cyberattacks and malware [9,10]. Especially, in the healthcare sector, cyber-attacks can lead to compromised sensitive personal patient data, while data tampering can lead to incorrect treatment, with irreparable damage to patients [11].

Today, computer programs and applications are developed at high speed. Malicious software (malware) has appeared and is growing in many formats and is becoming increasingly sophisticated. Computer criminals use them as a tool to infiltrate, steal or falsify information, causing huge damage to individuals, businesses and even threatening national security. A generic term generally used to describe all various types of unauthorized software programs is malware (malicious software), which includes viruses, worms, Trojans, spyware [12], Android malicious apps [13], bots, rootkits [14] and ransomware [15]. In achieving its objectives, malware has been used by cybercriminals as weapons. Malware has been used to conduct a wide variety of security threats, such as stealing confidential data, stealing cryptocurrency, sending spam, crippling servers, penetrating networks and overloading critical infrastructures. While large numbers of malware samples have been identified and blocked by cybersecurity service providers and antivirus software manufacturers, a significant number of malware samples have been created or mutated (e.g., "zero-day" malware [16]) and appear to evade conventional anti-virus scanning tools based on signatures. As these techniques are primarily based on modifications of signature-based models, this has caused the information security industry to reconsider their malware recognition techniques.

Malware detection methods can be classified into methods based on signatures and behavior. Currently, signature-based malware detectors can work effectively with previously known malware that has already been detected by some anti-malware vendors. However, it cannot detect polymorphic malware that can change its signatures, as well as new malware whose signatures have not yet been created. One solution to this problem is to use heuristic analysis in combination with machine learning techniques that provide higher detection efficiency. As practice has shown, the traditional approach to the field of malware detection, which is based on signature analysis [17], is not acceptable for detecting unknown computer viruses. To maintain the proper level of protection, users are forced to constantly and timely update anti-virus databases. However, the delay in the response from the anti-virus companies for the emergence of new malware (its detection and signature creation) can vary from several hours to several days. During this time, malicious new programs can cause irreparable damage.

To address this problem, in addition to the signature approach, heuristic analysis is used. At the same time, the file can be considered "potentially dangerous" with some probability based on its behavior (dynamic approach) or the analysis of its structure (static approach). Static analysis generally consists of two main stages: the training stage and the stage of using the results (detection of virus programs). At the training stage, a sample of infected (virus) and "clean" (legitimate) files is formed. In the structure of the files, some signs characterize each of them as viral or legitimate. As a result, a list of feature characteristics is compiled for each file. Next, the most significant (informative) features are selected, and redundant and irrelevant features are discarded. At the detection stage, feature characteristics are extracted from the scanned file. Heuristic algorithms developed specifically to detect unknown malware are characterized by a high error rate. Heuristic-based detection uses rules formulated by experts to distinguish between malicious and benign files. Additionally, behavior-based, model checking-based and cloud-based methods have performed effectively in malware detection [18].

Modern research in the area of information security aimed at creating such protection methods and algorithms that would be able to detect and neutralize unknown malware, and thus not only increase the computer security but also save the user from constant updates of antivirus software. The size of gray lists is constantly growing with the advancement of malware writing and production techniques. Intelligent methods for automatically detecting malware are, therefore, urgently required. As a result, several studies have been published on the development of smart malware recognition systems using artificial intelligence methods [19–22].

A prerequisite for creating effective anti-virus systems is the development of artificial neural network (ANN)-based technologies. The ability of such systems to learn and generalize results makes it possible to create smart information security systems. Artificial intelligence (AI) has several advantages when it comes to cybersecurity: AI can discover new previously unknown attacks; AI can handle a high volume of data; AI-based cybersecurity systems can learn over time to respond better to threats [23].

This study aims to implement an ensemble of neural networks for the detection of malware. The novel contributions of this paper are the following:

(1) The detailed experimental analysis and verification of machine learning and deep learning methods for malware recognition performed on the Classification of Malware with PE headers (ClaMP) dataset;
(2) A novel ensemble learning-based hybrid classification framework for malware detection with a heterogeneous batch of convolutional neural networks (CNNs) as base classifiers and a machine learning algorithm as a final-stage classifier, which allows us to achieve the improvement of malware detection accuracy;
(3) An extensive ablation study to select CNN model architectures and a machine learning algorithm for the best overall malware detection performance.

The other parts of this study are structured as follows. In Section 2, related works are discussed including the presentation of adequate criticism of existing methods and approaches. Section 3 describes the methodology used in this paper. Section 4 discusses the implementation and results obtained. Section 5 presents the conclusion of the study.

2. Related Works

Malware search algorithms are divided into two classes based on the method of collecting information—dynamic and static. In static analysis, suspicious objects are considered without starting them, based on the assembly code and attributes of executable files [24]. Dynamic analysis algorithms work either with already running programs or run them themselves in an isolated environment, exposing the information that has arisen in the course of work: they analyze the behavior of the program, sections of code and data and monitor resource consumption [25]. According to the type of objects detected, malware search algorithms are divided into signature and anomalous ones. Signature programs tend to highlight the signatures of malware. Anomaly detection algorithms seek to describe legitimate programs and learn to look for deviations from the norm.

At the same time, machine learning is also widely used as a powerful tool for security experts to identify malicious programs with high accuracy, when the number of malicious programs is high enough, and their options have become diverse. Among the main methods is the Windows Portable Executable 32-bit (PE32) file header analysis [26]. For example, Nisa et al. [27] transformed malware code into images and applied segmentation-based fractal texture analysis for feature extraction. Deep neural networks (AlexNet and Inception-v3) were used for classification. Previously, the use of ensemble methods, such as random forest and extremely randomized trees, allowed the improvement of the performances of machine learning models in detecting malware in internet of things (IoT) environments [28] and Wireless Sensor Networks (WSN) [29].

Many studies are being performed to analyze malware to curb the increase in malicious software [30]. The existing deep learning-based malware analysis methods include convolutional neural networks (CNN) [31], deep belief network (DBN) [32], graph convolutional network (GCN) [33], LSTM and Gated Recurrent Unit (GRU) [34], VGG16 [35] and generative adversarial networks (GAN) [36]. However, it is not possible to guarantee the generalization potential of artificial neural network-based models [37].

To solve the above-mentioned problems, more general and robust methods are, therefore, required. Researchers are creating numerous ensemble classifiers [38–42] that are less susceptible to malware feature collection. Ensemble methods [43] are a class of techniques that incorporate several learning algorithms to enhance the precision of overall prediction. To minimize the risk of overfitting in the training results, these ensemble classifiers integrate several classification models. In this way, the training dataset can be more effectively used, and generalization efficiency can be increased as a result. While several models of ensemble classification are already developed, there is still space for researchers to improve the accuracy of sample classification, which would be useful for improving malware detection.

Therefore, this paper proposes an ensemble earning-based approach for using fully connected and convolution neural networks as base learners for malware detection.

3. Materials and Methods

Malware developers are primarily focused on targeting computer networks and infrastructure to steal information, make financial demands or prove their potential. The standard approaches for detecting malware were effective in detecting known malware. Via these approaches, however, new malware can never be blocked. The latest machine learning platform [44] has significantly enhanced the identification capability of models used for malware detection. It is possible to detect malware using machine learning methods in two steps, namely, extracting features from input data and choosing important ones that best represent the data, and classifying/clustering. The technology proposed is focused on machine learning that can learn and discern malicious and benign files, as well as make reliable forecasts of new files that have not been seen before.

The phases involved in achieving the final solution are (1) data processing and feature selection and (2) model engineering, which includes the following steps: data selection and scaling, reduction in dimensionality, ANN model exploration and meta-learner classifier selection, ensemble model development, model testing and performance evaluation. Figure 1 indicates the flow to the model evaluation stage of the stages involved in the system methodology, beginning with data selection, which is described in more depth in the following subsections.

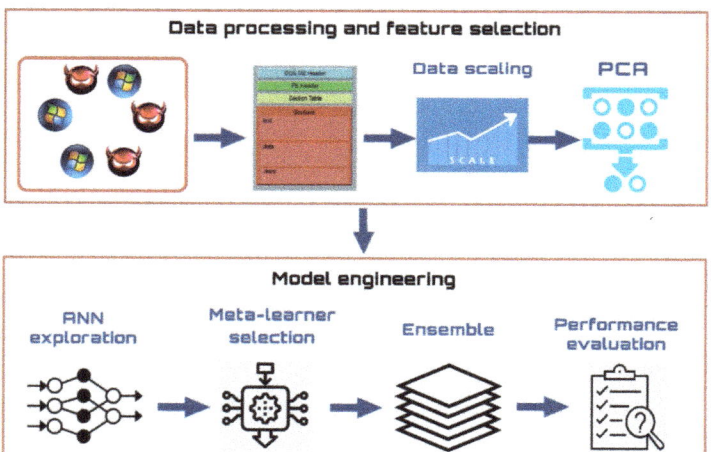

Figure 1. Outline of malware detection methodology.

3.1. Data Collection and Processing

For machine learning to be a success, the selection of a representative dataset is necessary. This is because it is important to train a machine learning algorithm on a dataset that correctly represents the conditions for the model's real-world applications.

For this model, the dataset gathered contains malicious and benign data from the Classification of Malware with PE headers (ClaMP) dataset, obtained from GitHub. We used the ClaMP_Integrated dataset, which has 2722 malware and 2488 Benign instances. The dataset has 69 features, which include, among others, the following features:

- DOS image header: e_cp–pages in file, e_cblp–bytes on the last page, e_cparhdr–size of header, e_crlc–number of relocations, e_cs–initial CS value, e_csum - checksum,

e_p–initial IP value, e_lfanewe_lfarlc, e_magic–Magic number, e_maxalloc–maximum extra paragraphs, e_minalloc–minimum number of extra paragraphs, e_oemid–OEM ID, e_oeminfo–OEM information, e_ovno–overlay number, e_res and e_res2–reserved words, e_sp–initial SP value, e_ss–initial SS value.
- File header features: CharacteristicsCreationYear, Machine, NumberOfSections, NumberOfSymbols, PointerToSymbolTable, SizeOfOptionalHeader.
- Other raw features: AddressOfEntryPoint, BaseOfCode, BaseOfData, CheckSum, DllCharacteristics, FileAlignment, ImageBase, LoaderFlags, Magic, MajorImageVersion, MajorLinkerVersion, MajorOperatingSystemVersion, MajorSubsystemVersion, MinorImageVersion, MinorLinkerVersion, MinorOperatingSystemVersion, MinorSubsystemVersion, NumberOfRvaAndSizes, SectionAlignment, SizeOfCode, SizeOfHeaders, SizeOfHeapCommit, SizeOfHeapReserve, SizeOfImage, SizeOfInitializedData, SizeOfStackCommit, SizeOfStackReserve, SizeOfUninitializedData, Subsystem.
- Derived features: sus_sections, non_sus_sections, packer, packer_type, E_text, E_data, filesize, E_file, fileinfo.

However, we used only 68 features (all numerical), because one feature "packer_type" is a string, which was not used. The numerical features were scaled using the standard scaling method. These features, along with the class label (0 for benign and 1 for malicious), were used to build the ensemble classification model.

3.2. Dimensionality Reduction

To fix a variety of estimation and classification questions, machine learning methods are commonly used. Bad machine learning output can be triggered by overfitting or underfitting the results. Removing the unimportant characteristics guarantees the algorithms' optimal efficiency and improves pace. Principal Component Analysis (PCA) was introduced to perform attribute dimensionality reduction. Based on previous studies, 40 features were chosen to be passed into the machine learning model (representing 95% of the total variability in the dataset), because these features are critical in neural network learning, whether a file is malicious or benign.

3.3. Deep Learning Models

As deep learning models, we considered fully connected (FC) multilayer perceptron (MLP) and one-dimensional convolutional neural networks (1D-CNN), which are discussed in detail below.

3.3.1. Multilayer Perceptron

As a baseline approach, we adopted a simple multilayer perceptron (MLP). Let the output of the MLP be known $y(t)$ at the input $X(t)$, where $X(t)$ is a vector with components $(x_1, x_2, \ldots, x_n)$, t is the number of the sequence value and $t = \overline{1,T}$ (T is predetermined).

To find model parameters $w = (w_0, w_1, \ldots, w_m)$ and $V_k = (V_{1k}, V_{2k}, \ldots, V_{nk})$, h_k, $k = \overline{1,m}$ such that the model output $F(X, V, w)$ and the real output of the MLP $y(t)$ would be as close as possible. The relationship between the input and output of a two-layer perceptron is established by the following relationships:

$$Z_k = \sigma(V_{1k}x_1 + V_{2k}x_2 + \ldots V_{nk}x_n - h_k), k = \overline{1,m} \qquad (1)$$

$$y = \sigma(w_1 Z_1 + w_2 Z_2 + \ldots w_m Z_m + w_0) \qquad (2)$$

The following expression describes a perceptron with one hidden layer, which is able to approximate any continuous function defined on a bounded set.

$$\sum_{k=1}^{m} w_k \cdot \sigma(V_{1k}x_1 + V_{2k}x_2 + \ldots V_{nk}x_n - h_k) + w_0 \qquad (3)$$

Training of MLP occurs by applying a gradient descent algorithm (such as error backpropagation) similar to a single-layer perceptron.

3.3.2. One-Dimensional Convolutional Neural Network (1D-CNN)

While CNN models have been developed for image processing, where an internal representation of a two-dimensional input (2D) is learned by the model, the same mechanism can be used in a process known as feature learning on one-dimensional (1D) data sequences, such as in the case of malware detection. The model understands how to extract features from observational sequences and how to map hidden layers to different types of software (malware or benign).

$$\hat{y} = \Phi([x_1, \ldots, x_N]), \tag{4}$$

where $X : x_1, \ldots, x_N$ indicates the input of the network, and $Y : \hat{y}$ is the output. Therefore, the network learns a mapping from the input space X to the output space Y.

The key block of the convolutional network is the convolutional layer. A group of trainable filters are the parameters of this layer (scan windows). Each filter operates in size through a tiny window. The scanning window sequentially traverses the whole picture during the forward propagation of the signal (from the first layer to the last layer) according to the tiling principle and measures the dot products of two vectors: the filter values and the outputs of the chosen neurons. Thus, a two-dimensional activation map is generated after passing all the shifts in the width and height of the input field, which gives the effect of applying a particular filter in each spatial area. The network uses filters that are enabled when there is an input signal of some kind. A series of filters are used for each convolutional layer, and each generates a different activation map.

$$x_j^l = f\left(\sum_{i=1}^{M} x_i^{l-1} \cdot k_{ij}^l + b_j^l\right), \tag{5}$$

where k is the convolution kernel, j is the size of kernels, M is the number of inputs x_i^{l-1}, b, is kernel bias, $f(\ )$ is the neuron activation function and $(\cdot)$ represents the convolution operator.

The sub-sampling layer is another feature of a convolutional neural network. It is usually positioned between successive convolution layers, so it may occur periodically. Its purpose is to reduce the spatial size of the vector gradually to reduce the number of network parameters and calculations, as well as to balance overfitting. The convolution layer resizes the feature map, using the max operation most frequently. If the output from the previous layer is to be fed to the fully connected layer, the flattening layer is used, and then it needs to be flattened. The layer of the Parametric Rectified Linear Unit (PReLU) is an activation function that complements the rectified unit with a negative value slope.

The dropout layer is used to regularize the network. It also makes it possible to be thinner for the network size. The neurons that are less likely to raise the weight of learning are randomly removed. The practical importance of dropout unit is to prevent overfitting [45]. This dropout layer, as we have two classes, is succeeded by a fully linked (dense) layer that reduces the final output vector to two classes, and we expect the program's behavior to be either malicious or benevolent. The final activation function is SoftMax, which shrinks the two outputs to one.

The output of each convolutional layer in 1D-CNN is also the input of the subsequent layer. It also represents the weights learned by the convolution kernel from the training samples.

A unique and essential part of CNNs is the fully connected (FC) layer, which outputs a final output. The output of the network's previous layers is reshaped into a single vector (flattened). Any of them reflects the probability that a class label is a special function. The final probabilities for each label are supplied by the output of the FC layer.

3.4. Network Model Optimization

Optimization of neural network hyper-parameters, which rule how the network operates and governs its accuracy and validity, is still an unsolved problem. Optimizers adjust the parameters of neural networks, such as weight and learning rate, to minimize loss. Known examples of neural network optimization algorithms are Stochastic Gradient Descent (SGD) [46], AdaGrad [47], RMSProp [48] and Adam [49], which usually show a tradeoff of optimization vs. generalization. This means that higher training speed and higher accuracy in the training may result in poorer accuracy on the testing dataset. Here, we adopted the Exponential Adaptive Gradients (EAG) optimization [50], which combines Adam and AdaBound [51]. During training, it exponentially sums the gradient in the past and adaptively adjusts the learning rate to address poor generalization of the Adam optimizer.

3.5. Ensemble Classification

The basic principle of ensemble methods is that training datasets are rearranged in several ways (either by resampling or reweighting) and by adding a base classifier to each rearranged training dataset, an ensemble of base classifiers is built. After that, a new ensemble classifier is developed using the stacked ensemble method by combining the prediction effects of all those base classifiers, where a new model learns how to better integrate predictions from multiple base models. We used the two-stage stacking technique [52]. First, several models are trained based on a dataset. Then, the output of each of the models is processed to create a new dataset. The actual value it is supposed to approximate is related to each instance in the current dataset. Second, the dataset with the meta-learning algorithm is used to provide the final output.

In the design of a stacking model (Figure 2), base models are often referred to as level-0 models, and a meta-learner (or generalizer) that integrates base model projections, referred to as a level-1 model, is involved. Models that fit into the training data and are compiled with forecasts are the base models. The meta-learner (level-1 model) is a classification model trained to combine the predictions of the base model. The meta-learner is informed by simple models on the choices made. To train the base models, a new batch of previously unused data is used and predictions are made, and the input and output value pairs of the training dataset are used to fit the meta-learner, along with projected outputs given by these predictions.

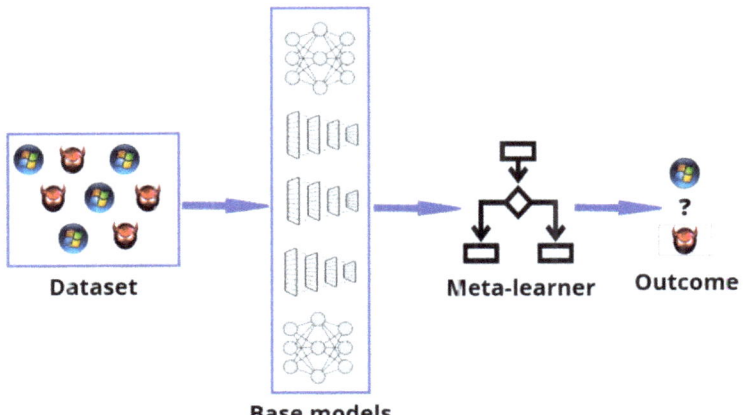

Figure 2. Schematics of ensemble classification approach.

The ensemble learner algorithm consists of three stages:

1. Set up the ensemble:
   (a) Select N base learners;
   (b) Select a meta-learning algorithm.
2. Train the ensemble:
   (a) Train each of the N base learners on the training dataset $\{X_1, X_2, \ldots, X_M\}$, where M is the number of samples;
   (b) Perform the k-fold cross-validation on each of the base learners and record the cross-validated predictions $\{y_1, y_2, \ldots, y_N\}$;
   (c) Combine cross-validated predictions from base learners to form a new feature matrix as follows. Train the meta-learner on the new data (features x predictions from base-level classifiers) $\{(X_1, X_2, \ldots, X_M, y_1), (X_1, X_2, \ldots, X_M, y_2), \ldots, (X_1, X_2, \ldots, X_M, y_N)\}$. Combine base learning models and the meta-learner to generate more accurate predictions on unknown data.
3. Test on new data:
   (a) Record output decisions from the base learners;
   (b) Send base-level decisions to the meta-learner to make ensemble decision.

On the training dataset, stacking capitalizes over every single best learner. Usually, the greatest gains are made when base classifiers used for stacking have high variability and uncorrelated outputs predicted values. As base models, we used the following neural networks: fully connected MLP with one hidden layer (Dense-1), fully connected MLP with two hidden layers (Dense-2) and one-dimensional CNN (1D-CNN). The configurations of neural networks are summarized in Table 1.

Table 1. Model configuration of neural networks with their parameters. FC—fully connected. Conv1D—one-dimensional convolution. PReLU—Parametric Rectified Linear Unit.

Dense-1 Network	Dense-2 Network	1D-CNN Network
Parameters		
X—number of neurons in 1st hidden layer	X—number of neurons in 1st hidden layer Y—number of neurons in 2nd hidden layer	F—number of filters in convolutional layers N—number of neurons in dense layer
Input layer of 40 × 1 features		
1 FC layer (X neurons)	1 FC layer (X neurons)	2 Conv1D layers (F 2 × 2 filters)
PReLU	PReLU	Max-pooling layer
Dropout layer (p = 0.3)	Dropout layer (p = 0.3)	2 Conv1D layers (F 2 × 2 filters)
1 FC layer (2 neurons)	1 FC layer (Y neurons)	Max-pooling layer
	PReLU	1 FC layer (N neurons)
Softmax output layer	Dropout layer (p = 0.3)	Dropout layer (p = 0.5)
	1 FC layer (2 neurons)	1 FC layer (2 neurons)
	Softmax output layer	Softmax output layer

The examples of neural network architectures are presented in Figure 3.

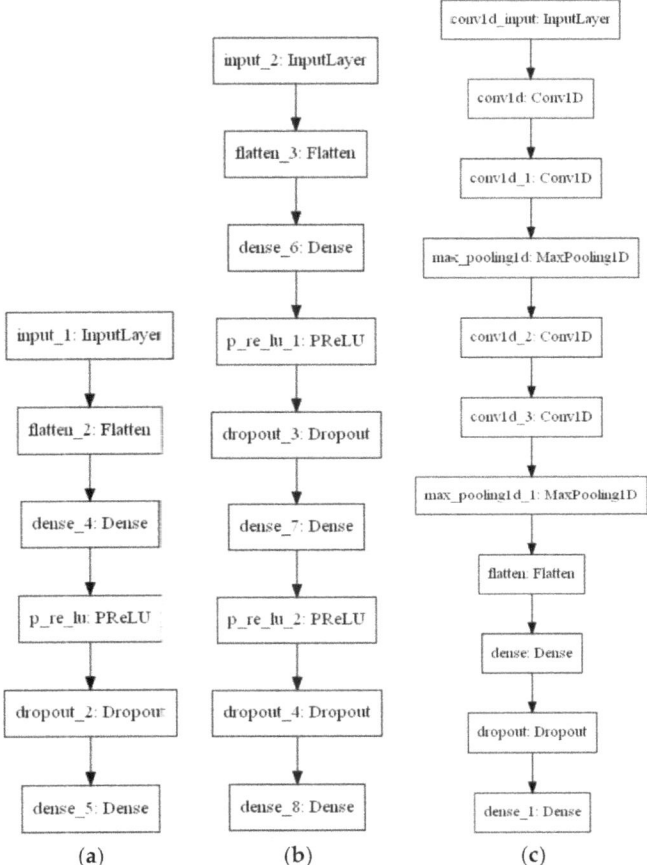

Figure 3. Example of architectures used as base learners: (a) Dense-1 network architecture, (b) Dense-2 network architecture and (c) 1D-CNN network architecture.

The role of the meta-learner is to find how best to aggregate the decisions of the base classifiers. As meta-learners, we explored K-Nearest Neighbors (KNN), Support Vector Machine (SVM) with linear kernel, SVM with radial basis function (RBF) kernel, Decision Tree (DT), Random Forest (RF), Multi-Layer Perceptron (MLP), AdaBoost Classifier, Extra-Trees (ET) classifier, Isolation Forest, Gaussian Naïve Bayes (GNB), Linear Discriminant Analysis (LDA), Quadratic Discriminant Analysis (QDA), Logistic Regression (LR), Ridge Classifier (RC) and stochastic gradient descent classifier (SGDC). Here, KNN is a model that classifies unknown input data based on having the most similarity (least distance) to known input data. SVM is a supervised learning method that constructs a higher dimensional hyperplane to separate input data belonging to various classes while maximizing the data input distance to the hyperplane. The DT classifier creates a decision tree by splitting according to the feature which has the highest information gain. RF fits many DT classifiers on different sub-samples of the dataset and uses averaging to improve the prediction accuracy. AdaBoost fits a classifier on the dataset and performs weighting of incorrectly classified instances to improve accuracy. Isolation Forest performs classification based on identified anomalies in data. GNB performs classification based on the probability distributions of features and classes. The ET Classifier [53] creates a meta-estimator that fits multiple decision trees on the training dataset sub-samples and uses averaging to improve precision and over-fitting management. The goal of LDA is to find a linear combination of input characteristics that distinguishes two or more input data groups. A quadratic decision surface is used by QDA to distinguish two or more groups of input data. LR is a linear regression-like statistical approach that predicts a result for a binary output variable from an input variable. RC converts the label data to $(-1,1)$ and fixes the regression method problem. As a target class, the greatest value of prediction is admitted. SGDC is a learning algorithm for stochastic gradient descent that finds the decision boundary with a linear hinge loss.

3.6. Evaluation of Malware Detection Results

To measure the classification potential of the proposed ensemble learning model, the performance of the proposed model was evaluated using the Leave-One-Out Cross-Validation (LOOCV) with a 10-fold cross-validation method.

The true labels were compared against the predicted labels and the true positive (TP), true negative (TN), false positive (FP) and false-negative (FN) values were calculated. The recall, precision, accuracy, error rate and F-score values were calculated (we assumed the binary classification problem, where a positive class is labeled by +1 and a negative class is labeled by -1):

False positive rate (FPR) (also specificity):

$$FPR = \frac{\sum_{i=1}^{m}[a(x_i) = +1][y_i = -1]}{\sum_{i=1}^{m}[y_i = -1]} \qquad (6)$$

here $[\cdot]$ is the Iverson bracket operator.

True positive rate (TPR) (also sensitivity and recall):

$$TPR = \frac{\sum_{i=1}^{m}[a(x_i) = +1][y_i = +1]}{\sum_{i=1}^{m}[y_i = +1]} \qquad (7)$$

False negative rate (FNR):

$$FNR = \frac{\sum_{i=1}^{m}[a(x_i) = -1][y_i = +1]}{\sum_{i=1}^{m}[y_i = +1]} \qquad (8)$$

Here, $a(x)$ is the classifier with inputs $X^m = (\ x_1, \ \ldots, \ x_m\ )$, and $(\ y_1, \ \ldots, \ y_m\ )$ are outputs.

Precision is calculated as:

$$Precision = \frac{TPR}{TPR + FPR} \qquad (9)$$

To compute F-score, the following equation is used:

$$F-score = 2\frac{Precision \times Recall}{Precision + Recall} \qquad (10)$$

The Matthews Correlation Coefficient (MCC) is calculated as:

$$MCC = \frac{TP \cdot TN - FP \cdot FN}{\sqrt{(TP+FP)(TP+FN)(TN+FP)(TN+FN)}} \qquad (11)$$

The Cohen's Kappa statistic (shortly, kappa) is

$$k = 1 - \frac{1-p_0}{1-p_e} \qquad (12)$$

where p_0 represents the ratio of correct agreement in the test dataset, and p_e is the ratio of agreement that is expected by random selection.

In this study, performance was calculated using 10-fold cross-validation. According to F1-score, instead of checking the performance of the model with accuracy alone, we selected the best model. The accuracy can be a confusing metric in datasets where a major class imbalance occurs. For a highly imbalanced sample, a model would correctly guess the value of the majority class for all predicted outcomes, and achieve a high performance in classification but making erroneous predictions in the minority and main classes. The F1-score discourages this type of action by computing the metrics for each mark and finding its unweighted average. We also consider area under curve (AUC) as a measure of binary classification consistency, which is known as a balanced metric that can be used even though there are classes of very different sizes in the dataset. Furthermore, the performance of the proposed model on a binary dataset is represented using the confusion matrix.

We used the performance outcomes achieved from the results from each fold of the 10-fold cross-validation for statistical analysis. We adopted the non-parametric Friedman test followed by the post-hoc Nemenyi test to compare the findings and measure their statistical value. Second, both strategies were ranked based on the selected performance measures (we used accuracy, AUC and F1-score). Then, each method's mean ranks were determined. If the difference between the mean ranks of the methods was less than the critical difference obtained from the Nemenyi test, the difference between method outputs was assumed not to be significant.

4. Implementation and Results

4.1. Experimental Settings

The machine learning models were trained on the features acquired from the dataset using Python's Scikit-learn libraries. All experiments were performed on a laptop computer with 64-bit Windows 10 OS with Intel® Core™ i5-8265U CPU @ 1.60 GHz 1.80 GHz with 8 GB RAM (Intel, Santa Clara, CA, USA).

4.2. Results of Machine Learning Methods

The results from using classical machine learning models are summarized in Table 2, while their confusion matrices are summarized in Figure 4. The best results were obtained by the ExtraTrees (ET) model, achieving an accuracy of 98.8%. As can be seen from Table 2 and Figure 3, the ET model generated very good results for the precision, recall, F1 and accuracy of the two classes. This agrees with the low FPR and FNR of 0.8% and 1.4% obtained by the ET model.

Table 2. Summary of results of machine learning models. Acc–Accuracy. Prec–Precision. Rec–Recall. Spec–Specificity. FPR–False Positive Rate. FNR–False Negative Rate. AUC–Area Under Curve. MCC–Matthews Correlation Coefficient. SVM–Support Vector Machine. RBF–Radial Basis Function. LDA–Linear Discriminant Analysis. SGDC–Stochastic Gradient Descent Classifier.

Meta-Learner	Acc	Prec	Rec	Spec	FPR	FNR	F1	AUC	MCC	Kappa
Nearest Neighbors	0.973	0.973	0.973	0.973	0.029	0.025	0.973	0.973	0.946	0.946
Linear SVM	0.954	0.954	0.954	0.954	0.045	0.047	0.954	0.954	0.908	0.908
RBF SVM	0.924	0.924	0.924	0.924	0.012	0.132	0.924	0.928	0.856	0.849
Decision Tree	0.933	0.933	0.933	0.933	0.107	0.032	0.933	0.93	0.866	0.864
Random Forest	0.931	0.931	0.931	0.931	0.08	0.059	0.931	0.93	0.861	0.861
Neural Net	0.977	0.977	0.977	0.977	0.021	0.025	0.977	0.977	0.954	0.954
AdaBoost	0.962	0.962	0.962	0.962	0.041	0.036	0.962	0.961	0.923	0.923
ExtraTrees	0.988	0.988	0.988	0.988	0.008	0.014	0.988	0.989	0.977	0.977
LDA	0.936	0.936	0.936	0.936	0.08	0.05	0.936	0.935	0.871	0.871
Logistic	0.959	0.959	0.959	0.959	0.039	0.043	0.959	0.959	0.917	0.917
Passive	0.933	0.933	0.933	0.933	0.037	0.094	0.933	0.935	0.867	0.866
Ridge	0.936	0.936	0.936	0.936	0.08	0.05	0.936	0.935	0.871	0.871
SGDC	0.958	0.958	0.958	0.958	0.035	0.049	0.958	0.958	0.915	0.915

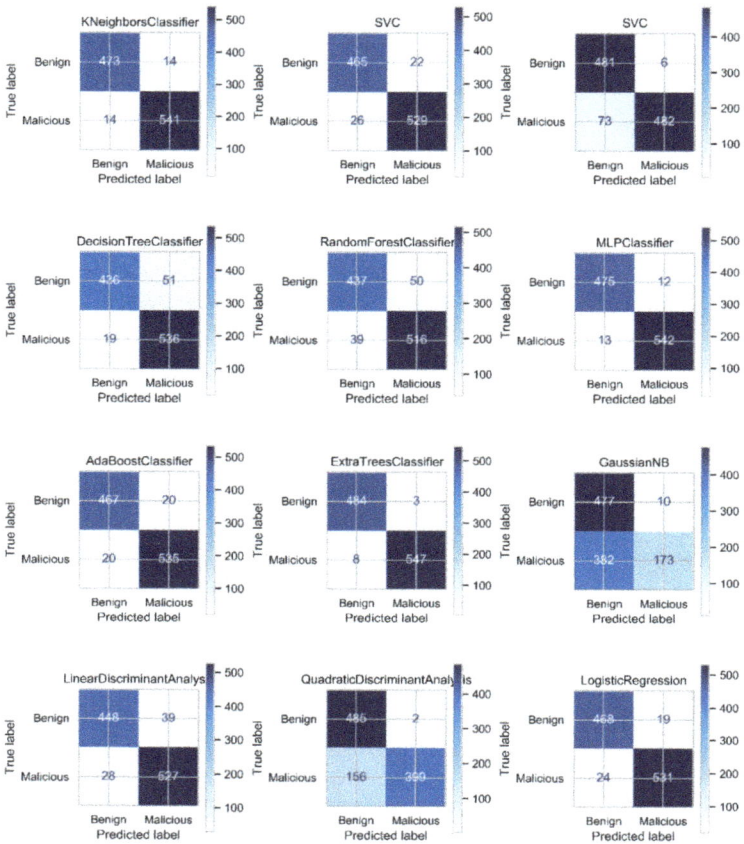

Figure 4. Confusion matrices of machine learning models.

4.3. Results of Neural Network Classifiers

To select the base classifiers, first, we performed an ablation study to find the best representatives of Dense-1, Dense-2 and 1D-CNN models in terms of their performance with respect with different values of hyperparameters. The results are presented in Tables 3–5. Note that in all cases, we used sparse categorical cross-entropy loss function and an Adam optimizer. For the training of Dense-1 and Dense-2 models, we used 100 epochs, while for the training of 1D-CNN models, we used 20 epochs. In all cases, 80% of data were used for training and 20% for testing.

Table 3. Malware detection performance with different number of neurons in hidden layer of Dense-1 model. Best models are shown in bold.

No. of Neurons in 1st Layer	Acc	Prec	Rec	Spec	FPR	FNR	F1	AUC	MCC	Kappa
5	0.956	0.956	0.956	0.956	0.035	0.052	0.956	0.956	0.912	0.912
10	0.962	0.962	0.962	0.962	0.043	0.033	0.962	0.962	0.924	0.924
15	0.965	0.965	0.965	0.965	0.034	0.036	0.965	0.965	0.929	0.929
20	0.971	0.971	0.971	0.971	0.026	0.033	0.971	0.971	0.941	0.941
25	0.977	0.977	0.977	0.977	0.023	0.023	0.977	0.977	0.954	0.954
30	0.977	0.977	0.977	0.977	0.022	0.024	0.977	0.977	0.954	0.954
35	**0.980**	**0.980**	**0.980**	**0.980**	**0.022**	**0.019**	**0.980**	**0.979**	**0.959**	**0.959**
40	0.979	0.979	0.979	0.979	0.023	0.019	0.979	0.979	0.958	0.958

Table 4. Malware detection performance with different number of neurons in hidden layers of Dense-2 model. Best models are shown in bold.

No. of Neurons in 1st Layer	No. of Neurons in 2nd Layer	Acc	Prec	Rec	Spec	FPR	FNR	F1	AUC	MCC	Kappa
5	5	0.954	0.954	0.954	0.954	0.057	0.036	0.954	0.953	0.908	0.908
5	10	0.958	0.958	0.958	0.958	0.046	0.039	0.958	0.958	0.915	0.915
5	15	0.960	0.960	0.960	0.960	0.032	0.047	0.960	0.960	0.919	0.919
5	20	0.964	0.964	0.964	0.964	0.039	0.033	0.964	0.964	0.928	0.928
5	25	0.961	0.961	0.961	0.961	0.042	0.036	0.961	0.961	0.922	0.922
5	30	0.965	0.965	0.965	0.965	0.038	0.033	0.965	0.965	0.929	0.929
5	35	0.963	0.963	0.963	0.963	0.040	0.034	0.963	0.963	0.926	0.925
10	5	0.963	0.963	0.963	0.963	0.042	0.033	0.963	0.963	0.926	0.925
10	10	0.969	0.969	0.969	0.969	0.031	0.032	0.969	0.969	0.937	0.937
10	15	0.965	0.965	0.965	0.965	0.036	0.033	0.965	0.965	0.931	0.931
10	20	0.968	0.968	0.968	0.968	0.030	0.034	0.968	0.968	0.936	0.935
10	25	0.971	0.971	0.971	0.971	0.028	0.030	0.971	0.971	0.941	0.941
10	30	0.971	0.971	0.971	0.971	0.028	0.030	0.971	0.971	0.941	0.941
10	35	0.972	0.972	0.972	0.972	0.024	0.030	0.972	0.973	0.945	0.945
15	5	0.958	0.958	0.958	0.958	0.055	0.029	0.958	0.958	0.917	0.917
15	10	0.972	0.972	0.972	0.972	0.026	0.030	0.972	0.972	0.944	0.944
15	15	0.972	0.972	0.972	0.972	0.034	0.023	0.972	0.972	0.944	0.944
15	20	0.969	0.969	0.969	0.969	0.028	0.034	0.969	0.969	0.937	0.937
15	25	0.971	0.971	0.971	0.971	0.024	0.033	0.971	0.971	0.942	0.942
15	30	0.977	0.977	0.977	0.977	0.026	0.021	0.977	0.977	0.954	0.954
15	**35**	**0.980**	**0.980**	**0.980**	**0.980**	**0.022**	**0.019**	**0.980**	**0.979**	**0.959**	**0.959**
20	5	0.967	0.967	0.967	0.967	0.035	0.032	0.967	0.967	0.933	0.933
20	10	0.976	0.976	0.976	0.976	0.016	0.032	0.976	0.976	0.951	0.951
20	15	0.972	0.972	0.972	0.972	0.028	0.027	0.972	0.972	0.945	0.945
20	20	0.973	0.973	0.973	0.973	0.027	0.027	0.973	0.973	0.946	0.945
20	**25**	**0.980**	**0.980**	**0.980**	**0.980**	**0.023**	**0.018**	**0.980**	**0.979**	**0.959**	**0.959**
20	30	0.978	0.978	0.978	0.978	0.023	0.022	0.978	0.978	0.955	0.955
20	35	0.978	0.978	0.978	0.978	0.022	0.022	0.978	0.978	0.956	0.955
25	5	0.970	0.970	0.970	0.970	0.030	0.030	0.970	0.970	0.940	0.940
25	10	0.974	0.974	0.974	0.974	0.024	0.027	0.974	0.974	0.949	0.949
25	15	0.974	0.974	0.974	0.974	0.022	0.030	0.974	0.974	0.947	0.947
25	20	0.975	0.975	0.975	0.975	0.028	0.022	0.975	0.975	0.950	0.950
25	**25**	**0.980**	**0.980**	**0.980**	**0.980**	**0.023**	**0.017**	**0.980**	**0.980**	**0.960**	**0.960**
25	30	0.980	0.980	0.980	0.980	0.023	0.018	0.980	0.979	0.959	0.959
25	35	0.980	0.980	0.980	0.980	0.024	0.017	0.980	0.979	0.959	0.959
30	5	0.976	0.976	0.976	0.976	0.031	0.017	0.976	0.976	0.953	0.952
30	10	0.978	0.978	0.978	0.978	0.015	0.029	0.978	0.978	0.955	0.955
30	15	0.974	0.974	0.974	0.974	0.024	0.027	0.974	0.974	0.949	0.949
30	20	0.979	0.979	0.979	0.979	0.024	0.018	0.979	0.979	0.958	0.958
30	25	0.980	0.980	0.980	0.980	0.024	0.017	0.980	0.979	0.959	0.959
30	**30**	**0.981**	**0.981**	**0.981**	**0.981**	**0.022**	**0.017**	**0.981**	**0.981**	**0.962**	**0.962**
30	**35**	**0.983**	**0.983**	**0.983**	**0.983**	**0.013**	**0.019**	**0.983**	**0.984**	**0.967**	**0.967**

Table 5. Malware detection performance with different number of filters in convolutional layers and neurons in the final fully connected layer of 1D-CNN model. Best models are shown in bold.

No. of Filters	No. of Neurons	Acc	Prec	Rec	Spec	FPR	FNR	F1	AUC	MCC	Kappa
32	10	0.957	0.957	0.957	0.957	0.045	0.041	0.957	0.957	0.914	0.914
32	15	0.960	0.960	0.960	0.960	0.055	0.027	0.960	0.959	0.919	0.919
32	20	0.964	0.964	0.964	0.964	0.036	0.036	0.964	0.964	0.927	0.927
32	25	0.960	0.960	0.960	0.960	0.053	0.028	0.960	0.960	0.921	0.920
32	30	0.961	0.961	0.961	0.961	0.058	0.022	0.961	0.960	0.922	0.922
32	35	0.964	0.964	0.964	0.964	0.049	0.026	0.964	0.963	0.927	0.927
32	40	0.966	0.966	0.966	0.966	0.039	0.029	0.966	0.966	0.932	0.932
32	45	0.967	0.967	0.967	0.967	0.042	0.026	0.967	0.966	0.933	0.933

Table 5. Cont.

No. of Filters	No. of Neurons	Acc	Prec	Rec	Spec	FPR	FNR	F1	AUC	MCC	Kappa
48	10	0.967	0.967	0.967	0.967	0.032	0.033	0.967	0.967	0.935	0.935
48	15	0.965	0.965	0.965	0.965	0.039	0.032	0.965	0.965	0.929	0.929
48	20	0.972	0.972	0.972	0.972	0.020	0.035	0.972	0.972	0.944	0.944
48	25	0.962	0.962	0.962	0.962	0.032	0.043	0.962	0.963	0.924	0.924
48	30	0.969	0.969	0.969	0.969	0.016	0.045	0.969	0.969	0.938	0.937
48	35	0.970	0.970	0.970	0.970	0.018	0.041	0.970	0.971	0.940	0.940
48	40	0.972	0.972	0.972	0.972	0.019	0.036	0.972	0.972	0.944	0.944
48	45	0.971	0.971	0.971	0.971	0.026	0.033	0.971	0.971	0.941	0.941
64	10	0.961	0.961	0.961	0.961	0.053	0.027	0.961	0.960	0.922	0.922
64	15	0.965	0.965	0.965	0.965	0.020	0.047	0.965	0.966	0.931	0.931
64	**20**	**0.980**	**0.980**	**0.980**	**0.980**	**0.019**	**0.022**	**0.980**	**0.980**	**0.959**	**0.959**
64	25	0.972	0.972	0.972	0.972	0.040	0.017	0.972	0.971	0.944	0.943
64	30	0.974	0.974	0.974	0.974	0.016	0.035	0.974	0.974	0.948	0.947
64	35	0.969	0.969	0.969	0.969	0.046	0.017	0.969	0.969	0.939	0.938
64	40	0.979	0.979	0.979	0.979	0.022	0.021	0.979	0.979	0.958	0.958
64	45	0.974	0.974	0.974	0.974	0.023	0.029	0.974	0.974	0.947	0.947
128	10	0.978	0.978	0.978	0.978	0.013	0.029	0.978	0.979	0.957	0.956
128	15	0.980	0.980	0.980	0.980	0.022	0.018	0.980	0.980	0.960	0.960
128	20	0.975	0.975	0.975	0.975	0.011	0.038	0.975	0.976	0.950	0.950
128	25	0.980	0.980	0.980	0.980	0.022	0.019	0.980	0.979	0.959	0.959
128	30	0.979	0.979	0.979	0.979	0.020	0.022	0.979	0.979	0.958	0.958
128	**35**	**0.985**	**0.985**	**0.985**	**0.985**	**0.018**	**0.013**	**0.985**	**0.985**	**0.969**	**0.969**
128	40	0.983	0.983	0.983	0.983	0.019	0.015	0.983	0.983	0.967	0.967
128	45	0.979	0.979	0.979	0.979	0.013	0.028	0.979	0.979	0.958	0.958

4.4. Results of Ensemble Learning

Based on the ablation study, we selected one Dense-1 (with 35 neurons) model, two Dense-2 (with (40,40) and (40,50) neurons) models and two 1D-CNN (with (25,25) and (30,35) neurons) models as base learners based on their kappa and F1-score performance. We performed classification with several different meta-learner classification algorithms. For KNN, the number of nearest neighbors was set to 3. For linear SVM, C was set to 0.025. For RBF SVM, the C parameter (which performs regularization by applying a penalty to reduce overfitting) was set to 1, and gamma was set to 2. For DT and RF, the max depth was set to 5. In all cases, 10-fold cross-validation was used, where each cross-validation fold was made by randomly selecting 80% of samples, and the remaining 20% were used for testing. The results are presented in Table 6.

Table 6. Ensemble learning results with different meta-learners: mean values from 10-fold cross-validation. Best values are shown in bold.

Meta-Learner	Acc	Prec	Rec	Spec	FPR	FNR	F1	AUC	MCC	Kappa
Nearest Neighbors	0.984	0.984	0.984	0.984	0.014	0.018	0.984	0.984	0.967	0.967
Linear SVM	0.974	0.974	0.974	0.974	0.029	0.023	0.974	0.974	0.948	0.948
RBF SVM	0.979	0.979	0.979	0.979	0.021	0.022	0.979	0.979	0.958	0.958
Decision Tree	0.987	0.987	0.987	0.987	0.002	0.023	0.987	0.987	0.973	0.973
Random Forest	0.991	0.991	0.991	0.991	0.008	0.009	0.991	0.991	0.983	0.983
Neural Net	0.974	0.974	0.974	0.974	0.029	0.023	0.974	0.974	0.948	0.948
AdaBoost	0.997	0.997	0.997	0.997	0.006	0	0.997	0.997	0.994	0.994
ExtraTrees	**0.999**	**0.999**	**0.998**	**1.000**	**0.000**	**0.002**	**0.999**	**0.999**	**0.999**	**0.999**
Naive Bayes	0.968	0.968	0.968	0.968	0.027	0.036	0.968	0.969	0.937	0.936
LDA	0.975	0.975	0.975	0.975	0.027	0.023	0.975	0.975	0.95	0.95
QDA	0.974	0.974	0.974	0.974	0.029	0.023	0.974	0.974	0.948	0.948

Table 6. Cont.

Meta-Learner	Acc	Prec	Rec	Spec	FPR	FNR	F1	AUC	MCC	Kappa
Logistic	0.973	0.973	0.973	0.973	0.031	0.023	0.973	0.973	0.946	0.946
Ridge	0.971	0.971	0.971	0.971	0.049	0.011	0.971	0.97	0.943	0.942
SGDC	0.975	0.975	0.975	0.975	0.027	0.023	0.975	0.975	0.95	0.95

The average performance results are visualized in Figures 5–7, whereas the results from the 10-fold cross-validation are shown as boxplots in Figures 8–10. The results demonstrate that the ExtraTrees meta-learner achieved the highest performance in terms of accuracy, AUC and F1-score measures.

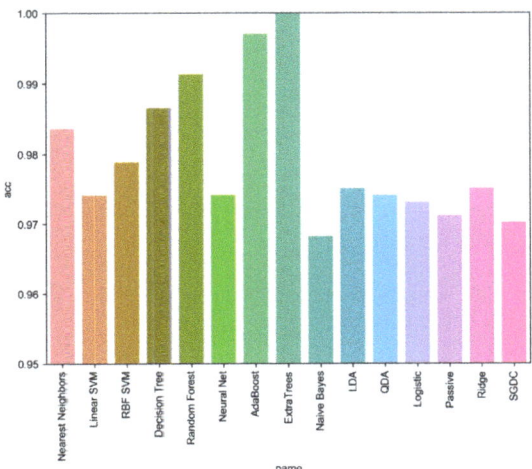

Figure 5. Malware detection performance of deep learning ensemble model by final stage meta-learner classifier: accuracy.

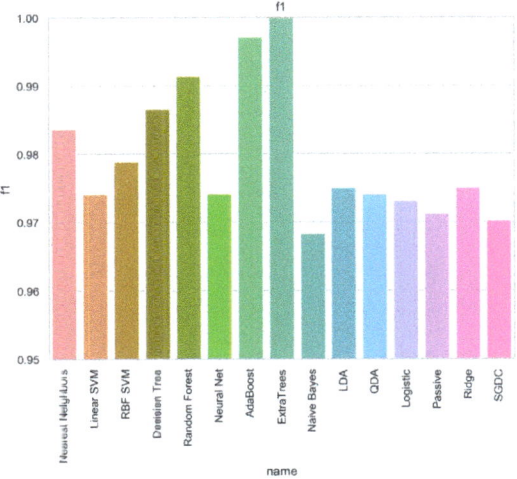

Figure 6. Malware detection performance of deep learning ensemble model by final stage meta-learner classifier: F1-score.

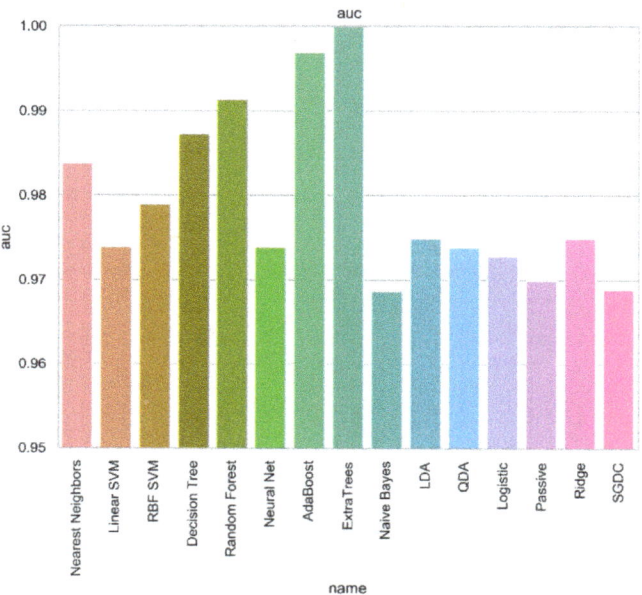

Figure 7. Malware detection performance of deep learning ensemble model by final stage meta-learner classifier: AUC.

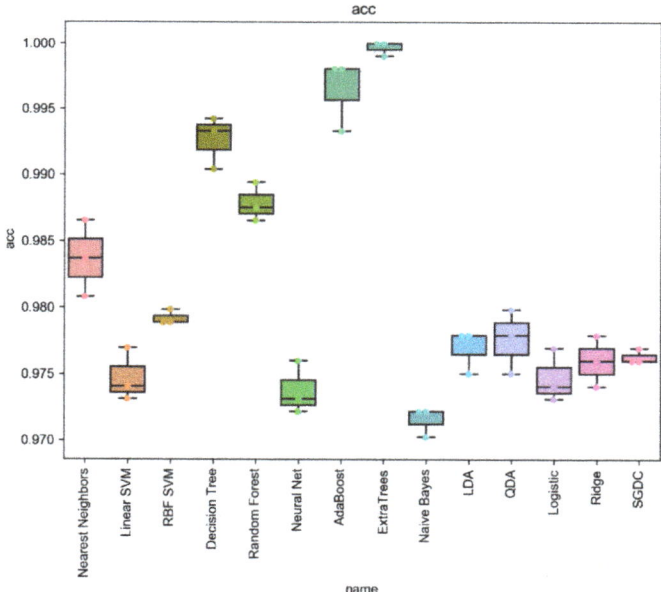

Figure 8. Malware detection performance of deep learning ensemble model by final stage meta-learner classifier: accuracy.

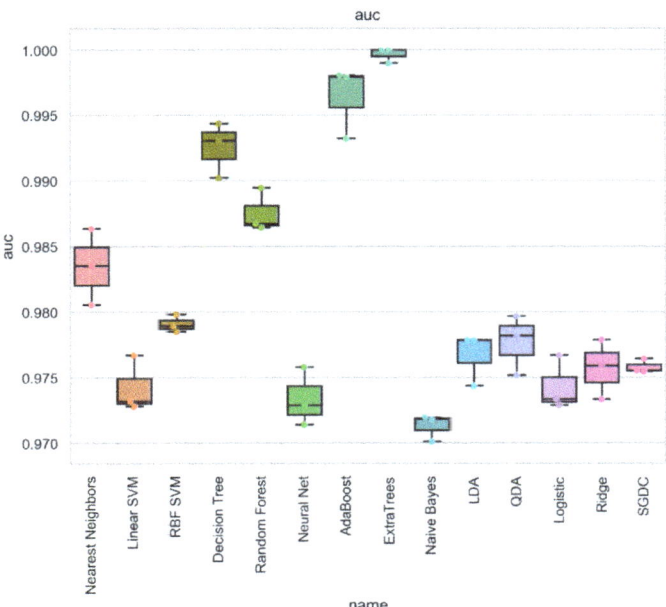

Figure 9. Malware detection performance of deep learning ensemble model by final stage meta-learner classifier: area under curve.

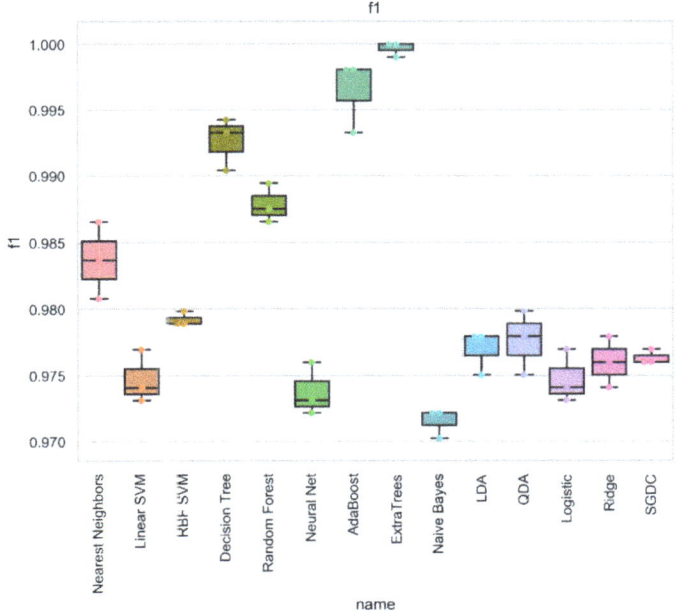

Figure 10. Malware detection performance of deep learning ensemble model by final stage meta-learner classifier: F1-score.

Finally, we present the confusion matrix of the best ensemble model (with the ET classifier as the meta-learner) in Figure 11.

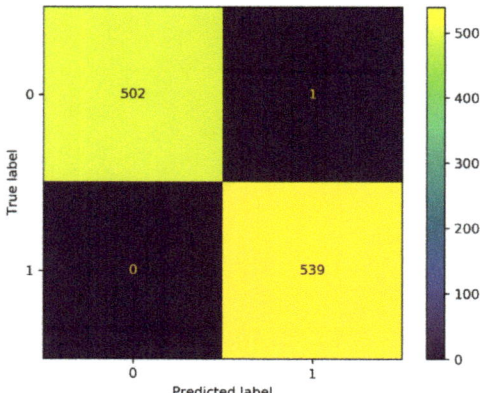

Figure 11. Confusion matrix of the best ensemble model (with the ET classifier as meta-learner).

4.5. Statistical Analysis

To perform the statistical analysis of the experimental results, we adopted the Friedman test and the Nemenyi test. The results are presented as critical difference (CD) diagrams in Figures 12–14. If the difference between the mean ranks of the meta-learners is smaller than the CD, then it is not statistically significant. The results of the Nemenyi test again show that the ExtraTrees meta-learner allows us to achieve the best performance; however, the performance of AdaBoost and Decision Tree meta-learners is not significantly different.

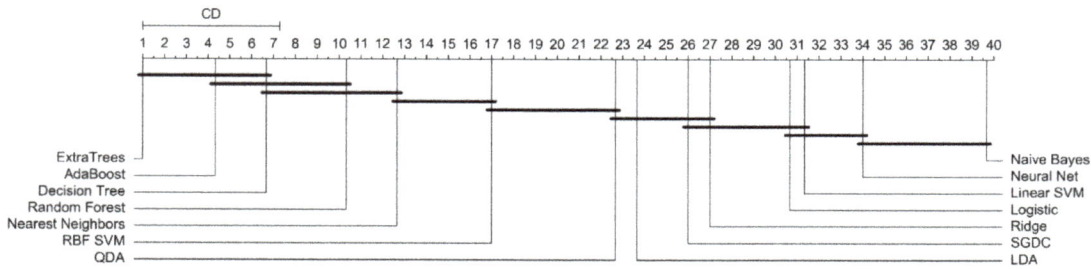

Figure 12. Comparison of mean ranks of meta-learners based on their accuracy performance: results of Nemenyi test.

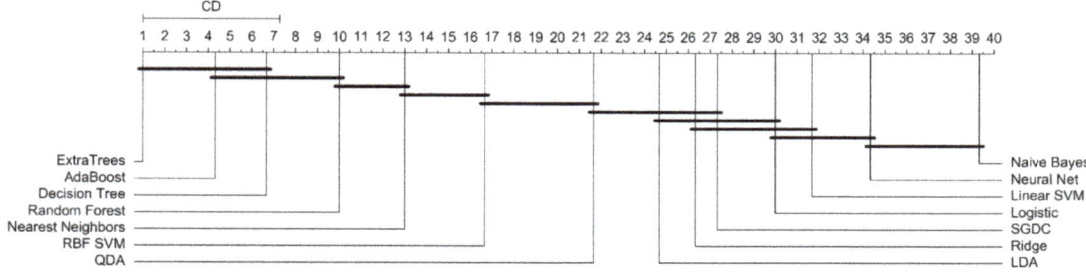

Figure 13. Comparison of mean ranks of meta-learners based on their AUC performance: results of Nemenyi test.

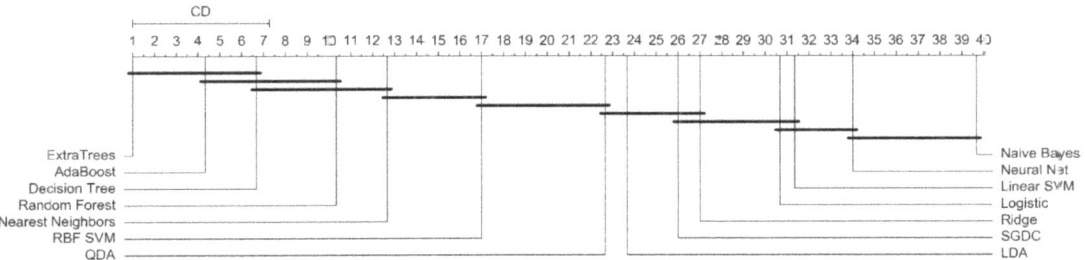

Figure 14. Comparison of mean ranks of meta-learners based on their F1-score performance: results of Nemenyi test.

4.6. Ablation Study of the Ensemble

We also conducted the ablation study to evaluate the contribution of the individual parts in the proposed ensemble classification base framework for malware recognition. We compared and analyzed the impact of the ensemble size of the classification results. We analyzed the following ensembles, consisting of a smaller number (4) of neural networks models:

1. Case ENSEMBLE1: two Dense-2 (with (40,40) and (40,50) neurons) models and two 1D-CNN (with (25,25) and (30,35) neurons) models;
2. Case ENSEMBLE2: one Dense-1 (with 35 neurons) model, one Dense-2 (with (40,40) neurons) model and two 1D-CNN ((25,25), and (30,35) neurons) models;
3. Case ENSEMBLE3: one Dense-1 (with 35 neurons) model, two Dense-2 (with (40,40) and (40,50) neurons) models and one 1D-CNN (with (30;35) neurons) model.

The results are summarized and compared in Table 7. In all cases, the ExtraTrees Classifier was used as a meta-learner. The Full Model here corresponds to the five-model ensemble with PCA scaling of data. The results show that the best performance was achieved by the full five-model ensemble with data scaling using PCA and ExtraTrees as the meta-learner.

Table 7. Comparison of ensemble models. Best values are shown in bold.

Case	Acc	Prec	Rec	Spec	FPR	FNR	F1	AUC	MCC	Kappa
ENSEMBLE1	0.989	0.988	0.987	0.979	0.011	0.012	0.989	0.989	0.968	0.968
ENSEMBLE2	0.985	0.983	0.985	0.983	0.017	0.014	0.984	0.984	0.967	0.967
ENSEMBLE3	0.985	0.984	0.984	0.985	0.013	0.016	0.986	0.988	0.971	0.970
Full Model	**0.999**	**0.999**	**0.998**	**1.000**	**0.000**	**0.002**	**0.999**	**0.999**	**0.999**	**0.999**

4.7. Comparison with Related Work.

Finally, we compare our results with some of the related work on classifying benign and malware files in Table 8 and explain in more detail below. Note that the methods working on different malware datasets were compared. Alzaylaee et al. [54] explored 2-, 3- and 4-layer fully connected neural networks on a dataset of 31,125 Android apps, with 420 static and dynamic features, while comparing the results to machine learning classifiers. The best results were achieved with a three-layer network with 200 neurons in each layer. Bakour and Ünver [55] suggested a visualization-based approach that converted software characteristics into grayscale images and then applied local and global image features as voters in an ensemble voting classifier Cai et al. [56] used information gain for feature selection and weight mapping functions derived by machine learning methods, which were optimized by the differential evolution algorithm. Chen et al. [57] used an attention network architecture based on CNN to classify apps based on their Application Programming Interface (API) call sequences. Farg et al. [58] used a DeepDetectNet deep learning model for static PE malware detection model, and an adversarial generation network RLAttackNet based on reinforcement learning, which was trained to bypass DeepDetectNet. The generated adversarial samples were used to retrain DeepDetectNet, which allowed the improvement of malware recognition accuracy.

Imtiaz et al. [59] proposed a deep multi-layer fully connected Artificial Neural Network (ANN) that has an input layer, few hidden layers and an output layer. The approach has been validated with the CICInvesAndMal2019 dataset of Android malware. Jeon and Moon [60] proposed a convolutional recurrent neural network (CRNN), which uses the opcode sequences of software as input. The front-end CNN performs opcode compression, and the back end dynamic recurrent neural network (DRNN) detects malware from the compressed sequence.

Jha et al. [61] proposed using RNN with feature vectors obtained by skip-grams of the Word2Vec embedding model for malware recognition. Namavar Jahromi et al. [62] proposed a modified Two-hidden-layered Extreme Learning Machine (TELM), which was tested on Ransomware, Windows, Internet of Things (IoT) and other malware datasets.

Narayanan and Davuluru [63] suggested using CNNs and Long Short-Term Memory (LSTM) networks for feature extraction and SVM or LR for the classification of malware based on their machine language opcodes. The approach was validated on Microsoft's Malware Classification Challenge (BIG 2015) dataset with nine malware classes. Song et al. [64] proposed a JavaScript malware detection based on the Bidirectional LSTM neural network. Wang et al. [65] suggested CrowdNet, a radial basis function network, as a malware predictor. Yen and Sun [66] extracted instruction code and applied hashing to extract features. Then, the features were transformed into images and used to train a CNN.

Table 8. Comparison with other known deep learning approaches for malware recognition. n/a—data were not provided.

Reference	Benign	Malware	Acc. (%)	Prec. (%)	Recall (%)	F-Score (%)
Alzaylaee et al. [54]	19,620	11,505	98.5	98.09	99.56	98.82
Bakour and Ünver [55]	-	4850	98.14	n/a	n/a	n/a
Cai et al. [56]	3000	3000	96.92	96.75	97.23	96.99
Chen et al. [57]	4596	4596	97.23	98.69	98.69	98.69
Fang et al. [58]	749	726	n/a	n/a	98.07	n/a
Imtiaz et al. [59]	5065	426	93.4	93.5	93.4	93.2
Jeon and Moon [60]	1000	1000	96	n/a	95	n/a
Jha et al. [61]	20,000	20,000	91.91	n/a	n/a	91.76
Namavar Jahromi et al. [62]	-	18,831	99.03	n/a	n/a	n/a
Narayanan and Davuluru [63]	-	7826	99.8	n/a	n/a	n/a
Song et al. [64]	30,487	29,893	97.71	n/a	n/a	98.29
Wang et al. [65]	6375	6375	n/a	94	n/a	n/a
Yen and Sun [66]	720	720	92.67	n/a	n/a	n/a
This paper	2488	2722	99.99	99.99	99.98	99.99

5. Conclusions

There is an increase in demand for smart methods that detect new malware variants, because the existing methods are time-consuming and vulnerable to many errors. This paper analyzed various machine learning algorithms and models of neural networks, which are smart approaches that can be used for malware detection. With neural networks used as base learners, we proposed an ensemble learning-based architecture and explored 14 machine learning algorithms as meta-learners. As baseline models, we used machine learning algorithms for comparison. We conducted our experiments on a dataset that included malware and benign files from Windows Portable Executables (PE).

In this paper, we analyzed and experimentally validated the use of ensemble learning to combine the malware prediction results given by different machine learning and deep learning models. The aim of this practice is to improve the recognition of Windows PE malware. With ensemble methods, it is not required to select any specific machine learning model. Instead, the prediction capability of each combination of the machine learning models is aggregated to create a learning procedure that achieves the best malware detection performance. We explored our proposed ensemble classification framework with lightweight fully connected and convolutional neural network architectures, and combined deep learning and machine learning techniques to learn effective and efficient malware detection models. We conducted extensive experiments on various lightweight deep learning architectures and machine learning models within the framework of ensemble learning under the same conditions for a fair comparison.

The results achieved show that the malware detection ability of ensemble stacking exceeds the ability of other machine-learning methods, including neural networks. We showed that the ensemble learning framework based on lightweight deep models could successfully tackle the problem of malware detection. The results obtained indicate that ensemble learning methods can be implemented and used as intelligent techniques for the identification of malware. The classification system with the Extra Trees algorithm as a meta-learner and an ensemble of dense ANN and 1-D CNN models obtained the best accuracy value for the classification procedure, outperforming other machine learning classification methods. Our proposed framework can lead to highly accurate malware detection models that are adapted for real-world Windows PE malware.

The application of explanatory artificial intelligence (XAI) [67] strategies to interpret the outcomes of deep learning models for malware detection will be carried out in the future to provide useful information for malware analysis researchers. We also intend to explore ensemble learning architectures and run further tests with larger databases of malware. We strive to improve the classification ability and accuracy of the ensemble learning model by refining the model architecture and validating it for multiple malware datasets in future work.

Author Contributions: Conceptualization and methodology, R.D. and A.V.; software, R.D.; validation, R.D., A.V. and J.T.; formal analysis, R.D. and J.T.; investigation, R.D., A.V. and Š.G.; resources, A.V. and J.T.; writing—original draft preparation, R.D., A.V., J.T. and Š.G.; writing—review and editing, R.D. and A.V.; visualization, R.D. and J.T.; supervision, A.V. All authors have read and agreed to the published version of the manuscript.

Funding: This paper was supported in part by European Union's Horizon 2020 research and innovation programme under Grant Agreement No. 830892, project "Strategic programs for advanced research and technology in Europe" (SPARTA).

Data Availability Statement: The dataset is available from https://github.com/urwithajit9/ClaMP.

Conflicts of Interest: The authors declare no conflict of interest. The funders had no role in the design of the study; in the collection, analyses, or interpretation of data; in the writing of the manuscript, or in the decision to publish the results.

References

1. Lallie, H.S.; Shepherd, L.A.; Nurse, J.R.C.; Erola, A.; Epiphaniou, G.; Maple, C.; Bellekens, X.J.A. Cyber Security in the Age of COVID-19: A Timeline and Analysis of Cyber-Crime and Cyber-Attacks during the Pandemic. *arXiv* **2020**, arXiv:2006.11929.
2. Anderson, R.; Barton, C.; Böhme, R.; Clayton, R.; Van Eeten, M.J.G.; Levi, M.; Moore, T.; Savage, S. Measuring the Cost of Cybercrime. In *The Economics of Information Security and Privacy*; Springer: Berlin/Heidelberg, Germany, 2013; pp. 265–300.
3. Bissell, K.; la Salle, R.; Dal, C.P. The 2020 Cyber Security Report. 2020. Available online: https://pages.checkpoint.com/cyber-security-report-2020 (accessed on 4 May 2020).
4. Chebyshev, V. Mobile Malware Evolution. 2018 Available online: https://securelist.com/mobile-malware-evolution-2018/89617 (accessed on 18 June 2020).
5. Kaspersky Lab. The Great Bank Robbery. 2015. Available online: https://www.kaspersky.com/about/press-releases/2015_the-great-bank-robbery-carbanak-cybergang-steals--1bn-from-100-financial-institutions-worldwide (accessed on 17 February 2015).
6. Kingsoft. 2015–2016 Internet Security Research Report in China. 2016. Available online: https://cn.cmcm.com/news/media/2016-01-14/60.html (accessed on 14 January 2016).
7. Bissell, K.; la Salle, R.M.; Dal, C.P. The Cost of Cybercrime—Ninth Annual Cost of Cybercrime Study. Technical Report, Ponemon Institute LLC, Accenture. 2019. Available online: https://www.accenture.com/_acnmedia/pdf-96/accenture-2019-cost-of-cybercrime-study-final.pdf (accessed on 3 March 2020).
8. Cybersecurity Ventures. Cybercrime Damages Will Cost the World $6 Trillion Annually by 2021. 2017. Available online: https://cybersecurityventures.com/cybercrime-damages-6-trillion-by-2021/ (accessed on 21 December 2020).

9. Williams, C.M.; Chaturvedi, R.; Chakravarthy, K. Cybersecurity Risks in a Pandemic. *J. Med. Internet Res.* **2020**, *22*, e23692. [CrossRef]
10. Hakak, S.; Khan, W.Z.; Imran, M.; Choo, K.-K.R.; Shoaib, M. Have You Been a Victim of COVID-19-Related Cyber Incidents? Survey, Taxonomy, and Mitigation Strategies. *IEEE Access* **2020**, *8*, 124134–124144. [CrossRef]
11. Seh, A.H.; Zarour, M.; Alenezi, M.; Sarkar, A.K.; Agrawal, A.; Kumar, R.; Khan, R.A. Healthcare Data Breaches: Insights and Implications. *Healthcare* **2020**, *8*, 133. [CrossRef]
12. Pierazzi, F.; Mezzour, G.; Han, Q.; Colajanni, M.; Subrahmanian, V.S. A Data-driven Characterization of Modern Android Spyware. *ACM Trans. Manag. Inf. Syst.* **2020**, *11*, 1–38. [CrossRef]
13. Odusami, M.; Abayomi-Alli, O.; Misra, S.; Shobayo, O.; Damasevicius, R.; Maskeliunas, R. Android Malware Detection: A Survey. In *Communications in Computer and Information Science*; Springer International Publishing: Cham, Switzerland, 2018; Volume 942, pp. 255–266. [CrossRef]
14. Subairu, S.O.; Alhassan, J.; Misra, S.; Abayomi-Alli, O.; Ahuja, R.; Damasevicius, R.; Maskeliunas, R. An Experimental Approach to Unravel Effects of Malware on System Network Interface. In *Lecture Notes in Electrical Engineering*; Springer International Publishing: Cham, Switzerland, 2019; Volume 612, pp. 225–235. [CrossRef]
15. Alsoghyer, S.; Almomani, I. Ransomware Detection System for Android Applications. *Electronics* **2019**, *8*, 868. [CrossRef]
16. Hindy, H.; Atkinson, R.; Tachtatzis, C.; Colin, J.-N.; Bayne, E.; Bellekens, X. Utilising Deep Learning Techniques for Effective Zero-Day Attack Detection. *Electronics* **2020**, *9*, 1684. [CrossRef]
17. Martín, I.; Hernández, J.A.; Santos, S.D.L. Machine-Learning based analysis and classification of Android malware signatures. *Futur. Gener. Comput. Syst.* **2019**, *97*, 295–305. [CrossRef]
18. Aslan, O.; Samet, R. A Comprehensive Review on Malware Detection Approaches. *IEEE Access* **2020**, *8*, 6249–6271. [CrossRef]
19. Souri, A.; Hosseini, R. A state-of-the-art survey of malware detection approaches using data mining techniques. *Hum. Cent. Comput. Inf. Sci.* **2018**, *8*, 3. [CrossRef]
20. Ye, Y.; Li, T.; Adjeroh, D.; Iyengar, S.S. A Survey on Malware Detection Using Data Mining Techniques. *ACM Comput. Surv.* **2017**, *50*, 1–40. [CrossRef]
21. Ucci, D.; Aniello, L.; Baldoni, R. Survey of machine learning techniques for malware analysis. *Comput. Secur.* **2019**, *81*, 123–147. [CrossRef]
22. Liu, K.; Xu, S.; Xu, G.; Zhang, M.; Sun, D.; Liu, H. A Review of Android Malware Detection Approaches Based on Machine Learning. *IEEE Access* **2020**, *8*, 124579–124607. [CrossRef]
23. Truong, T.C.; Diep, Q.B.; Zelinka, I. Artificial Intelligence in the Cyber Domain: Offense and Defense. *Symmetry* **2020**, *12*, 410. [CrossRef]
24. Ngo, Q.-D.; Nguyen, H.-T.; Le, V.-H.; Nguyen, D.-H. A survey of IoT malware and detection methods based on static features. *ICT Express* **2020**, *6*, 280–286. [CrossRef]
25. Egele, M.; Scholte, T.; Kirda, E.; Kruegel, C. A survey on automated dynamic malware-analysis techniques and tools. *ACM Comput. Surv.* **2012**, *44*, 1–42. [CrossRef]
26. Ye, Y.; Wang, D.; Li, T.; Ye, D.; Jiang, Q. An intelligent PE-malware detection system based on association mining. *J. Comput. Virol.* **2008**, *4*, 323–334. [CrossRef]
27. Nisa, M.; Shah, J.H.; Kanwal, S.; Raza, M.; Khan, M.A.; Damaševičius, R.; Blažauskas, T. Hybrid Malware Classification Method Using Segmentation-Based Fractal Texture Analysis and Deep Convolution Neural Network Features. *Appl. Sci.* **2020**, *10*, 4966. [CrossRef]
28. Yong, B.; Wei, W.; Li, K.; Shen, J.; Zhou, Q.; Wozniak, M.; Połap, D.; Damaševičius, R. Ensemble machine learning approaches for webshell detection in Internet of things environments. *Trans. Emerg. Telecommun. Technol.* **2020**. [CrossRef]
29. Wei, W.; Woźniak, M.; Damaševičius, R.; Fan, X.; Li, Y. Algorithm research of known-plaintext attack on double random phase mask based on WSNs. *J. Internet Technol.* **2019**, *20*, 39–48. [CrossRef]
30. Berman, D.S.; Buczak, A.L.; Chavis, J.S.; Corbett, C.L. A Survey of Deep Learning Methods for Cyber Security. *Information* **2019**, *10*, 122. [CrossRef]
31. Ren, Z.; Wu, H.; Ning, Q.; Hussain, I.; Chen, B. End-to-end malware detection for android IoT devices using deep learning. *Ad Hoc Netw.* **2020**, *101*, 102098. [CrossRef]
32. Yuxin, D.; Siyi, Z. Malware detection based on deep learning algorithm. *Neural Comput. Appl.* **2017**, *31*, 461–472. [CrossRef]
33. Pei, X.; Yu, L.; Tian, S. AMalNet: A deep learning framework based on graph convolutional networks for malware detection. *Comput. Secur.* **2020**, *93*, 101792. [CrossRef]
34. Čeponis, D.; Goranin, N. Investigation of Dual-Flow Deep Learning Models LSTM-FCN and GRU-FCN Efficiency against Single-Flow CNN Models for the Host-Based Intrusion and Malware Detection Task on Univariate Times Series Data. *Appl. Sci.* **2020**, *10*, 2373. [CrossRef]
35. Huang, X.; Ma, L.; Yang, W.; Zhong, Y. A Method for Windows Malware Detection Based on Deep Learning. *J. Signal Process. Syst.* **2020**, 1–9. [CrossRef]
36. Martins, N.; Cruz, J.M.; Cruz, T.; Abreu, P.H. Adversarial Machine Learning Applied to Intrusion and Malware Scenarios: A Systematic Review. *IEEE Access* **2020**, *8*, 35403–35419. [CrossRef]
37. Zador, A.M. A critique of pure learning and what artificial neural networks can learn from animal brains. *Nat. Commun.* **2019**, *10*, 1–7. [CrossRef] [PubMed]

38. Idrees, F.; Rajarajan, M.; Conti, M.; Chen, T.M.; Rahulamathavan, Y. PIndroid: A novel Android malware detection system using ensemble learning methods. *Comput. Secur.* **2017**, *68*, 36–46. [CrossRef]
39. Feng, P.; Ma, J.; Sun, C.; Xu, X.; Ma, Y. A Novel Dynamic Android Malware Detection System with Ensemble Learning. *IEEE Access* **2018**, *6*, 30996–31011. [CrossRef]
40. Wang, W.; Li, Y.; Wang, X.; Liu, J.; Zhang, X. Detecting Android malicious apps and categorizing benign apps with ensemble of classifiers. *Futur. Gener. Comput. Syst.* **2018**, *78*, 987–994. [CrossRef]
41. Yan, J.; Qi, Y.; Rao, Q. Detecting Malware with an Ensemble Method Based on Deep Neural Networks. *Secur. Commun. Netw.* **2018**, *2018*, 1–16. [CrossRef]
42. Gupta, D.; Rani, R. Improving malware detection using big data and ensemble learning. *Comput. Electr. Eng.* **2020**, *86*, 106729. [CrossRef]
43. Sagi, O.; Rokach, L. Ensemble learning: A survey. *Wiley Interdiscip. Rev. Data Min. Knowl. Discov.* **2018**, *8*, e1249. [CrossRef]
44. Basu, I. Malware detection based on source data using data mining: A survey. *Am. J. Adv. Comput.* **2016**, *3*, 18–37.
45. Srivastava, N.; Hinton, G.; Krizhevsky, A.; Sutskever, I.; Salakhutdinov, R. Dropout: A simple way to prevent neural networks from overfitting. *J. Mach. Learn. Res.* **2014**, *15*, 1929–1958.
46. Ruder, S. An overview of gradient descent optimization algorithms. *arXiv* **2016**, arXiv:1609.04747.
47. Duchi, J.; Hazan, E.; Singer, Y. Adaptive subgradient methods for online learning and stochastic optimization. *J. Mach. Learn. Res.* **2011**, *12*, 2121–2159.
48. Tieleman, T.; Hinton, G. Lecture 6.5-rmsprop: Divide the gradient by a running average of its recent magnitude. *Coursera Neural Netw. Mach. Learn.* **2012**, *4*, 26–31.
49. Kingma, D.P.; Ba, J. Adam: A method for stochastic optimization. In Proceedings of the International Conference on Learning Representation (ICLR), San Diego, CA, USA, 5–8 May 2015.
50. Ragab, M.; Abdulkadir, S.; Aziz, N.; Al-Tashi, Q.; Alyousifi, Y.; Alhussian, H.; Alcushaibi, A. A Novel One-Dimensional CNN with Exponential Adaptive Gradients for Air Pollution Index Prediction. *Sustainability* **2020**, *12*, 10090. [CrossRef]
51. Luo, L.; Xiong, Y.; Liu, Y.; Sun, X. Adaptive gradient methods with dynamic bound of learning rate. *arXiv* **2019**, arXiv:1902.09843.
52. Van der Laan, M.J.; Polley, E.C.; Hubbard, A.E. Super Learner. *Stat. Appl. Genet. Mol. Biol.* **2007**, *6*. [CrossRef] [PubMed]
53. Geurts, P.; Ernst, D.; Wehenkel, L. Extremely randomized trees. *Mach. Learn.* **2006**, *63*, 3–42. [CrossRef]
54. Alzaylaee, M.K.; Yerima, S.Y.; Sezer, S. DL-Droid: Deep learning based android malware detection using real devices. *Comput. Secur.* **2020**, *89*, 101663. [CrossRef]
55. Bakour, K.; Ünver, H.M. VisDroid: Android malware classification based on local and global image features, bag of visual words and machine learning techniques. *Neural Comput. Appl.* **2020**, *2020*, 1–21. [CrossRef]
56. Cai, L.; Li, Y.; Xiong, Z. JOWMDroid: Android malware detection based on feature weighting with joint optimization of weight-mapping and classifier parameters. *Comput. Secur.* **2021**, *100*, 102086. [CrossRef]
57. Chen, J.; Guo, S.; Ma, X.; Li, H.; Guo, J.; Chen, M.; Pan, Z. SLAM: A Malware Detection Method Based on Sliding Local Attention Mechanism. *Secur. Commun. Netw.* **2020**, *2020*, 1–11. [CrossRef]
58. Fang, Y.; Zeng, Y.; Li, B.; Liu, L.; Zhang, L. DeepDetectNet vs RLAttackNet: An adversarial method to improve deep learning-based static malware detection model. *PLoS ONE* **2020**, *15*, e0231626. [CrossRef]
59. Imtiaz, S.I.; Rehman, S.U.; Javed, A.R.; Jalil, Z.; Liu, X.; Alnumay, W.S. DeepAMD: Detection and identification of Android malware using high-efficient Deep Artificial Neural Network. *Futur. Gener. Comput. Syst.* **2021**, *115*, 844–856. [CrossRef]
60. Jeon, S.; Moon, J. Malware-Detection Method with a Convolutional Recurrent Neural Network Using Opcode Sequences. *Inf. Sci.* **2020**, *535*, 1–15. [CrossRef]
61. Jha, S.; Prashar, D.; Long, H.V.; Taniar, D. Recurrent neural network for detecting malware. *Comput. Secur.* **2020**, *99*, 102037. [CrossRef]
62. Jahromi, A.N.; Hashemi, S.; Dehghantanha, A.; Choo, K.-K.R.; Karimipour, H.; Newton, D.E.; Parizi, R.M. An improved two-hidden-layer extreme learning machine for malware hunting. *Comput. Secur.* **2020**, *89*, 101655. [CrossRef]
63. Narayanan, B.N.; Davuluru, V.S.P. Ensemble Malware Classification System Using Deep Neural Networks. *Electronics* **2020**, *9*, 721. [CrossRef]
64. Song, X.; Chen, C.; Cui, B.; Fu, J. Malicious JavaScript Detection Based on Bidirectional LSTM Model. *Appl. Sci.* **2020**, *10*, 3440. [CrossRef]
65. Wang, X.; Li, C.; Song, D.; Wang, C. CrowdNet: Identifying Large-Scale Malicious Attacks Over Android Kernel Structures. *IEEE Access* **2020**, *8*, 15823–15837. [CrossRef]
66. Yen, Y.-S.; Sun, H.-M. An Android mutation malware detection based on deep learning using visualization of importance from codes. *Microelectron. Reliab.* **2019**, *93*, 109–114. [CrossRef]
67. Zanni-Merk, C. On the Need of an Explainable Artificial Intelligence. In Proceedings of the 40th Anniversary International Conference on Information Systems Architecture and Technology, Wroclaw, Poland, 15–17 September 2019; p. 3.

Article

Deep Learning Techniques for Android Botnet Detection

Suleiman Y. Yerima [1,*], Mohammed K. Alzaylaee [2], Annette Shajan [3] and Vinod P [4]

1. Cyber Technology Institute, De Montfort University, Leicester LE1 9BH, UK
2. College of Computing in Al-Qunfudhah, Umm Al-Qura University, Mecca 21955, Saudi Arabia; mkzaylaee@uqu.edu.sa
3. RV College of Engineering, Bengaluru 560059, India; annetteshajan@gmail.com
4. Department of Computer Applications, Cochin University of Science and Technology, Cochin 682022, India; vinod.p@cusat.ac.in
* Correspondence: syerima@dmu.ac.uk

Abstract: Android is increasingly being targeted by malware since it has become the most popular mobile operating system worldwide. Evasive malware families, such as Chamois, designed to turn Android devices into bots that form part of a larger botnet are becoming prevalent. This calls for more effective methods for detection of Android botnets. Recently, deep learning has gained attention as a machine learning based approach to enhance Android botnet detection. However, studies that extensively investigate the efficacy of various deep learning models for Android botnet detection are currently lacking. Hence, in this paper we present a comparative study of deep learning techniques for Android botnet detection using 6802 Android applications consisting of 1929 botnet applications from the ISCX botnet dataset. We evaluate the performance of several deep learning techniques including: CNN, DNN, LSTM, GRU, CNN-LSTM, and CNN-GRU models using 342 static features derived from the applications. In our experiments, the deep learning models achieved state-of-the-art results based on the ISCX botnet dataset and also outperformed the classical machine learning classifiers.

Keywords: botnet detection; deep learning; Android botnets; convolutional neural networks; dense neural networks; recurrent neural networks; long short-term memory; gated recurrent unit; CNN-LSTM; CNN-GRU

1. Introduction

The increase in Android's popularity worldwide has made it a continuous target for malware authors. The volume of malware targeting Android has continued to grow in the last few years [1,2]. Android has been attacked by numerous malware families aimed at infecting mobile devices and turning them into bots. These bots become parts of larger botnets that are usually under the control of a malicious user or group of users known as botmasters. The Android botnets may be used to launch various types of attacks such as distributed denial of service (DDoS) attacks, phishing, click fraud, theft of credit card details or other credentials, generation and distribution of spam, etc. Nowadays, malicious Android botnets have become a serious threat. Additionally, their increasing use of sophisticated evasive techniques such as self-protection or multi-staged payload execution [3], calls for more effective approaches to detect them.

The Chamois malware family [3–5], which was discovered on Google Play in August 2016 is one example of the emerging sophisticated Android botnet threats. By March 2018, Chamois had infected over 20 million devices, which were commandeered into a botnet that received instructions from a remote command and control server [5]. The botnet was used to serve malicious advertisements and to direct victims to premiums Short Message Service (SMS) scams. The early version of Chamois disguised as benign apps that tricked users into downloading it on their devices, and this was detected and almost completely eradicated by the Android security team. Later versions of Chamois appeared which were

distributed by tricking developers and device manufacturers into incorporating the botnet code directly into their apps. Chamois was sold to developers as a legitimate software development kit, and to the device manufacturers as a mobile payment solution [5].

The emergence of evasive and technically complex families like Chamois has driven interest in adopting machine learning based techniques as a means to improve existing detection systems. In the past few years, several works have investigated traditional machine learning techniques such as Support Vector Machines (SVM), Random Forest, Decision Trees, etc., for Android botnet detection. Some of the more recent machine learning based Android botnet detection work, such as ref. [6] and ref. [7] have focused on deep learning. Nevertheless, empirical studies that extensively investigate various deep learning techniques to provide insight into their relative performance for Android botnet detection are currently lacking. Hence, in this paper, we present a comparative analysis of deep learning models for Android botnet detection using the publicly available ISCX botnet dataset. Our approach is based on classification of unknown applications into 'clean' or 'botnet' using 342 static features extracted from the apps. We evaluate the performance of several deep learning models on 6802 apps consisting of 1929 botnet apps from the ISCX botnet dataset. The models investigated include Convolutional Neural Networks (CNN), Dense Neural Networks (DNN), Gated Recurrent Units (GRU), Long Short-Term Memory (LSTM), as well as more complex networks like CNN-LSTM and CNN-GRU.

The rest of the paper is organized as follows: Section 2 contains related works. Section 3 gives an overview of the overall system for deep learning-based Android botnet detection, while Section 4 provides brief background discussions of the deep learning models that were built for this study. Section 5 discusses the methodology and experimental approach, while Section 6 presents the results and discussion of results. Finally, the conclusions and future work are outlined in Section 7.

2. Related Work

Kadir et al. in their paper [8], studied several families of Android botnets aiming to gain a better understanding of the botnets and their communication characteristics. They presented a deep analysis of the Command and Control channels and built-in URLs of the Android botnets. They provided insights into each malicious infrastructure underlying the families, and uncovered the relationships between the botnet families by using a combination of static and dynamic analysis with visualization. From their work, the ISCX Android botnet dataset consisting of 1929 samples from 14 Android botnet families emerged. Since then, several works on Android botnet detection have been based on the dataset which is available from ref. [9].

Anwar et al. [10] proposed a mobile botnet detection method based on static features. They combined permissions, MD5 signatures, broadcast receivers, and background services to obtain a comprehensive set of features. They then utilized these features to implement machine learning based classifiers to detect mobile botnets. Having performed experiments using 1400 botnet applications of the ISCX dataset, combined with an extra 1400 benign applications, they recorded an accuracy of 95.1%, a recall of 0.827, and a precision of 0.97 as their best results.

Android Botnet Identification System (ABIS) was proposed in [11] to detect Android botnets. The method is based on static and dynamic features consisting of API calls, permissions, and network traffic. ABIS was evaluated with several machine learning techniques. In the end, Random Forest was found to perform better than the other algorithms by achieving 0.972 precision and 0.96 recall.

In ref. [12], machine learning was used to detect Android botnets using permissions and their protection levels as features. Initially, 138 features were utilized and then increased to 145 after protection levels were added as novel features. In total, four machine learning models (i.e., Random Forest, multilayer perceptron (MLP), Decision Trees, and Naive Bayes) were evaluated on 3270 applications containing 1635 benign and 1635 botnets from the ISCX dataset. Random Forest was found to have the best results yielding 97.3%

accuracy, 0.987 recall, and 0.985 precision. The authors of [13] also utilized only the requested permissions' as features and applied Information Gain to reduce the features and select the most significant requested permissions. They evaluated their approach using Decision Trees, Naive Bayes, and Random Forest. In their experiments, Random Forest performed best, with an accuracy of 94.6% and false positive rate of 0.099%.

Karim et al. in [14], proposed DeDroid, a static analysis approach to investigate properties that are specific to botnets that can be used in the detection of mobile botnets. In their approach, 'critical features' were first identified by observing the coding behavior of a few known malware binaries that possess Command and Control features. These 'critical features' were then compared with features of malicious applications from Drebin dataset [15]. The comparison with 'critical features' suggested that 35% of the malicious applications in the Drebin dataset could be classed as botnets. However, according to their study, a closer examination confirmed 90% of the apps as botnets.

Jadhav et al. [16], present a cloud-based Android botnet detection system that leverages dynamic analysis by using a virtual environment with cluster analysis. The toolchain for the dynamic analysis process is composed of strace, netflow, logcat, sysdump, and tcpdump within the botnet detection system. However, in the paper there were no experimental results provided to evaluate the effectiveness of the proposed cloud-based solution. Moreover, the virtual environment can easily be evaded by the botnets using different fingerprinting techniques. In addition, being a dynamic-analysis based approach, the systems effectiveness could be degraded by the lack of complete code coverage [17,18].

In ref. [19], a method was proposed by Bernardeschia et al. to identify Android botnets through model checking. Model checking is an automated technique used in verifying finite state systems. This is achieved by checking whether a structure representing a system satisfies a temporal logic formula describing their expected behavior. In particular, static analysis is used to derive a set of finite state automata from the Java byte code that represents approximate information about the run-time behavior of an app. Afterwards, the botnet malicious behavior is formulated using temporal logic formulae [20]; then by adopting a model checker, it can be automatically checked whether the code is malicious and identify where the botnet code is located within the application. These properties are checked using the CAAL (Concurrency Workbench, Aalborg Edition) [21] formal verification environment. The authors evaluated their approach on 96 samples from the Rootsmart botnet family, 28 samples from the Tigerbot botnet family, in addition to 1000 clean samples. The results obtained on the 1124 app samples showed perfect (100%) accuracy, precision, and recall.

Alothman and Rattadilok [22] proposed a source code mining approach based on reverse engineering and text mining techniques to identify Android botnet applications. Dex2Jar was used to reverse engineer the Android apps to Java source code. Natural Language Processing techniques were applied to the obtained Java source code. They also evaluated a 'source code metrics (SCM)' approach of classifying the apps into 'botnet' or 'clean'. In the SCM approach, statistical measures, such as total number of code lines, code to comment ratio, etc., were extracted from the source code and the metrics were used as features for training machine learning classifiers. The Java source code was extracted from 9 apps from 9 ISCX botnet families, as well as 12 normal apps. The TextToWordVector filter within WEKA (Waikato Environment for Knowledge Analysis), together with TF-IDF, was then applied to the code. They also applied WEKA's StringToWordVector with TF-IDF filter while varying the numbers of the 'words to keep' parameter. SubSetEval feature selection method was used to reduce the features. The features were applied to Naive Bayes, KNN, J48, SVM, and Random Forest algorithms, were KNN obtained the best performance.

In ref. [23], a real-time signature-based detection system is proposed to combat SMS botnets by first applying pattern-matching detection approaches for incoming and outgoing SMS text messages. In the second step, rule-based techniques are used to label unknown SMS messages as suspicious or normal. Their method was evaluated with over 12,000 test messages, where all 747 malicious SMS messages were detected in the dataset.

However, the system produced some false positives where 349 SMS messages were flagged as suspicious. In ref. [24], a botnet detection technique called 'Logdog' is proposed for mobile devices using log analysis. The approach relies on analyzing the logs of mobile devices to find evidence of botnet activities. Logdog writes logcat messages to a text file in the background while the Android user continues to use their device. The system targets HTTP botnets looking for events or series of events that indicate botnet activities and was tested manually on a botnet and a normal app.

In ref. [6], Android botnet detection based on CNN and using permissions as features was proposed. In the proposed method, apps are represented as images that are constructed based on the co-occurrence of permissions used within the applications. The images were then used to train a CNN-based binary classifier. The binary classifier was evaluated using 5450 apps containing 1800 botnet apps from the ISCX dataset. They obtained an accuracy of 97.2%, with a recall of 0.96, precision of 0.955, and f-measure of 0.957. Similarly, ref. [7] proposes an Android botnet detection approach based on CNN, where not only permissions were used as features but also API calls, Commands, Intents, and Extra Files. Unlike in ref. [6], 1D CNN was used and the model was evaluated with the 1929 ISCX botnet apps and 4873 benign apps resulting in 98.9% accuracy, 0.978 recall, 0.983 precision, and 0.981 F1-score.

Different from the aforementioned earlier works, this paper aims to investigate the performance of several deep learning techniques to gain insight into their effectiveness in detecting Android botnets based on the extraction of 342 static features from the applications. To this end, we implemented CNN, DNN, LSTM, GRU, CNN-LSTM, and CNN-GRU models and evaluated the models using 1929 ISCX botnet apps and 4873 benign apps. The deep learning models developed in the study are discussed in Section 4 and the results of the experiments with the models are presented in Section 6.

3. Deep Learning-Based Android Botnet Detection System

At a fundamental level, our botnet detection system is designed to distinguish between clean apps and botnet apps. As a result, it may sometimes fail to correctly classify an unknown app by mistakenly identifying a benign app as botnet or vice-versa. The various accuracy metrics used in the experiments presented in Section 6 will enable us to capture the extent to which a given deep learning model used as a classifier can be relied upon to correctly predict which category an unknown app should belong to. The classification system is implemented by extracting static features from thousands of applications consisting of both botnet and clean examples. A bespoke tool that we developed in Python for automated reverse engineering of Android Package files (APKs) was utilized in the process. Using the tool, we extracted a total of 342 features from 5 different categories shown in Table 1.

Table 1. The five types of features used in developing the deep learning models.

Feature Types	Number
API calls	135
Permissions	130
Commands	19
Extra executables	5
Intents	53
Total	342

The five feature types include: (1) API calls (2) commands (3) permissions (4) Intents (5) extra (binary or executable) files. Most of the features were from the 'API calls' and 'permissions' category as shown in Table 1. A selection of the features is shown in Table 2. These features are represented as vectors of binary numbers with each feature in the vector represented by a '1' or '0'. Each feature vector (corresponding to one application) is labelled with its class. The feature vectors are loaded into the deep learning model during the training phase. After training, the model can then be used to predict the class (clean or botnet) of an unknown application using its extracted feature vector. Figure 1 gives a high level overview of the overall botnet detection system.

Table 2. Examples of features extracted for the deep learning models.

Feature Name	Type
TelephonyManager.*getDeviceId	API
TelephonyManager.*getSubscriberId	API
abortBroadcast	API
SEND_SMS	Permission
DELETE_PACKAGES	Permission
PHONE_STATE	Permission
SMS_RECIVED	Permission
Ljava.net.InetSocketAddress	API
READ_SMS	Permission
Android.intent.action.BOOT_COMPLETED	Intent
io.File.*delete(	API
Chown	Command
Chmod	Command
Mount	Command
.apk	Extra File
.zip	Extra File
.dex	Extra File
.jar	Extra file
CAMERA	Permission
ACCESS_FINE_LOCATION	Permission
INSTALL_PACKAGES	Permission
android.intent.action.BATTERY_LOW	Intent
.so	Extra File
android.intent.action.POWER_CONNECTED	Intent
System.*LoadLibrary	API

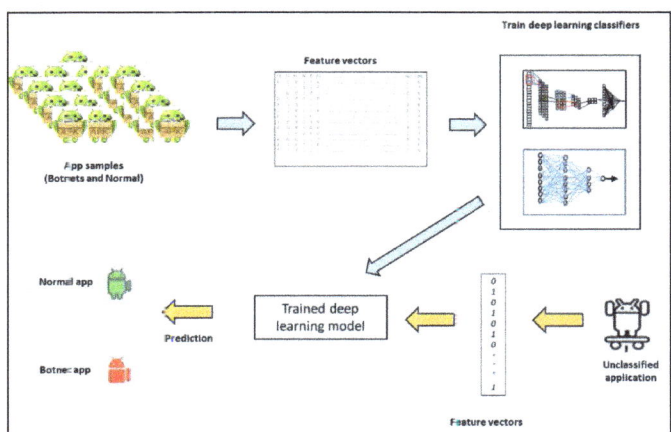

Figure 1. Overview of the deep learning-based detection system for Android botnets.

4. Deep Learning Techniques Applied to Android Botnet Detection

4.1. Convolutional Neural Networks

A CNN is a feedforward neural network whereby the information moves only in the forward direction from the input node, through the hidden nodes to the output nodes with no loops or cycles. Such (feedforward) networks are primarily used for pattern recognition. CNN generally works well for identifying simple patterns in data which will then be used to form more complex patterns in the higher or deeper layers. CNNs typically consists of convolutional layers and pooling layers. The role of the convolutional layer is to detect local conjunctions of features from the previous layer, while the role of the pooling layer is to merge semantically similar features into one [25]. CNNs combine concepts such as shared weights, local receptive fields and spatial subsampling [26]. They take advantage of many parallel and cascaded convolutional filters to solve high dimensional non-convex problems such as regression, image classification, semantic segmentation, object detection, etc. Due to weight sharing in each layer and by processing limited dimensions, a CNN requires fewer parameters than a traditional neural network and is much easier to train.

Datasets that possess a one-dimensional structure can be processed using a one-dimensional convolutional neural network (1D CNN). A 1D CNN is quite effective when you expect to derive interesting features from shorter (fixed-length) segments of the overall feature set, and where the location of the feature within the segment is not important. The use of 1D CNN can be commonly found in Natural Language Processing (NLP) applications. Similarly, 1D CNN is applicable in problems where vectorized data are used to represent the characteristics of the items whose state or category is being predicted (e.g., an Android application). The 1D CNN could be used to extract potentially more discriminative feature representations that describe any existing patterns or relationships within segments of the vectors characterizing each entity in the dataset. These new features are then fed into a classifier (e.g., LSTM, GRU or a fully connected layer) which will in turn process the derived features to produce a set of outputs that will contribute towards a final classification decision. Hence, CNNs can be employed as feature extraction layers for a given classifier which then eliminates the need to apply separate feature ranking and selection outside of the deep learning model.

Figure 2 depicts a 1D CNN model made up of two convolutional layers and two max pooling layers. The output of last pooling layer is flattened and connected to a dense (fully connected) layer of N units. The N-unit dense layer is in then connected to a final output layer containing a single neuron with a sigmoid activation function which is given by: $S = \frac{1}{1+e^{-x}}$.

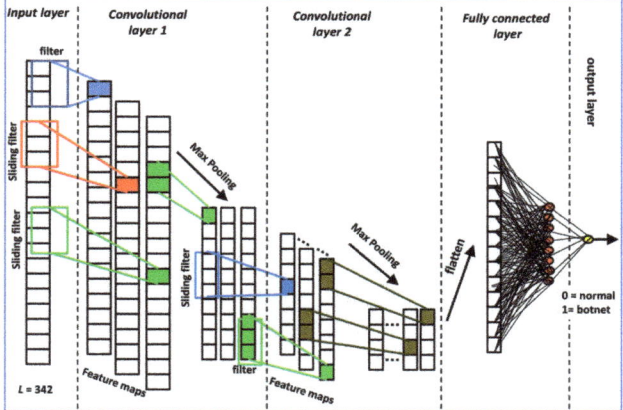

Figure 2. 1D CNN model with 2 convolutional and max pooling layers feeding a dense (fully connected) layer. The model is designed for botnet detection by classifying Android applications into 'normal' or 'botnet'.

The output layer performs the final classification into one of two classes, i.e., 'botnet' or 'normal'. The convolutional layers utilize Rectified Linear Units (ReLU) with activation function given by: $f(x) = max(0, x)$. ReLU helps to mitigate vanishing and exploding gradient issues [27].

4.2. Long-Short Term Memory

LSTM [28,29] is a type of recurrent neural network (RNN) which, unlike feedforward networks, utilizes feedback and is able to 'memorize' parts of the input and use them in making predictions. RNNs are designed to handle sequential data and thus have found popular application in areas such as speech recognition and machine translation. Different from traditional artificial neural networks that fully connect all nodes, or CNN that explore nodes from local to global layer by layer, RNNs use state neurons to explore the relationship in context. Traditional RNNs have a known problem of vanishing gradients which hinders their ability to have long term memory and thus can only make predictions based on the most recent information in the sequence. LSTM solves the vanishing gradient problem and is therefore able to process longer sequences (long term memory). LSTM is a recurrent neural network that can understand contextual information from a sequence of features. It has the ability to add or remove information from the hidden state vector with the aid of a gate function, thereby retaining important information in the hidden layer vectors.

As shown in Figure 3a, LSTM consists of three gate functions. These include: the forget gate, the input gate, and the output gate. The forget gate is used to control the amount of information in C_{t-1} is retained in the process of computing C_t and it (the forget vector) can be expressed as:

$$f_t = \sigma\left(U^f x_t + W^f h_{t-1} + b_f\right) \quad (1)$$

where U^f, W^f, and b_f constitute the parameters of the forget gate and x_t is the input vector in step t, while h_{t-1} is the hidden state vector in the previous step $t-1$. The input gate determines how much information of x_t is added to C_t and can be expressed as:

$$f_t = \sigma\left(U^f x_t + W^f h_{t-1} + b_f\right) \quad (2)$$

where U^i, W^i, and b_i are the parameters of the input gate and hence C_t can be calculated by relying on the forget gate vector f_t as well as the input gate vector i_t as follows:

$$f_t = \sigma\left(U^f x_t + W^f h_{t-1} + b_f\right) \quad (3)$$

where $\tilde{C}_t = tanh(U^c x_t + W^c h_{t-1} + b_C)$ denotes the information represented in the hidden layer vector. Note that $*$ denotes the Hadamard (element-wise) product. The output gate controls the output in C_t, and we have:

$$o_t = \sigma(U^o x_t + W^o h_{t-1} + b_o), \ h_t = o_t * tanh(C_t) \quad (4)$$

where U^o, W^o, and b_o are the parameters of the output gate and C_t is the internal state in step t.

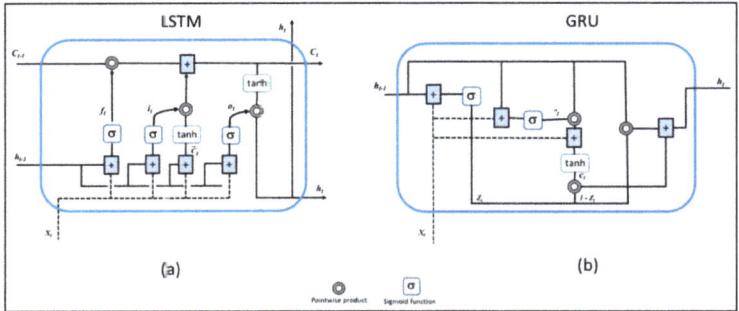

Figure 3. Recurrent neural networks. (a) LSTM; (b) GRU.

4.3. Gated Recurrent Units

A GRU [30] is also a kind of RNN model and a variant of LSTM. However, unlike LSTM which has three gates, GRU has only two gates, i.e., reset gate and update gate, as shown in Figure 3b. This makes GRU less complicated and therefore faster to train than LSTM. The gates are two vectors that decide which information should be passed to the output. The update gate enables the model to determine how much of the past information needs to be passed along to the future. The update gate Z_t is calculated for step t using the formula given by:

$$z_t = \sigma(U^z x_t + W^z h_{t-1}) \quad (5)$$

where U^z, W^z are the parameters (weights) of the update gate and h_{t-1} holds information for the previous $t - 1$ units. The reset gate is used to decide how much of the past information to forget which can be calculated using:

$$r_t = \sigma(U^r x_t + W^r h_{t-1}) \quad (6)$$

where U^r, W^r are the parameters (weights) of the reset gate and h_{t-1} holds information for the previous $t - 1$ units. The current memory content will use the reset gate to store the relevant information from the past as follows:

$$c_t = tanh(U^c x_t + r_t * W^c h_{t-1}) \quad (7)$$

where U^c, W^c are the parameters (weights). Note that $*$ denotes the Hadamard (element-wise) product. At the last step, the vector h_t which holds the information for the current unit and passes it down the network will be calculated by:

$$h_t = z_t * h_{t-1} + (1 - z_t) * c_t \quad (8)$$

A GRU network obtains a long spatial (or temporal) sequence with lower computational complexity compared to traditional encoder-decoder architecture. With its gating mechanisms, GRU can overcome the vanishing gradient problem and is therefore capable of processing longer sequences than standard RNN. Both GRU and LSTM can be applied to sequences of spatial features to determine the extent of dependencies or establish context between features that are located several places apart.

4.4. Dense Neural Networks

The DNN model is a regular deeply connected neural network with several layers. In a DNN model, each neuron in a layer receives an input from all the neurons present in the previous layer. The layers are known as the dense layers and constitute the hidden layers of the network. Such neural networks are also known as Multilayer perceptron (MLP). It is composed of an input layer, an output layer that makes a decision or prediction about the input, and an arbitrary number of hidden layers in between. The model is often trained on a set of input–output pairs and learns to model the correlation (or dependencies) between those inputs and outputs. The basic unit (a perceptron) of the model produces a single output based on several real-valued inputs by forming a linear combination using its input weights. The output is typically passed through a non-linear activation function ϑ:

$$y = \vartheta \left(\sum_{1}^{n} w_i x_i + b \right) = \vartheta \left(\mathbf{W}^T \mathbf{X} + b \right) \quad (9)$$

where W denotes the vector of weights, X is the vector of inputs, b is the bias and ϑ is the non-linear activation function.

The sigmoid or the hyperbolic tangent functions were the non-linear activation functions typically used in the past due to their ability to map complex relationships within data. However, these two non-linear activation functions do not perform well in networks with many layers due to the vanishing gradient problem. Nowadays, rectified linear activation function ReL (and its variants) is the preferred function used in training dense neural networks. Hence, the neurons in a network employing ReL activation are known as ReLU (Rectified Linear activated Units). ReL is a piecewise linear function given by: $f(x) = max(0, x)$. It will output the input directly if positive, and will output a zero if negative. ReLU overcomes the vanishing gradient problem and enable models to learn faster and perform better. Hence, it is used as the default activation function when developing the DNN and the CNN networks. In our study, we have experimented with different numbers of hidden layers for the DNN, and numbers of units per layer and recorded the performance of each configuration.

4.5. Hybrid Models

In this paper, we refer to hybrid models as those combining different deep learning techniques to leverage the unique capabilities of each of the techniques. For example in CNN-LSTM or CNN-GRU depicted in Figure 4, the model will utilize CNN to extract local n-gram features (where n is set by the length of the filters). The CNN's max pooling layer downsamples the output to reduce the dimensionality, which also contributes to the reduction in overfitting. The LSTM or GRU layers are then used to capture long-range dependencies that may be present within the features encoded by the CNN layers. The vectors output by the LSTM-GRU layer with the context and dependencies information will then be transmitted to dense layers for further processing before the final classification by the sigmoid activated output layer consisting of a single unit.

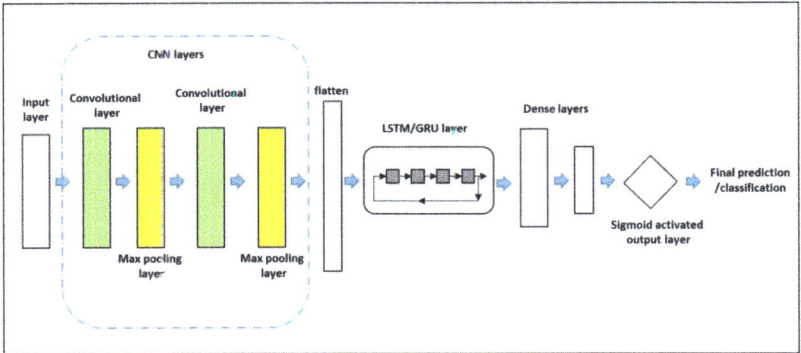

Figure 4. Overview of the CNN-LSTM and CNN-GRU hybrid model architecture.

5. Methodology and Experiments

In this section, we further detail our approach and outline the experiments undertaken to evaluate the deep learning models implemented in this paper. The models were developed in Python using the Keras library with TensorFlow backend. Other libraries utilized include Scikit Learn, Seaborn, Pandas, and Numpy. The experiments were performed on a Ubuntu Linux 16.04 64-bit Machine with 8GB RAM.

5.1. Problem Definition

Let $A = \{a_1, a_2, \ldots a_n\}$ be a set of applications where each a_i is represented by a vector containing the values of n features (where $n = 342$). Let $a = \{f_1, f_2, f_3, \ldots f_n, cl\}$ where $cl \in \{botnet, normal\}$ is the class label assigned to the app. Thus, A can be used to train a model to learn the behaviors of botnet and normal apps, respectively. The goal of a trained model is then to classify a given unlabeled app $A_{unknown} = \{f_1, f_2, f_3, \ldots f_n, ?\}$ by assigning a label cl, where $cl \in \{botnet, normal\}$.

5.2. Dataset Used for the Investigation

As mentioned earlier, the ISCX Android botnet dataset from [9] was utilized for the experiments in this paper. This dataset contains 1929 botnet apps and has been employed in previous works including [6–8,10–13,22]. Table 3 shows the distribution of samples within the 14 different botnet families present in the dataset. To complement the ISCX dataset, we obtained 4873 clean from Google Play store. These apps were cross-checked for maliciousness using Virus Total (https://www.virustotal.com (accessed on 20 December 2020)). Thus, a total of 6802 apps were used in our experiments.

Table 3. Botnet dataset composition.

Botnet Family	Number of Samples
Anserverbot	244
Bmaster	6
Droiddream	363
Geinimi	264
Misosms	100
Nickyspy	199
Notcompatible	76
Pjapps	244
Pletor	85
Rootsmart	28
Sandroid	44
Tigerbot	96
Wroba	100
Zitmo	80
Total	1929

5.3. Experiments to Evaluate the Deep Learning Techniques on the Android Dataset

In order to investigate the performance of the deep learning models, we performed several experiments with different configurations of the models to enable us observe the optimum performance that is possible with each model architecture. The models are designed to exploit the capabilities of the constituent neural network types as discussed in Section 4. The following metrics are used in measuring the performance of the models: Accuracy, precision, recall, and F1-score. Given TP as true positives, FP as false positives, FN as false negatives, and TN as true negatives (all with respect to the botnet class), the metrics are defined as follows (taking the botnet class as positive):

- Accuracy: the ratio between correctly predicted outcomes and the sum of all predictions expressed as: $\frac{TP+TN}{TP+TN+FP+FN}$
- Precision: All true positives divided by all positive predictions, i.e., was the model right when it predicted positive? Expressed as: $\frac{TP}{TP+FP}$
- Recall: All true positives divided by all actual positives. That is, how many positives did the model identify out of all possible positives? Expressed as: $\frac{TP}{TP+FN}$
- F1-score: This is the weighted average of precision and recall, given by: $\frac{2 \times Recall \times Precision}{Recall+Precision}$

All the results of the experiments are from 10-fold cross validation where the dataset is divided into 10 equal parts with 10% of the dataset held out for testing, while the models are trained from the remaining 90%. This is repeated until all of the 10 parts have been used for testing. The average of all 10 results is then taken to produce the final result. Additionally, during the training of all the deep learning models (for each fold), 10% of the training set was used for validation.

6. Results and Discussions

This section will present the results of investigating CNN-GRU, CNN-LSTM, CNN, and DNN models. Subsequently, a comparative performance evaluation of the models and how they measure against traditional machine learning models will be discussed.

Finally, we will examine how these models have performed compared to results reported in previous works on Android botnet detection.

6.1. CNN-GRU Model Results

Here, we present the results obtained from CNN-GRU model where the configurations of both the CNN layer and the GRU layer were varied. A summary of the results is presented in Table 4. In the top half of the table, the configuration had 1 convolutional layer and 1 max pooling layer in the CNN part. These models are named as CNN-1-layer-GRU-X where X stands for the number of hidden units in the GRU layer. The Convolutional layer receives input vector of dimension 342 from the input layer, and it consists of 32 filters each of size = 4. The max pooling layer has its parameter set to 2, which means it would reduce the output of the convolutional layer by half. The outputs from the max pooling layer are concatenated into a flat vector before sending to the GRU layer. As depicted in Figure 4, the output from the CNN-GRU layers are forwarded to 2 dense layers. The first dense layer had 128 units, while the second one had 64 units. The 64-unit layer is finally connected to a sigmoid-activated single-unit output layer where the final classification decision into 'clean' or 'botnet' is made. The model can be summarized in the following sequence:

Input [342] -> Conv [32 filters, Size=4] -> max pooling -> flatten -> GRU[X] -> Dense [128, ReLU] -> Dense[64, ReLU]-> Dense[1, Sigmoid] where X is the number of GRU hidden units taken as 5, 10, 25, and 50, respectively.

Table 4. Results from the CNN-GRU models of various configurations using the architecture depicted in Figure 4.

	Accuracy	Precision	Recall	F1-Score	Number of Parameters
CNN-1-layer-GRU-5	0.988	0.982	0.976	0.979	90,442
CNN-1-layer-GRU-10	**0.989**	**0.986**	**0.975**	**0.980**	**172,457**
CNN-1-layer-GRU-25	0.988	0.981	0.978	0.979	419,402
CNN-1-layer-GRU-50	**0.989**	**0.980**	**0.980**	**0.980**	**833,977**
CNN-2-layer-GRU-5	0.988	0.985	0.974	0.979	53,528
CNN-2-layer-GRU-10	0.989	0.986	0.974	0.980	93,993
CNN-2-layer-GRU-25	0.989	0.982	0.978	0.980	217,098
CNN-2-layer-GRU-50	**0.991**	**0.988**	**0.979**	**0.984**	**425,273**

The results of the bottom half of Table 4 are from the same CNN-GRU architecture described above, but with the CNN part having 2 convolutional layers and 2 max pooling layers. The model can be summarized in the following sequence:

Input [342] -> Conv [32 filters, Size=4] -> max pooling -> Conv [32 filters, Size=4] -> max pooling -> flatten -> GRU[X]->Dense [128, ReLU] -> Dense[64, ReLU]->Dense[1, Sigmoid] where X is the number of GRU hidden units taken as 5, 10, 25, and 50, respectively.

Note that a dropout = 0.25 is incorporated between each of the layers in the models to reduce overfitting.

From Table 4, we can see that the model with the 1-layer CNN had higher overall accuracy of 98.9% when the number of GRU hidden units were set at 10 or at 50. The corresponding F1-score were also the highest at 0.980. The recall of the GRU-50 model was 0.980 compared to that of the GRU-10 model, which was 0.975. This means that the GRU-50 model was better at detecting botnet apps than the GRU-10 model in the top half of Table 4. Note that the GRU-5 model from the 1-layer CNN batch (top-half) which had 5 hidden units actually did perform well also by obtaining an overall accuracy of 98.8%, with an F1-score of 0.979 and a botnet detection rate (recall) of 97.6%. It had the least numbers of parameters to train, i.e., 90,422.

From the bottom half of Table 4, the 2-layer CNN models with the best overall accuracy was the one with 50 units in the GRU layer (i.e., CNN-2-GRU-50). It obtained 99.1%

accuracy, and the best F1 score of 0.984. The recall (botnet detection rate) was 97.9% while the precision was 98.8%, the highest in all of the CNN-GRU models. From these set of results, we can conclude the following:

- The best overall performance for the CNN-GRU models was from the model with 2 convolutional layers and 50 hidden units in the GRU layer.
- Very good results can be obtained by CNN-GRU model with only 1 convolutional layer and few hidden units (5) in the GRU layer. The accuracy observed was 98.8%, and the F1 score was 0.979. The lower the number of hidden units, the faster it is to train the model.

6.2. CNN-LSTM Model Results

This section presents the results of the CNN-LSTM models with different configurations in both the CNN layer and the GRU layer. The results are presented in Table 5. Similar to the results of CNN-GRU in Table 4, the top half is for the models with 1 convolutional layer and 1 max pooling layer in the CNN part, while the bottom half (of Table 5) shows the results of the models having 2 convolutional layers and 2 max pooling layers in the CNN part. The models are named with the convention CNN-1-layer-LSTM-X in the top half, or CNN-2-layer-LSTM-X in the bottom half, where X stands for the number of hidden units in the LSTM layer. As depicted in Figure 4, the output from the CNN-LSTM layers are forwarded to 2 dense layers. The first dense layer had 128 units, while the second one had 64 units. The 64-unit layer is finally connected to a single unit sigmoid activated output layer where the final classification decision into 'clean' or 'botnet' is made. The model can be summarized in the following sequence:

Input [342] -> Conv [32 filters, Size=4] -> max pooling -> flatten -> LSTM[X] -> Dense [128, ReLU] -> Dense [64, ReLU]-> Dense[1, Sigmoid]

where X is the number of LSTM hidden units taken as 5, 10, 25, and 50, respectively.

Table 5. Results from the CNN-LSTM models of various configurations using the architecture depicted in Figure 4.

	Accuracy	Precision	Recall	F1-Score	Number of Parameters
CNN-1-layer-LSTM-5	0.990	0.987	0.977	0.982	117,497
CNN-1-layer-LSTM-10	**0.990**	**0.986**	**0.980**	**0.983**	**226,617**
CNN-1-layer-LSTM-25	0.990	0.985	0.979	0.982	555,117
CNN-1-layer-LSTM-50	**0.990**	**0.983**	**0.983**	**0.983**	**1,106,777**
CNN-2-layer-LSTM-5	0.989	0.986	0.973	0.980	66,553
CNN-2-layer-LSTM-10	0.989	0.986	0.974	0.980	120,633
CNN-2-layer-LSTM-25	0.990	0.984	0.979	0.981	284,073
CNN-2-layer-LSTM-50	0.989	0.984	0.976	0.980	560,473

For the bottom half of Table 5, the models can be summarized in the following sequence:

Input [342] -> Conv [32 filters, Size=4] -> max pooling -> Conv [32 filters, Size=4] -> max pooling -> flatten -> LSTM[X]->Dense [128, ReLU] -> Dense [64, ReLU]->Dense [1, Sigmoid]

where X is the number of LSTM hidden units taken as 5, 10, 25, and 50, respectively.

Note that a dropout = 0.25 is incorporated between each of the layers in the models to reduce overfitting.

Table 5 (top half), it can be seen that the all the CNN-LSTM models with only 1 layer in the CNN part had overall accuracy of 99%. The models with 10 and 50 hidden units, respectively, in the LSTM layer obtained identical F1-score of 0.983, compared to the ones with 5 and 25, respectively, which had F1-score of 0.982. The best recall (or botnet detection rate) of 98.3% was recorded with the LSTM-50 model. However, having more than 1.1 million parameters, the LSTM-50 model will be longer to train than the LSTM-10 model which has only 226,617 parameters.

From the bottom half of Table 5, the 2-layer CNN models with the best overall accuracy was the one with 25 units in the LSTM layer (i.e., CNN-LSTM-25). It obtained 99% accuracy, and the best F1 score of 0.981. The recall (botnet detection rate) was 97.9% while the precision was 98.4%. From the results of Table 5, we can conclude the following:

- The best overall performance for the CNN-LSTM models was from the model with 1 convolutional layer and 25 hidden units in the LSTM layer.
- Very good results can be obtained by CNN-LSTM model with only 1 convolutional layer and few hidden units (5) in the LSTM layer. This is evident from the results of the CNN-1-layer-LSTM-5 where the accuracy observed was 99%, and the F1 score was 0.982, precision was 98.7%, and recall (botnet detection rate) was 97.7%.
- Comparing Tables 4 and 5, the results of CNN-LSTM were generally better than those of CNN-GRU even though a CNN-GRU model obtained the highest F1-score of 0.984 with an overall accuracy of 99.1%.

6.3. CNN Model Results

In this section we discuss the results of the CNN model which is summarized in Table 6. The CNN model consists of 2 convolutional layers and 2 max pooling layers. The resulting vectors are 'flattened' and fed into a dense layer containing 8 units. The model's sequence can be summarized as follows:

Input [342] -> Conv [32 filters, Size=4] -> max pooling -> Conv [32 filters, Size=4] -> max pooling -> flatten -> Dense [8, ReLU] -> Dense [1, Sigmoid]

Table 6. Results from a 2-layer CNN model obtained by varying the number of filters, with length of filters = 4 in both convolutional layers.

	Accuracy	Precision	Recall	F1-Score	Number of Parameters
CNN-2-layer-4-filters	0.986	0.978	0.974	0.976	2,777
CNN-2-layer-8-filters	0.988	0.980	0.977	0.978	5,657
CNN-2-layer-16-filters	0.988	0.980	0.976	0.978	11,801
CNN-2-layer-32-filters	**0.989**	**0.983**	**0.978**	**0.981**	**25,625**
CNN-2-layer-64-filters	0.987	0.980	0.975	0.977	59,419

In our preliminary study presented in [24], this particular configuration of the model has been determined to yield the best performance on the same features extracted from the same app dataset used for the other models presented in this paper. More extensive performance evaluation of the CNN model has been presented in [24], where the effect of varying the other parameters, such as filter length, number of layers, and max pooling parameter has been investigated.

As shown in Table 6, the CNN model with 32 filters yielded the best results with 98.9% overall accuracy, precision = 0.983, recall = 0.978, and F1-score = 0.981. When compared to the results in Tables 4 and 5, it can be observed that most of the CNN-LSTM configurations and some of the CNN-GRU configurations achieved higher results than the CNN-only model. This suggests that the LSTM and GRU were able to capture some dependencies amongst the features thus improving the performance of the model.

6.4. DNN Model Results

The results obtained from the Dense Neural Network model is presented in this section. The naming convention used to describe the models is DNN-Y-layer-N as shown in Table 7, where Y stands for the number of hidden layers and N is the number of units in the layer. For example, the sequence of the DNN-2-layer-200 model can be summarized as follows:

Input [342] -> Dense [200, ReLU] -> Dense [200, ReLU] -> Dense [1, Sigmoid]

Table 7. Results from the DNN model with various numbers of layers and units per layer.

	Accuracy	Precision	Recall	F1-Score	Number of Parameters
DNN-1-layer-100	0.990	0.984	0.982	0.983	34,401
DNN-2-layer-100	**0.991**	**0.990**	**0.979**	**0.984**	**44,501**
DNN-3-layer-100	**0.991**	**0.988**	**0.980**	**0.984**	**54,601**
DNN-1-layer-200	0.990	0.982	0.982	0.982	68,801
DNN-2-layer-200	0.990	0.981	0.983	0.982	109,001
DNN-3-layer-200	0.989	0.982	0.980	0.981	149,201
DNN-1-layer-300	0.990	0.985	0.981	0.983	103,201
DNN-2-layer-300	0.989	0.984	0.978	0.981	193,501
DNN-3-layer-300	0.989	0.979	0.984	0.981	283,801

Note that a dropout = 0.25 is incorporated between each of the layers in the models to reduce overfitting. Additionally, in all of the DNN models and the previous models in Sections 6.1–6.3, the optimization algorithm used was 'Adam' and 'Binary cross entropy' was used for the loss function. Furthermore, all the models were configured to automatically terminate the training after the validation loss is observed to have not changed for a specific number of K training epochs, where K was set to 20.

From Table 7, it can be observed that the DNN models with a single hidden layer did not result in the best outcomes. Likewise, in most cases, using 3 hidden layers as observed with the DNN-3-layer-200 and DNN-3-layer-300 also did not give the best outcomes. The best performance was obtained from the model with 2 hidden layers and 100 units in each layer, where the overall accuracy is 99.1% and F1-score = 0.984. The model with 3 hidden layers and 100 units in each layer also gave identical results. This shows that increasing the number of units in each layer is unlikely to improve the performance any further.

6.5. Best Deep Learning Results vs. Classical Non-Deep Learning Classifiers

In Table 8, we juxtapose the best results from our investigation of the deep learning classifiers with the results from the classical machine learning techniques. The DNN and the CNN-GRU models achieved the best results as depicted in the table. The highest accuracy achieved by both models were 99.1% which also corresponds to the highest F1-score of 0.984. These results are followed closely by the CNN-LSTM model which achieved 99% overall accuracy and F1-score of 0.983. Next, was the CNN-only model with 98.9% accuracy and F1-score of 0.981. All of these models outperformed the classical machine learning classifiers where the best two were SVM and Random Forest. SVM had 98.7% overall accuracy and F1-score of 0.976, while Random Forest obtained 98.5% accuracy and F1-score of 0.973. These results suggest that with the static based features extracted for detecting Android botnets, the deep learning models will perform beyond the limits of the classical machine learning classifiers.

In the table, the GRU-only model is shown as having the least accuracy results compared to all the other models. This GRU model consisted of 200 hidden units and obtained an overall accuracy of 82.9%. Similarly, with LSTM-only models, overall accuracies below 75% were observed (results not shown in the table). This confirmed our initial expectation that pattern recognition (e.g., with convolutional layers or dense layers) was more important for the type of feature vectors used in the study, rather than context or dependencies. However, the results of Sections 6.1 and 6.2 for the hybrid models suggests that a combination of methods that can capture both characteristics is promising.

Table 8. A summary of the best results of each technique compared to popular non-deep learning classifiers.

	Accuracy	Precision	Recall	F1-Score	Number of Parameters
DNN-2-layer-100	0.991	0.990	0.979	0.984	44,501
DNN-3-layer-100	0.991	0.988	0.980	0.984	54,601
CNN-2-layer-GRU-50	0.991	0.988	0.979	0.984	425,273
CNN-1-layer-LSTM-10	0.990	0.986	0.980	0.983	226,617
CNN-1-layer-LSTM-50	0.990	0.983	0.983	0.983	1,106,777
CNN-2-layer-32-filters	0.989	0.983	0.978	0.981	25,625
SVM	0.987	0.980	0.973	0.976	-
Random Forest	0.985	0.982	0.965	0.973	-
Simple Logistic	0.984	0.983	0.963	0.973	-
Decision Tree (J48)	0.981	0.974	0.958	0.966	-
Naïve Bayes	0.872	0.728	0.874	0.795	-
Bayes Net	0.867	0.736	0.832	0.781	-
GRU-200	0.829	0.677	0.766	0.718	122,001

6.6. Model Training Times

When training the deep learning models, the number of epochs has a major influence on the overall model training time. In our experiments we utilized a stopping criterion based on minimum validation loss rather than specifying a fixed number of training epochs. For this reason, the number of training epochs varied between the different configurations of a given model. Hence, the longest CNN-GRU model to train was the CNN-2-layer-GRU-25 which took 145 s, and the testing time was 0.482 s. Whereas the shortest CNN-GRU model to train was the CNN-1-layer-GRU-10 model which took 84.4 s with a testing time of 0.399 s. The longest CNN-LSTM model to train was the CNN-2-layer-LSTM-5 which took 141 s with a testing time of 0.468 s. The shortest CNN-LSTM model to train was the CNN-1-layer-LSTM-25 model which took 83.6 s with a testing time of 0.419 s. Compared to the other models, the DNN was the fastest to train with training times ranging from 10 to 26 s and an average testing time of 0.15 s.

6.7. Comparison with Previous Works

The results obtained in our study improves the performance beyond the reported results in previous papers that also used the ISCX botnet dataset in their work. This can be observed in Table 9. The second column shows the numbers of the botnet and benign samples used in each of the referenced paper. Note that in some papers, some of the metrics were not reported. Even though the complete datasets and techniques used were different in each of the previous works, Table 9 shows that the models developed in this paper achieved state-of-the-art results with the ISCX botnet dataset compared to the others

Table 9. Performance comparisons with previous works that utilize ISCX botnets samples.

Paper	Botnets/Benign	ACC (%)	Prec.	Rec.	F1-Score
Hojjatinia et al. [6]	1800/3650	97.2	0.955	0.960	0.957
Tansettanakorn et al. [11]	1926/150	-	0.972	0.969	-
Anwar et al. [10]	1400/1400	95.1	0.970	0.827	-
Abdullah et al. [13]	1505/850	-	0.931	0.946	-
Alqatawna and Faris [12]	1635/1635	97.3	0.987	0.957	-
Yerima and Alzaylaee [7]	1929/4873	98.9	0.983	0.978	0.981
This paper	**1929/4873**	**99.1**	**0.990**	**0.979**	**0.984**

7. Conclusions and Future Work

In this paper, we presented an extensive evaluation of various deep learning techniques for Android botnet detection using 342 static features consisting of 5 different types. The deep learning models investigated include: CNN, DNN, GRU, LSTM, as well as CNN-LSTM and CNN-GRU. The experiments were undertaken using 6802 apps consisting of 1929 botnet apps from the ISCX botnet dataset which has been utilized in several previous works. The outcomes of our experiments showed that with optimum configuration, the deep learning models performed quite well yielding high accuracies that were beyond the limits of the classical machine learning classifiers. DNN showed the best overall performance, while CNN-GRU and CNN-LSTM showed promising results that were much better than GRU-only or LSTM-only models. In future work, we plan to further investigate the performance of the deep learning models for botnet detection using alternative static and dynamic features. Another possible direction is to explore alternative network architectures such as those consisting of parallel rather than purely sequential integrations of the deep learning model components.

Author Contributions: Conceptualization, S.Y.Y.; Data curation, M.K.A.; Investigation, S.Y.Y., M.K.A. and A.S.; Methodology, S.Y.Y.; Resources, V.P.; Supervision, V.P.; Validation, M.K.A., A.S. and V.P.; Writing—original draft, S.Y.Y.; Writing—review & editing, M.K.A. and V.P. All authors have read and agreed to the published version of the manuscript.

Funding: This research received no external funding.

Data Availability Statement: The data presented in this study are publicly available in FigShare at https://doi.org/10.6084/m9.figshare.14079581 (accessed on 10 December 2020).

Conflicts of Interest: The authors declare no conflict of interest.

References

1. McAfee Mobile Threat Report Q1. 2020. Available online: https://www.mcafee.com/en-us/consumer-support/2020-mobile-threat-report.html (accessed on 5 December 2020).
2. Yerima, S.Y.; Khan, S. Longitudinal Performance Analysis of Machine Learning based Android Malware Detectors. In Proceedings of the 2019 International Conference on Cyber Security and Protection of Digital Services (Cyber Security), Oxford, UK, 3–4 June 2019.
3. Grill, B.B.; Ruthven, M.; Zhao, X. "Detecting and Eliminating Chamois, a Fraud Botnet on Android" Android Developers Blog. March 2017. Available online: https://android-developers.googleblog.com/2017/03/detecting-and-eliminating-chamois-fraud.html (accessed on 10 December 2020).
4. Chris Brook "Google Eliminates Android Adfraud Botnet Chamois" Threat Post. March 2017. Available online: https://threatpost.com/google-eliminates-android-adfraud-botnet-chamois/124311/ (accessed on 10 December 2020).
5. Fahmida, Y. Rashid "Chamois: The Big Botnet You Didn't Hear about" April 2019 Decipher, by Duo Security. Available online: https://duo.com/decipher/chamois-the-big-botnet-you-didnt-hear-about (accessed on 10 December 2020).
6. Hojjatinia, S.; Hamzenejadi, S.; Mohseni, H. Android Botnet Detection using Convolutional Neural Networks. In Proceedings of the 2020 28th Iranian Conference on Electrical Engineering (ICEE), Tabriz, Iran, 4–6 August 2020.
7. Yerima, S.Y.; Alzaylaee, M.K. Mobile Botnet Detection: A Deep Learning Approach Using Convolutional Neural Networks. In Proceedings of the 2020 International Conference on Cyber Situational Awareness (Cyber SA 2020), Dublin, Ireland, 15–19 June 2020.
8. Kadir, A.F.A.; Stakhanova, N.; Ghorbani, A.A. Android botnets: What urls are telling us. In Proceedings of the International Conference on Network and System Security, New York, NY, USA, 3–5 November 2015; Springer: New York, NY, USA, 2015; pp. 78–91.
9. ISCX Android Botnet Dataset. Available online: https://www.unb.ca/cic/dataset/android-botnet.html (accessed on 23 December 2020).
10. Anwar, S.; Zain, J.M.; Inayat, Z.; Haq, R.U.; Karim, A.; Jabir, A.N. A static approach towards mobile botnet detection. In Proceedings of the 2016 3rd International Conference on Electronic Design (ICED), Phuket, Thailand, 11–12 August 2016; pp. 563–567.
11. Tansettanakorn, C.; Thongprasit, S.; Thamkongka, S.; Visoottiviseth, V. ABIS: A prototype of android botnet identification system. In Proceedings of the 2016 Fifth ICT International Student Project Conference (ICT-ISPC), Nakhon Pathom, Thailand, 27–28 May 2016; pp. 1–5.
12. Alqatawna, J.F.; Faris, H. Toward a Detection Framework for Android Botnet. In Proceedings of the 2017 International Conference on New Trends in Computing Sciences (ICTCS), Amman, Jordan, 11–13 October 2017; pp. 197–202.

13. Abdullah, Z.; Saudi, M.M.; Anuar, N.B. ABC: Android botnet classification using feature selection and classification algorithms. *Adv. Sci. Lett.* **2017**, *23*, 4717–4720. [CrossRef]
14. Karim, A.; Rosli, S.; Syed, S. DeDroid: A Mobile Botnet Detection Approach Based on Static Analysis. In Proceedings of the 7th International Symposium on UbiCom Frontiers—Innovative Research Systems and Technologies, Beijing, China, 10–14 August 2015. [CrossRef]
15. The Drebin Dataset. Available online: https://www.sec.cs.tu-bs.de/~{}danarp/drebin/index.html (accessed on 22 December 2020).
16. Jadhav, S.; Dutia, S.; Calangutkar, K.; Oh, T.; Kim, Y.H.; Kim, J.N. Cloud-based android botnet malware detection system. In Proceedings of the 2015 17th International Conference on Advanced Communication Technology (ICACT), PyeongChang, Korea, 1–3 July 2015; pp. 347–352.
17. Yerima, S.Y.; Alzaylaee, M.K.; Sezer, S. Machine learning-based dynamic analysis of Android apps with improved code coverage. *EURASIP J. Inf. Secur.* **2019**. [CrossRef]
18. Alzaylaee, M.K.; Yerima, S.Y.; Sezer, S. Improving dynamic analysis of android apps using hybrid test input generation. In Proceedings of the 2017 International Conference on Cyber Security and Protection Of Digital Services (Cyber Security 2017), London, UK, 19–20 June 2017.
19. Bernardeschia, C.; Mercaldo, F.; Nardonec, V.; Santoned, A. Exploiting Model Checking for Mobile Botnet Detection. *Procedia Comput. Sci.* **2019**, *159*, 963–972. [CrossRef]
20. Clarke, E.; Emerson, E.; Sistla, A. Automatic verification of finite-state concurrent systems using temporal logic specifications. *ACM Trans. Program. Lang. Syst.* **1986**, *8*, 244–263. [CrossRef]
21. Andersen, J.R.; Andersen, N.; Enevoldsen, S.; Hansen, M.M.; Larsen, K.G.; Olesen, S.R.; Srba, J.; Wortmann, J.K. CAAL: Concurrency workbench, aalborg edition. In Proceedings of the Theoretical Aspects of Computing—ICTAC 2015—12th International Colloquium, Cali, Colombia, 29–31 October 2015; Springer: Cham, Switzerland, 2015; pp. 573–582.
22. Alothman, B.; Rattadilok, P. Android botnet detection: An integrated source code mining approach. In Proceedings of the 12th International Conference for Internet Technology and Secured Transactions (ICITST), Cambridge, UK, 11–14 December 2017; pp. 111–115.
23. Alzahrani, A.J.; Ghorbani, A.A. Real-time signature-based detection approach for sms botnet. In Proceedings of the 2015 13th Annual Conference on Privacy, Security and Trust (PST), Izmir, Turkey, 21–23 July 2015; pp. 157–164.
24. Girei, D.A.; Shah, M.A.; Shahid, M.B. An enhanced botnet detection technique for mobile devices using log analysis. In Proceedings of the 2016 22nd International Conference on Automation and Computing (ICAC), Colchester, UK, 7–8 September 2016; pp. 450–455.
25. LeCun, Y.; Bengio, Y.; Hinton, G. Deep learning *Nature* **2015**, *521*, 436–444. [CrossRef] [PubMed]
26. LeCun, Y.; Bottou, L.; Bengio, Y.; Haffner, P. Gradient-based learning applied to document recognition. *Proc. IEEE* **1998**, *86*, 2278–2324. [CrossRef]
27. Glorot, X.; Bordes, A.; Bengio, Y. Deep sparse rectifier neural networks. In Proceedings of the 4th International Conference on Artificial Intelligence and Statistics, Ft. Lauderdale, FL, USA, 11–13 April 2011; pp. 315–323.
28. Hochreiter, S.; Schmidhuber, J. Long short-term memory. *Neural Comput.* **1997**, *9*, 1735–1780. [CrossRef] [PubMed]
29. Graves, A. *Long Short-Term Memory*; Springer: Berlin/Heidelberg, Germany, 2012; pp. 37–45.
30. Chung, J.; Gulcehre, C.; Cho, K.; Bengio, Y. Empirical evaluation of gated recurrent neural networks on sequence modeling. *arXiv* **2014**, arXiv:1412.3555.

Article

Detection of Malicious Software by Analyzing Distinct Artifacts Using Machine Learning and Deep Learning Algorithms

Mathew Ashik [1], A. Jyothish [1], S. Anandaram [1], P. Vinod [2], Francesco Mercaldo [3,4,*], Fabio Martinelli [3] and Antonella Santone [4]

[1] Department of Computer Science and Engineering, SCMS School of Engineering and Technology, Ernakulam 682013, India; ashik.mathew@scmsgroup.org (M.A.); jyothish@scmsgroup.org (A.J.); anandaram.s@scmsgroup.org (S.A.)
[2] Department of Computer Science and Engineering, Cochin University of Science and Technology, Cochin 682001, India; vinodp@scmsgroup.org
[3] Institute for Informatics and Telematics, National Research Council of Italy, 56124 Pisa, Italy; fabio.martinelli@iit.cnr.it
[4] Department of Medicine and Health Sciences "Vincenzo Tiberio", University of Molise, 86100 Campobasso, Italy; antonella.santone@unimol.it
* Correspondence: francesco.mercaldo@iit.cnr.it

Abstract: Malware is one of the most significant threats in today's computing world since the number of websites distributing malware is increasing at a rapid rate. Malware analysis and prevention methods are increasingly becoming necessary for computer systems connected to the Internet. This software exploits the system's vulnerabilities to steal valuable information without the user's knowledge, and stealthily send it to remote servers controlled by attackers. Traditionally, anti-malware products use signatures for detecting known malware. However, the signature-based method does not scale in detecting obfuscated and packed malware. Considering that the cause of a problem is often best understood by studying the structural aspects of a program like the mnemonics, instruction opcode, API Call, etc. In this paper, we investigate the relevance of the features of unpacked malicious and benign executables like mnemonics, instruction opcodes, and API to identify a feature that classifies the executable. Prominent features are extracted using Minimum Redundancy and Maximum Relevance (mRMR) and Analysis of Variance (ANOVA). Experiments were conducted on four datasets using machine learning and deep learning approaches such as Support Vector Machine (SVM), Naïve Bayes, J48, Random Forest (RF), and XGBoost. In addition, we also evaluate the performance of the collection of deep neural networks like Deep Dense network, One-Dimensional Convolutional Neural Network (1D-CNN), and CNN-LSTM in classifying unknown samples, and we observed promising results using APIs and system calls. On combining APIs/system calls with static features, a marginal performance improvement was attained comparing models trained only on dynamic features. Moreover, to improve accuracy, we implemented our solution using distinct deep learning methods and demonstrated a fine-tuned deep neural network that resulted in an F1-score of 99.1% and 98.48% on Dataset-2 and Dataset-3, respectively.

Keywords: malware; machine learning; deep learning; static analysis; dynamic analysis; hybrid analysis; security

1. Introduction

Malware or malicious code is harmful code injected into legitimate programs to perpetrate illicit intentions. With the rapid growth of the Internet and heterogeneous devices connected over the network, the attack landscape has increased and has become a concern, affecting the privacy of users [1]. The primary source of infection, causing malicious programs to enter the systems without users' knowledge. Mostly freely downloadable software's are a primary source of malware, which include freeware comprising of games, web browsers, free antivirus, etc. Largely financial transactions are performed using the Internet, these have caused huge financial losses for organizations and individuals. Malware writing has transformed into profit-making industries, thus attracting a large number of hackers. Current malware is broadly classified as polymorphic or metamorphic, and they remain undetected by a signature-based detector [2].

Malware writers employ diverse techniques to generate new variants that commonly include (a) instruction permutation, (b) register re-assignment, (c) code permutation using conditional instructions, (d) no-operation insertion, etc. Malware analysis is the process aimed to inspect and understand a malicious behavior [3]. Normally malware are analyzed by extracting strings, opcodes, sequence of bytes, APIs/system call, and the network trace.

In this paper, we conduct a comprehensive analysis using multiple datasets by exploiting machine learning and deep learning approaches. Classifiers are trained independently using the static, dynamic feature, and their combinations. We employ dynamic instrumentation tools like Ether [4], a sandbox approach for analyzing malware. In addition, we also make use of sandbox [5]. The motivation behind using the aforesaid sandboxes are to stop side-effects induced to the host environment and to permit malware to exhibit its capabilities, which can be used as features for developing detection models. Ether in particular is based on the application of a hardware virtualization extension, such as Intel VT [6] and resides entirely outside of the target OS environment. In addition to providing anti-debugging facilities, Ether can also be used for software de-armoring dynamically.

Starting from these considerations, we propose a malware detector, exploiting machine learning and deep learning techniques. The experiments were conducted on malware and benign Portable Executables (PE), Android applications, and metamorphic samples created using virus kits. The motivation for using these types of files was arrived at by monitoring the submissions received over the Virus Total [7], a service that performs online scanning of malicious samples. In particular, we consider a set of features obtained from benign and malicious executables like mnemonics, instruction opcodes, and API/system calls for automatically discriminating legitimate and malicious samples. In summary, we list below the contributions of our proposal:

- Comprehensive analysis of machine learning and deep learning-based malware detection system using four datasets comprising of PE files, collection of ransomware, Android apps, and metamorphic samples;
- We show that information-theoretic and statistical feature selection methods improve the detection rate of traditional machine learning algorithms. However, the former approach exhibited better results in all cases comparing the statistical approach;
- Evaluation of classification models on different types of features such as the opcode sequence and API/system calls. Here, we investigate the performance of models trained on independent attribute categories and unifying static and dynamic features. We show that combining static features with dynamic attributes does not significantly improve classifier outcomes;
- Exhaustive analysis demonstrates an enhanced F1 score generated by deep learning methods on comparing machine learning algorithms. Furthermore, a detailed analysis of code obfuscation on samples developed using virus kits was performed. We conclude that malware kits generate metamorphic variants which employ simple obfuscation transformation easily identified using the local sequence alignment approach. Besides, we show that machine learning algorithms can precisely separate instances generated through virus kits using generic features like an opcode bigram.

The rest of the paper is organized as follows: In the next section we provide an overview about the current state of the art in the malware detection context; in Section 3 we present the proposed method for malware detection; experimental analysis is discussed in Section 4; and, finally, in Section 5 a conclusion and future research plan are presented.

2. Related Work

To highlight the novelty of our work, we examine malware detection techniques topics for which the proposed method is related: The technique for malware detection and classification through machine learning and deep learning algorithms, and other techniques.

2.1. Machine Learning-Based Malware Detection Techniques

Krugel et al. [8] used dynamic analysis to detect obfuscated malicious code using a mining algorithm. Authors in [9] proposed a hybrid model for the detection of malware using different features like byte n-gram, assembly n-gram, and library functions to classify an executable as malware or benign. The work [10] considers the system call subsequence as an element and regards the co-occurrence of system calls as features to describe the dependent relationship between system calls.

Furthermore, the work in [11] extracted 11 types of static features and employed multiple classifiers in a majority vote fusion approach where classifiers such as SVM, k-NN, naive Bayes, Classification and Regression tree (CART), and Random Forest were used. Nataraj et al. [12] consider the Gabor filter and evaluated it on 25×86 malicious families. Thus, they built a model using the k-nearest Neighbors approach with Euclidean distance.

2.2. Deep Learning-Based Malware Detection Techniques

Recently in [13], applications were represented in the form of an image to discriminate between malicious and benign applications. The solution considered static features extracted by reverse-engineering the malicious code and encoding it by SimHash. The DroidDetector tool [14] discriminates between legitimate and malicious samples in an Android environment by exploiting a deep learning network, relying on required permissions, sensitive APIs, and dynamic behaviors features. A deep convolutional neural network for malware detection is proposed by McLaughlin et al. [15], starting from the analysis of raw opcode sequence obtained by a reverse engineering Android applications. MalDozer [16] is a tool aimed at Android malware detection and family identification by analyzing API method calls. Furthermore, the study in [17] proposes a malware detector focused on the Android environment, aimed to discriminate between malicious and legitimate samples and to identify malware belonging to the family.

2.3. Malware Detection Using Other Techniques

API calls have been used in the past for modeling program behavior [18,19] and for detecting malware [20,21]. This paper relies on the fact that the behavior of malicious programs in a specific malware class differs considerably from programs in other malware classes and benign programs. Sathyanarayan et al. [22] used static extraction to extract API calls from known malware to construct a signature for an entire class. In [23], authors use static analysis to detect system call locations and run-time monitoring to check all system calls made from a location identified during static analysis.

Damodaran et al. [24] compared malware detection techniques based on static, dynamic, and hybrid analysis. Authors in [25] used Hidden Markov Models (HMMs) to represent the statistical properties of a set of metamorphic virus variants. The metamorphic virus data set was generated from metamorphic engines: Second Generation virus generator (G2), Next Generation Virus Construction Kit (NGVCK), Virus Creation Lab for Win32 (VCL32), and Mass Code Generator (MPCGEN). Vinod et al. [26] proposed a method to find the metamorphism in malware constructors like NGVCK, G2, IL_SMG, and MPCGEN by executing each malware sample in a controlled environment like QEMU and monitoring API calls using STraceNTX. Suarez-Tangil et al. [27] focus their efforts to discern malicious components from the legitimate ones in repackaged Android malware. They consider control flow graphs generated from code fragments of the application under analysis. They highlight that most research papers on Android malware detection are focused on outdated repositories, such as the MalGenome project [28] and the Drebin [29] datasets.

DroidScope [30] uses a customized Android kernel to reconstruct semantic views to collect detailed application execution traces. An approach aimed at detecting Android malware families was presented in [10,31]. The method is based on the analysis of system calls sequences and is tested obtaining an accuracy of 97% in mobile malware identification using a 3-gram syscall as a feature. Android malware detection exploiting a set of static features was addressed in [32]. Unsupervised machine learning techniques were used to build models with the considered feature set, statically obtained from permission invocations, strings, and code patterns. Furthermore, the Alde [33] framework employs static analysis and dynamic analysis to detect the actions of users collected by analytics libraries. Moreover, Alde analyses gives insight into what private information can be leaked by apps that use the same analytics library. Casolare et al. [34] also focused on the Android environment by proposing a model checking-based approach for detecting colluding between Android applications. A comparison of existing techniques is given in Table 1.

Table 1. Comparison of existing techniques.

Machine Learning-Based Techniques		
Author	Approach	Drawback
M. Christodorescu et al. [8]	Mine malicious behavior present in a known malware.	The impact of test program choices on the quality of mined malware behavior was not clear.
M. K. Alzaylaee et al. [9]	Deep learning system that detects malicious Android applications through dynamic analysis using stateful input generation.	Investigation on recent intrusion detection systems were not available.
G. Canfora et al. [10]	Android malware detection method based on sequences of system calls.	Assumption that malicious behaviors are implemented by specific system calls sequences.
W. Wang et al. [11]	Framework to effectively and efficiently detect malicious apps and benign apps.	Require datasets of features extracted from malware and harmless samples in order to train their models.
L. Nataraj et al. [12]	Effective method for visualizing and classifying malware using image processing techniques.	Path to a broader spectrum of novel ways to analyze malware was not fully explored.

Deep Learning-Based Techniques		
Author	Approach	Drawback
S. Ni et al. [13]	Classification algorithm that uses static features called Malware Classification using SimHash and CNN.	Time required for malware detection and classification was comparatively more.
Z. Yuan et al. [14]	An online deep learning-based Android malware detection engine (DroidDetector).	The semantic-based features of Android malware were not considered.
N. McLaughlin et al. [15]	A novel Android malware detection system that uses a deep convolutional neural network.	The same network architecture cannot be applied to malware analysis on different platforms.
E. B. Karbab et al. [16]	Android malware detection using deep learning on API method sequences.	Less affected by the obfuscation techniques because they only consider the API method calls.
G. Iadarola et al. [17]	Deep learning model for mobile malware detection and family identification.	Model needs to be trained with large sets of labeled data.

Other Techniques		
Author	Approach	Drawback
B. Zhang et al. [18]	Detect unknown malicious executables code using fuzzy pattern recognition.	Fuzzy pattern recognition algorithm suffers from a low detection and accuracy rate.
H. M. Sun et al. [19]	Detecting worms and other malware by using sequences of WinAPI calls.	Approach is limited to the detection of worms and exploits the use of hard-coded addresses of API calls.

Table 1. Cont.

Author	Approach	Drawback
	Other Techniques	
J. Bergeron et al. [20]	Proposed a slicing algorithm for disassembling binary executables.	Graphs created are huge in size, thus the model is not computationally feasible.
Q. Zhang et al. [21]	Approach for recognizing metamorphic malware by using fully automated static analysis of executables.	Absence of analysis of the parameters passed to library or system functions.
V. S. Sathyanarayan et al. [22]	Static extraction to extract API calls from known malware in order to construct a signature for an entire class.	Detection of malware families does not work for packed malware.
J. C. Rabek et al. [23]	Host-based technique for detecting several general classes of malicious code in software executables.	Not applicable for detecting all malicious code in executable.
A. Damodaran et al. [24]	Comparison of malware detection techniques based on static, dynamic, and hybdrid analysis.	Disadvantage is that it is thwarted easily by obfuscation techniques.
W. Wong et al. [25]	Method for detecting metamorphic viruses using Hidden Markov models.	Model can be defeated by inserting a sufficient amount of code from benign files into each virus
V. P. Nair et al. [26]	Tracing malware API calls via dynamic monitoring within an emulator to extract critical APIs.	Applicability of the method to detection of new malware families are limited.
G. Suarez-Tangi et al. [27]	Differential analysis to isolate software components that are irrelevant to study the behavior of malicious riders.	Study is vulnerable to update attacks since the payload is stored in a remote host.
Y. Zhou et al. [28]	Android platform with the aim to systematize or characterize existing Android malware.	Detection of Android malware that shows the rapid development and increased sophistication, posing significant challenges to this system.
L. K. Yan et al. [30]	DroidScope, an Android analysis platform that continues the tradition of virtualization-based malware analysis.	Overall performance of the model is less, compared to others.
G. Canfora et al. [31]	Method for detecting malware based on the occurrences of a specific subset of system calls, a weighted sum of a subset of permissions.	Precise patterns and sequences of system calls that could be recurrent in malicious code fragments were ignored.
D. Su et al. [32]	Automated community detection method for Android malware apps by building a relation graph based on their static features.	This method for malware app family classification is not as precise as supervised learning approaches.
X. Liu et al. [33]	Collect and analyze users action (API) on an Android platform to detect privacy leakage.	Fail when tracking APIs used by other apps that are not listed in the configure file.
R. Casolare et al. [34]	Static approach, based on formal methods, which exploit the model checking technique.	Limitations of method is that the generation of the first heuristic is not automatic.

3. Proposed Methodology

In the following subsections, we discuss our proposed methods for detecting malicious files. We prepare four datasets (a) the first dataset (dataset-I) comprises malicious executables collected from VX-Heavens [35] along with legitimate files, (b) the second dataset (dataset-II), which is the collection of malicious files including ransomware's downloaded from virusshare [36] along with goodware gathered from diverse sources finally, (c) malicious Android applications acquired from the Drebin project [37] and benign apks (dataset-III), and (d) synthetic malware samples created using virus generation kits. To improve readability, we present the expansion of abbreviations and meaning of symbols in abbreviations and mathematical symbols.

To predict unknown samples, we used malwares from a collection of sources such as the VX Heavens repository [35], ransomware downloaded from virusshare [36], synthetic malware samples created using a virus kit, and malicious Android apps. Additionally, we gathered legitimate samples from diverse sources. The generation of feature space of features like mnemonic, instruction opcode, API calls [38], and 4-gram mnemonic are extracted after unpacking the files. The basic idea of dynamic analysis is to monitor the program while in execution. Dynamic analysis of malware needs a virtual environment to avoid infection on the host system. We thus used different types of sandboxes each for a different dataset. For VX-Heavens samples, the executable files were made to run on a hardware virtualized machine such as Xen [39]. The advantage of using an emulator is that the actual host machine is not infected by the viruses during the dynamic API tracing step. Ransomware dataset were analyzed in the Parsa sandbox [5] which hooks the API calls to provide the requested resources to the executable matching an environment condition. Finally, malicious Android apps were executed in an emulator and system call traces were logged using strace utility and each application was subjected to random events such as clicks, swipes, change of battery level, update of geo-location, etc.

API Call tracing requires that the samples are unpacked or unarmored, as explained earlier since the packers generally try to destroy the import table [40] of the malware or benign program. To unpack samples, we used Ether patched XEN. Ether patched XEN is transparent to malware. Hence the anti-debugging techniques like Virtual Machine Detection [4], Debugger Detection (IsDebuggerPresent() API Call, EFLAGS bitmask) and Timing Attacks (analyzed values of RDTSC before and after) could be avoided due to a hardware virtualized environment. We have used XEN as a virtual environment running on top of Debian Lenny (Debian 5.0.8). Xen is a generic and open source virtualizer. XEN achieves near native performances by executing the guest code directly on the host CPU. In our process, we followed these steps:

- Disk Image Creation: We created a disk image, this disk image works as a separate hard disk;
- Windows XP Installation: Once the disk image was created, we installed Windows XP on that disk image. We chose Windows XP Service Pack 2 as most of the viruses, written for Windows environments only. Another reason to choose Windows XP Service Pack 2 is that the ether patched version of Xen has been tested with XP SP2 as guest OS and Debian Lenny as host OS;
- Running Ether Patched Xen: Once we have installed the Operating System on Ether patched XEN, we run the machine using the vncviewer [41];
- Unpacking using Ether Patched Xen: To analyze the malware dynamically, the malware is executed on the DomU machine (XP SP2) and its footprints are recorded on the Dom0 system (Debian Lenny).

Ether dumps the sample by finding the Original Entry Point using the memory writes a program does. The dumped sample could be found in the images directory of ether Once we have unpacked malware samples, they execute in an emulated environment and API tracing achieved using Veratrace.

3.1. Software Armoring

Software armouring or executable packing, as shown in Figure 1, is the process of compressing/encrypting an executable file and prepending a stub that is responsible for decompressing/decrypting the executable for execution [42,43]. When execution starts, the stub will unpack the original executable code and transfer control to it. Today most malware authors use packed executables to hide from detection. Due to software armoring, malware writers can defeat malicious applications from detection.

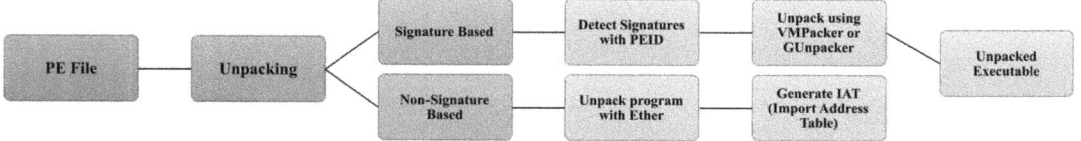

Figure 1. Software de-armoring.

Before beginning with the analysis of the malware, we should check whether the malware is armored or not. We use Ether as a tool for de-armoring since it is not signature-based and also due to its transparency to malware. Ether detects all the writes to memory a program does and dumps the program back in the binary executable form. It creates a hash-table for all the memory maps, and whenever there is a write to a slot in the hash table, it reports that as the Original Entry Point, which is the starting point of execution of a packed executable.

3.2. Feature Extraction

In our approach, we used API calls (dynamic malware analysis), and mnemonic/opcode, instruction opcode, and 4-gram mnemonic (static malware analysis). The process of feature extraction is briefly explained in Figure 2. To extract these features the various open-source tools used are listed below:

- API Call tracing using Veratrace. Veratrace is an Intel PIN-based API call tracing program for Windows. This program can trace API calls of only de-armored programs obtained from Ether. The output of Veratrace is parsed, and each executable is represented in the form of a vector. The collection of vectors of all applications is represented in the form of a two-dimensional matrix referred to us as the Frequency Vector Table (FVT) of API traces;
- Mnemonics and instruction opcode are extracted using ObjDump [44]. A custom-developed parser transforms the sample to FVT of mnemonics and instruction opcode.

In the following paragraphs, we briefly introduce the features extracted from malware and legitimate executables.

3.3. API Calls Tracing

The Windows API, informally WinAPI, is Microsoft's core set of application programming interfaces (APIs) available in the Microsoft Windows operating systems. In Windows, an executable program to perform its assigned work needs to make a set of API calls. For example, for file management, some of the API calls are OpenFile: Creates, opens, reopens, or deletes a file; DeleteFile: Deletes an existing file; FindClose: Closes a file search handle that is opened by FindFirstFile, FindFirstFileEx, or FindFirstStreamW function; FindFirstFile: Searches a directory for a file or subdirectory name that matches a specified name; and GetFileSize: Retrieves the size of a specified file, in bytes. Thus no executable program can run without the API calls. Hence, the API calls made by an executable is a good measure to record its behavior.

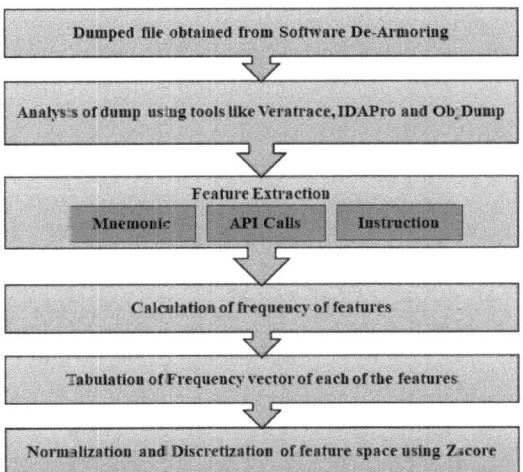

Figure 2. Feature extraction.

To extract APIs, we use Veratrace an API Call Tracer for Windows. It can trace all calls made by a process to the imported functions from a DLL. For extracting APIs, Veratrace mandates unpacking the samples. If packed, the import table would be populated with API calls like GetProcAddress() and LoadLibrary(), which are also common to legitimate executables. We have designed a parser to parse all the traces and filter out API names without argument, which is considered a feature in our work.

3.4. Mnemonic, Instruction Opcode, and 4-Gram Mnemonic Trace

We performed static analysis using the open-source ObjDump tool to obtain assembly language code. From these files, mnemonics, instruction opcode, and 4-gram mnemonic are extracted. An independent parser is developed to filter out mnemonics and instruction opcodes.

Each file is represented in the form of a vector, where the elements of the vector are the occurrence of an attribute. Since attribute values have different ranges, we normalize the data to a common scale. In our approach, we utilized a standard scalar approach "Z–Score" Besides, the normalized feature space is then discretized into three bins and used as an input to the Minimum Redundancy Maximum Relevance (mRMR) feature selection algorithm.

3.5. Feature Selection

In earlier studies, it has been reported that the feature selection is an integral component [45] in a machine learning pipeline. Many feature selection algorithms have been designed specifically for the application domain, furthermore every algorithm uses different criteria (such as information gain, Gini index, etc.) for extracting prominent attributes In the presence of irrelevant features, the detection model learns complex hypothesis functions, and learning models cannot generalize in identifying a new sample. Fundamentally, the role of a feature selection approach is to extract a prominent subset of attributes to improve classifier performance. The advantages of feature selection are listed below:

- Dimension reduction to haul down the computational cost;
- Reduction of noise to boost the classification accuracy;
- Introduce more interpretable features that can help identify and monitor the unknown sample.

We observe that the initial feature space obtained contained irrelevant attributes. By irrelevance, we mean set of feature which cannot identify a class and can never influence detection. In particular, these attributes appear equally in all samples of the target class. As a result, we selected discriminant features using maximal statistical dependency criterion based on mutual information known as Minimum Redundancy Maximum Relevance (mRMR), and by comparing means of two or more features using the ANOVA [46] as shown in Figure 3.

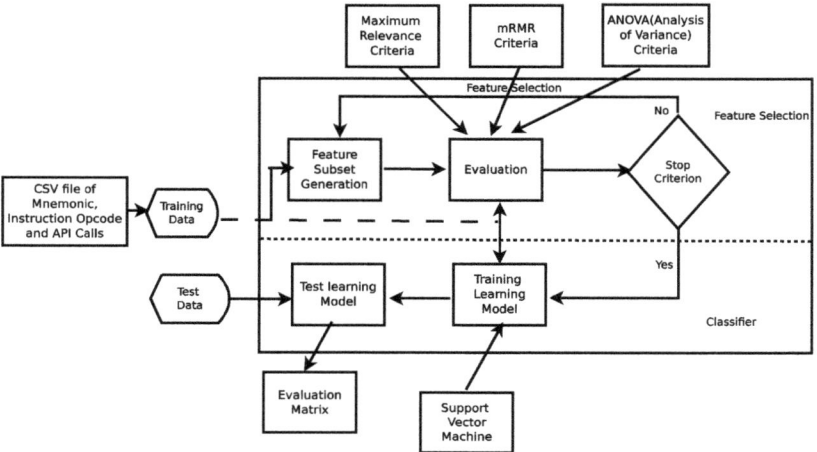

Figure 3. Feature selection process.

The training process can be either supervised, unsupervised, and semi-supervised. Supervised feature selection determines feature relevance by evaluating the correlation of attributes with the class. As our training data is labeled, we used supervised feature selection algorithms and filter methods to determine the correlation of the features with the class label. Using Filter methods, features are selected based on their intrinsic characteristics, their relevance, or controlling power concerning the target class. Such methods are based on mutual information, statistical test (t-test, F-test). A feature can become redundant due to the existence of other large volumes of relevant attributes in the feature space.

3.5.1. Minimum Redundancy Maximum Relevance

Maximum relevance criteria select features that highly correlated to the target class. mRMR is a filter method demanding the feature space to be discretized into states. However, this feature set is not a comprehensive representation of the characteristics of the target variable due to two essential aspects, as cited in [47]:

1. Efficiency: If a feature set of 50 samples contains many mutually highly correlated features, the representative features are very few, say 30, which means that 20 features are redundant, increasing the computational cost;
2. Broadness: According to their discriminative powers, we select some attributes, but such feature space is not maximally representative of the original space covered by the entire data set. The feature set may represent one or several dominant characteristics of an unknown sample, but it could also be a narrow region of relevant space. Thus the generalization ability could be limited.

To expand the representative power of the attribute set features while maintaining minimum pair-wise correlation, the minimum redundancy criterion supplements the maximum relevance criteria such as mutual information with target class. The mutual information of two features x and y is defined as the joint probabilistic distribution $P(x,y)$ and their respective marginal probabilities $P(x)$ and $P(y)$ (refer to Equation (1)).

$$I(x,y) = \sum_{i,j \in S} P(x_i, y_j) \log \frac{P(x_i, y_j)}{P(x_i)P(y_j)}, \qquad (1)$$

where x, y is the feature, namely mnemonic, instruction opcode, api call, 4-gram mnemonic, $P(x_i, y_j)$ is the joint probabilistic distribution of feature x and y, $P(x_i)$, $P(y_j)$ are the marginal probabilities, $I(x, y)$ is the mutual information between feature x and y, i indicates the level or state of feature x and j indicates the state of feature y, and S is the set obtained from cross product of set of states of x and y. Subsequently, we compute the relevance and redundancy value of attributes discussed below.

- Relevance value of an attribute x, $V(x)$ is computed using Equation (2):

$$V(x) = I(h, x), \qquad (2)$$

where h is the target variable or class, $I(h, x)$ is the mutual information between class and feature x.

- Redundancy value, $W(x)$ of feature x is obtained using Equation (3):

$$W(x) = \sum_{j \in N} I(y_j, x), \qquad (3)$$

where N is the total number of attributes, $I(y_j, x)$ is the mutual information of features y_j and x respectively.

Using Equations (2) and (3), minimum redundancy and maximum relevance of an attribute is computed, which is discussed below:

- Mutual Information Difference (MID): Is defined as the difference between the relevance value ($V(x)$) and the redundancy value ($W(x)$). To optimize the minimum redundancy and maximum relevance criteria, the difference between the relevance and redundancy value (see Equation (4)) was computed.

$$MID(x) = V(x) - W(x), \qquad (4)$$

Hence, the feature with maximum MID value indicates the mRMR feature;

- Mutual Information Quotient (MIQ): Is obtained by dividing the relevance value with the redundancy value, thus optimizing the mRMR criteria (refer to Equation (5)):

$$MIQ(x) = \frac{V(x)}{W(x) + 0.001}. \qquad (5)$$

Hence, the feature with a maximum MIQ value indicates the mRMR feature. Our approach use both these criteria, i.e., MID and MIQ, for selecting features, and compare classifier performance trained on the set of MID and MIQ attributes.

3.5.2. Analysis of Variance

Analysis of Variance (ANOVA) is a statistical method to compare the means of two or more groups. Depending upon the features and the level of features, ANOVA can be classified as follows:

- One way ANOVA: Requires one feature with at least two levels such that the levels are independent;

- Repeated Measures ANOVA: It commands one feature with at least two levels such that the levels are dependent;
- Factorial ANOVA: This approach demands two or more features, each of which with at least two levels either dependent, independent, or mixed.

Our proposed approach uses factorial ANOVA criteria for feature selection. In doing so, attributes highly correlated to the target class are determined. In particular, using ANOVA we estimate the impact of one or more independent variables on the dependent variable (i.e., class label). Feature influence is computed using variance, furthermore, it indicates separability between the class. Specifically, if the variance of an attribute is low then it has less impact on the target class. Using ANOVA, we choose a subset of independent variables having a stronger affinity towards classes. Generally, Post Hoc tests such as "F" statistics is performed to analyze the results of experiments. "F–Statistic" has its tailed distribution and is always positive. Variation in data can be due to two critical aspects (a) variation within the group and (b) variation between the group. Prominent features are derived using the procedure discussed below:

$$SS_T^p = SS_B^p + SS_W^p, \qquad (6)$$

where SS_T is the total sum of squares of feature p.

$$SS_T^p = \sum_{i=1}^{k}\sum_{j=1}^{l}(X_{ij} - \mu_p)^2, \qquad (7)$$

Here, k is the number of classes (malware/benign), l is the number of states of feature p,

$$\mu_p = \frac{1}{k*l}\sum_{i=1}^{k}\sum_{j=1}^{l}X_{ij}, \qquad (8)$$

μ_p is the mean of frequencies of feature p.

$$SS_W^p = \sum_{i=1}^{l}\sum_{j=1}^{k}(X_{ji} - \mu_i^p)^2, \qquad (9)$$

where SS_W^p is the sum of squares of within the group of feature p, and μ_i^p is the mean of frequencies of feature p in i^{th} discretization state.

$$DF_W^p = (k*l)^p - l^p, \qquad (10)$$

where DF_W^p is the degree of freedom of feature p within the group, and $(k*l)^p$ is the number of observations of feature p, l^p is the number of samples of feature p:

$$DF_B^p = l^p - 1, \qquad (11)$$

where DF_B^p is the degree of freedom of feature p between the group. Finally, F–Score is defined as:

$$F(DF_B^p, DF_W^p) = \frac{(SS_B^p/DF_B^p)}{(SS_W^p/DF_W^p)}. \qquad (12)$$

Eventually a feature p, with the highest F–Score is selected as a candidate member of the feature set.

3.6. Classification

A classification is a form of data analysis that can be used to extract models describing classes. It predicts categorical (discrete, unordered) labels. In our work, we utilized various machine learning and deep learning algorithms, such as Support Vector Machine

(SVM) [45,48], Naïve Bayes [49], J48 [50], Random Forest (RF) [51], and XGBoost [52]. In addition, we also evaluate the performance of the collection of deep neural networks like the Deep Dense network, One-Dimensional Convolutional Neural Network (1D-CNN) and CNN-LSTM in classifying unknown samples. The hyperparameters of all deep neural networks were tuned using the random search cross-validation approach. The above-mentioned classification algorithms were chosen as they have been extensively used in prior research work, and a subset of these classifiers have demonstrated to produce improved detection of unknown malware files [53–55].

In real-world applications, the size of the dataset is massive, data appears in a different form. The shallow network has a limited generalization capability. For obtaining better results, the shallow networks must be presented with features that are handpicked or suitably chosen after several iterations of the feature selection algorithms. Thus, the entire process is computationally expensive, also error-prone if attributes are extracted by humans. In contrast, deep neural networks employ a myriad of hidden layers, with each layer consisting of many neurons. Each neuron act as a processing unit to output complex features of input data. The lower layers extract features that are gradually amplified in the subsequent layers (higher layers). A deeper layer derives important aspects of the input data by omitting irrelevant details needed for classification. Thus, deep networks does not require feature extraction from scratch. In general, classification is a two step process as discussed below:

1. In the first step, we built a classifier describing a predetermined set of classes or concepts, also known as the learning step (or training phase). In this stage, a classification algorithm builds the classifier by analyzing or learning from a training set and their associated class labels. A tuple or feature X, is represented by an n-dimensional attribute vector, $X = (A_1 \& A_2 \& ... \& A_n)$, where, n depicts measurements made on the tuple from n database attributes, respectively, $A_1, A_2, ..., A_n$. Each tuple, X, is assumed to belong to a predefined class determined by the class label. The labels corresponding to the class attribute is discrete valued and unordered. The individual tuples making up the training set are referred to as training tuples and are randomly selected from the dataset. The class label of each training tuple is known to the classifier already, thus this approach is known as supervised learning;

2. In the second step, we use the classification model and predict the test data. This reserved set of samples is never used in the training phase. Eventually, the performance of the model on a given test set is estimated, generally evaluated as the percentage of test set tuples that are correctly classified by the classifier.

4. Experimental Evaluation and Results

4.1. Evaluation Metrics

We used following evaluation metrics:

- True Positive Rate (TPR) or Recall (R) = $\frac{TP}{TP+FN}$;
- Precision (P) = $\frac{TP}{TP+FP}$

Using Recall and Precision F-measure is estimated;

- F-measure = $2 \times \frac{P \times R}{P+R}$

Finally accuracy can be computed as shown below;

- Accuracy (A) = $\frac{TP+TN}{TP+FN+TN+FP}$

True Positive (TP) is number of samples correctly identified as malware. True Negative (TN) is the count of files identified as legitimate. False Negative (FN) is the number of malicious files misclassified as benign. False Positive (FP) is the number of benign files wrongly labeled as malware by the classifier.

4.2. Experiments Results

In this section, we discuss the experiment's setup, results obtained and the analysis of the result. The primary objective of this work is to perform analysis on different types of a dataset using various machine learning algorithms. For this purpose we created four datasets discussed below:

- Dataset-1 (VX-Dataset): A total of 2000 Portable Executables were collected which consists of 1000 malware samples gathered from sources VxHeaven (650) [35], User Agency (250), and Offensive Computing (100), and benign samples were collected from Windows XP System32 Folder (450), Windows7 System32 Folder (100), MikTex/Matlab Library (400), and Games (50);
- Dataset-2 (Virusshare-Dataset): A total of 622 executables were downloaded from virusshare [36], these belong to malware families like Mediyes, Locker. Intallcore, CryptoRansom, Citadel Zeus, and APT1_293. In addition, we collected 118 benign files from the freshly installed Windows operating system. These samples were used to evaluate the classifier performance on multi-label classification;
- Dataset-3 (Android-Dataset): A total 4000 applications were considered, out of which 2000 malicious samples were randomly chosen from the Derbin project [37], and 2000 legitimate applications were downloaded from the Google Playstore. Each benign application was submitted to the VirusTotal service to validate its genuinity;
- Dataset-4 (Synthetic Samples): VX Heavens reports nearly 152 synthetic kits and a few metamorphic engines to generate functionally equivalent malware code. Phalcon/Skin Mass-Produced Code generator (PS-MPCs), Second Generation virusgenerator (G2), Mass Code Generator (MPCGEN), Next Generation Virus Creation Kit (NGCVK), and Virus Creation Lab for Win32 (VCL32) are widely used to generate synthetic malware. A total of 320 viruses were generated with virus constructors and used as training samples. A separate test set is considered which includes 95 viruses (20 viruses from each generator and 15 real metamorphic) and 20 benign samples.

For experimenting on Dataset-1 and Dataset-4, we used a machine installed with Debian Lenny (Debian 5.08) as the host operating system, Windows XP Service Pack 2 as the guest operating system, i7 processor with 8GB RAM and 1TB HDD. Experiments on Dataset-2 and Dataset-3 were performed on Intel core i7, 10th generation with 16GB RAM, and 1TB HDD. Before executing samples in the system, we freshly installed the operating system and a snapshot of virtual environment was taken. After executing the sample, we restore the sandbox to its clean state, otherwise it would have a negative impact on the feature extraction phase.

4.3. Investigation of Relevant Feature Type-Dataset-1

We extracted mnemonics from 2000 samples. The experimental results obtained from feature reduction using mRMR (MID and MIQ) and ANOVA are as shown in Figure 4. We obtained these outcomes after classifying the samples using SVM, AdaBoost, Random Forest, and J48. Five mnemonic-based models were constructed at a variable length, starting from 40 to 120 at an interval of 20. Among these five models, ANOVA provides the best result with a strong positive likelihood ratio of 16.38 for the feature length of 120 mnemonics using AdaBoostM1 (J48 as base classifier). The main advantages of this model are its low error rate and speed. However, mnemonic-based features can be easily modified using code obfuscation techniques.

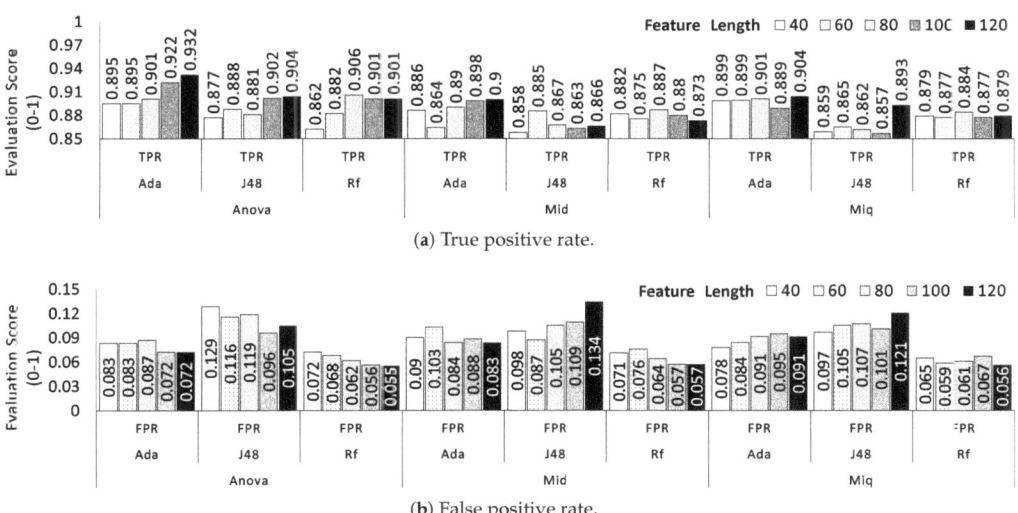

(a) True positive rate.

(b) False positive rate.

Figure 4. Performance of classifiers on mnemonic features.

The dynamic API features initially had a feature length of 4480. We use reduction algorithms like mRMR (MID), mRMR (MIQ), and ANOVA to obtain a reduced feature-length of 40, 60, 80, and 120 as illustrated in Figure 5. When we compare the different feature lengths, we observe that the likelihood for returning positive values is the highest in the case of mRMR (MIQ) at a feature-length of 120 prominent APIs using Random Forest.

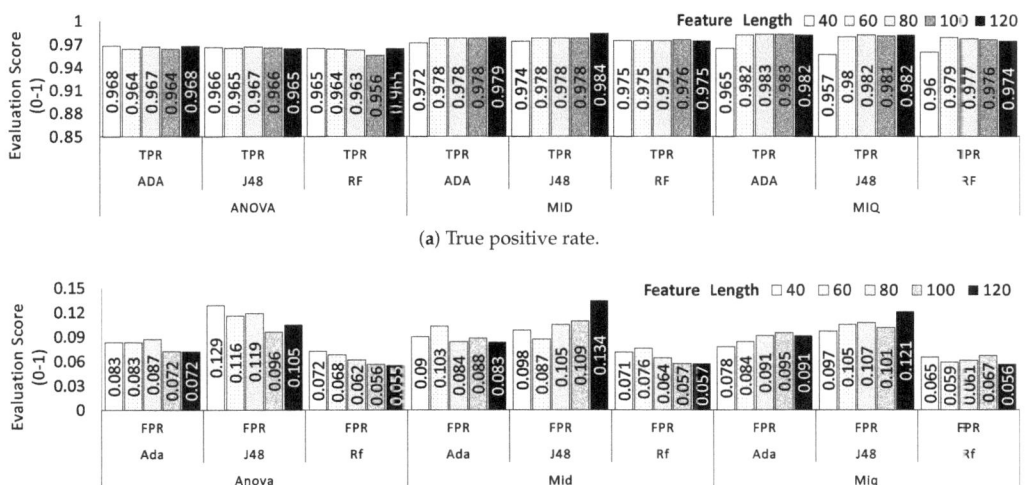

(a) True positive rate.

(b) False positive rate.

Figure 5. Performance of classifiers trained on API features.

Next, we derived 4-grams from a total feature space consisting of 1249 features. The features effectively reduced with mRMR (MID) methods, mRMR (MIQ), and ANOVA. The classification model's performance is estimated over variable feature-length starting from 40 until 120 in steps of 20 as shown in Figure 6. For the above-mentioned five-feature length, mRMR(MIQ) produces the best result with over 96% accuracy for feature-length 120 with Random Forest. However, the limitation stated in [56] is applicable in the current scenario. Generation of 4-grams are computationally expensive, exhibit diminishing returns with more data, are prone to over-fitting, and do not seem to carry vital information from discriminating samples. At the same time, 4-grams do exhibit some merits as it partially depicts the behavioral snapshot of a program and sometimes produces comparable results to other approaches.

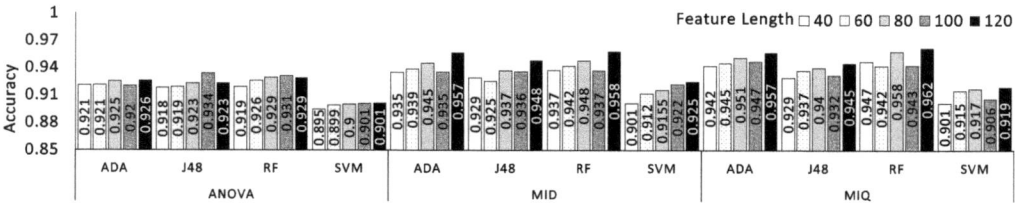

Figure 6. Performance of 4-gram features.

Finally, we derived the opcode-based feature set and reduced these features with mRMR (MID), mRMR (MIQ), and ANOVA, where the performance of the model is evaluated over a feature-length between 40 and 120 in increments of 20 as shown in Figure 7. Among these five-feature lengths, we observed that ANOVA attains the highest performance with a positive likelihood ratio of 19 for feature-length 100 using Random Forest. However, the results obtained with mRMR is very close to the ANOVA features.

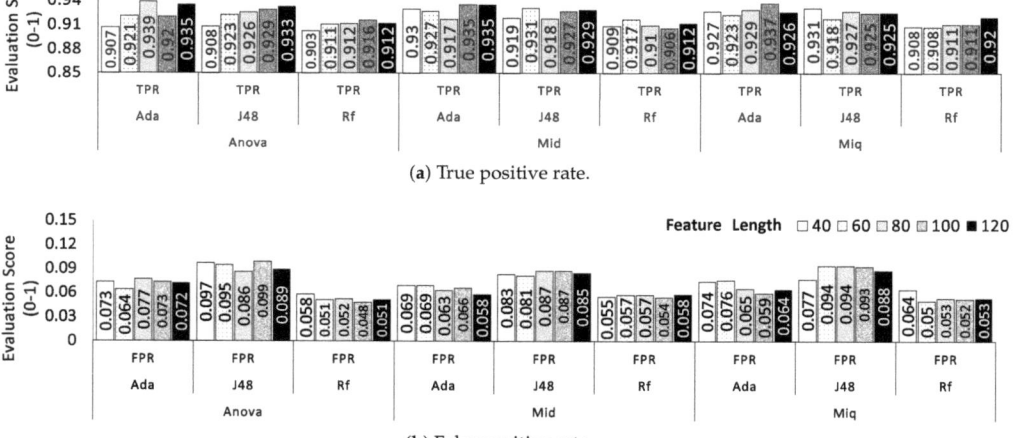

Figure 7. Performance of classifier trained on opcode features.

Hence, from the results we obtained, we observe that API features had a higher detection rate of 97.4% with only a fallout of 1.7% as against 4-gram's 93.15% accuracy. Again, when we compare the results obtained from API features as opposed to the results gathered from mnemonic features and opcode features, we see that opcode features had the highest likelihood ratio of 19 as against mnemonic features, and API features having a $LR+$ ratio of 16.38 and 15.69, respectively.

To summarize experiments on Dataset-1 considering each feature independently on four classification algorithms, namely J48, Support Vector Machine (SVM), AdaBoostM1 (with J48 as a base classifier), and Random Forest (RF), we observe that Random Forest and AdaBoost produced the best results. We can attribute this accuracy of Random Forest as it is an ensemble-based technique that derives its output from the sample by accumulating votes from multiple forests. We can credit the boosting technique that AdaBoost employs for improving its accuracy. AdaBoost cascades multiple weak classifiers to give a strong learner, ensuring a high degree of precision. The J48 classifier comes next in terms of the results produced in comparison to the other classifiers. The output produced by J48 is close to the best classifiers in some cases but is consistently inferior compared to the other two classifiers. SVM produces poor results among the four classifiers, which can be explained by SVM's tendency for over-fitting when the number of features is higher than the number of samples.

We further evaluate the performance of machine learning models generated by combining different feature categories. We consider such a feature space as a multimodal attribute set. The term modality means the particular mode in which something is expressed. In this context, it refers to the various features obtained with feature extraction, as shown in Figure 2. In the unimodal architecture, we perform classification based on a single modality and thus, this framework is limited to operating on a single attribute type. To investigate if blending different features from diverse feature categories could improve classification accuracy, we furthered our experiment using multimodal architecture.

Multimodal architecture involves learning based on multiple modalities. This solution is based on utilizing the relationship existing between the various features of the data available. This network can be used in converting data from one modality to another or in using one attribute set to assist the learning of another attribute set etc. We have achieved multimodal fusion in our experiment by carrying out feature selection (as shown in Figure 3) on the relevant attributes from diverse categories (4-gram, mnemonics, API, and opcodes) and then fusing them as shown in Figure 8.

As each feature has a different representation and correlation structure, the fusion of all these relevant features helps to extract maximum performance. Furthermore, after fusing these features, we were able to obtain a new feature space comprising of promising attributes. Additionally, we considered the new feature space for creating diverse classification models.

The presence of irrelevant features or redundancy in the data set might degrade the performance of the multimodal classification. Since we present the feature sets through various feature selection methods before performing feature fusion, our classifier is less susceptible to problems induced due to redundancy and extraneous features

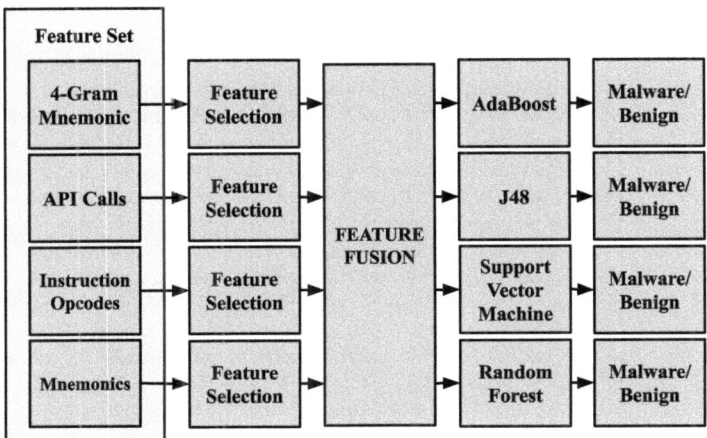

Figure 8. Multimodal architecture for feature fusion classification.

The ensemble classifier demonstrated the maximum accuracy of 97.98% with a feature-length of 240 using Random Forest, as shown in Figure 9. Among the unimodal classifiers, the API features demonstrated the highest detection rate of 97.4% with a FPR of 1.7%. Moreover, the opcode features displayed a detection rate of 91.6% and 0.48% FPR. By analyzing the results of both the unimodal and multimodal architectures, the results obtained using the multimodal architecture illustrate significant improvement compared to the results gained from the unimodal classifiers (as shown in Figures 4–7). Since the ensemble classifier was developed by concatenating prominent features from various feature sets, it is evident from the results that each modality considered for fusion has contributed to the overall performance of the classifier. Furthermore, this demonstrates that multimodal learning can be promising for increasing the detection in the malware detection task.

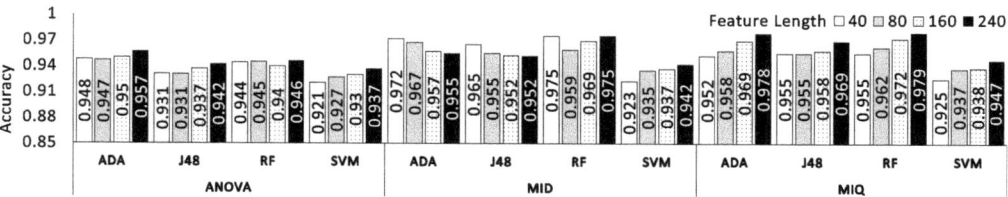

Figure 9. Performance of feature fusion.

Summary: Experiments on VX-Dataset demonstrates that combining prominent mRMR features results in improved results on comparing individual features. The highest detection rate is obtained with Random Forest and AdaBoost models, due to ensemble, bagging, and boosting strategies. APIs play a significant role in predicting examples, with poor outcomes obtained using opcodes. Another important trend noticed is that the results of multimodal feature space and API unimodal classifier marginally differ. This is because the opcode attribute in combined attribute space does not contribute towards classification, as they introduce more sparsity in feature vectors. Hence, we conclude that dynamic feature, i.e., API plays a critical role in discriminating malware and benign files.

4.4. Evaluation on Virusshare Samples Dataset-2

In this experiment, we perform a comprehensive evaluation on samples downloaded from Virusshare. As we saw inferior performance using static features, we performed analysis on APIs by running samples in the Parsa sandbox. The transition from Ether to an alternative sandbox (Parsa sandbox) arrived as many executables crashed while running in Ether. We observed that Parsa sandbox provides the requested resources to the executing samples by logging the API calls. This sandbox delivers the resources by matching the API used by executable with an API list which corresponds to a distinct set of operations corresponding to a mouse event, browser activities, file operations, etc. In this, the program is given an illusion of being run in a real environment as opposed to the virtual environment. While a program is executing, we log all APIs, extract call names, and select prominent calls using mRMR to create a machine learning model. In addition, we perform experiments using different deep learning models without feature engineering and compare the outcomes of both ML models and deep neural networks. Table 2 compares the average of the results of different models. We observe that the best results are obtained with a deep neural network, followed by one-dimensional convolutional neural network and XGBOOST. Table 3 exhibit the network topology and hyperparameters of deep neural network models. In all intermediate layers, we use ReLU activation function and randomly drop some neurons (i.e., dropout) to attain the best outcome for a particular neural network configuration.

Table 2. Results of models on Virusshare dataset.

Model	Precision (%)	Recall (%)	F1-Score (%)	Accuracy (%)
DNN	99.1	99.1	99.1	99.1
1D-CNN	97.9	97.9	97.9	97.9
CNN-LSTM	69.4	79.6	73.4	79.6
XGBOOST	97.8	97.8	97.5	97.8
Random Forest	97.3	97.25	96.89	97.35
AdaBoostM1	96.8	97.8	96.4	97.0
SVM	89.28	88.24	88.72	86.8
J48	87.54	88.08	87.6	86.53

Table 3. Network architecture and hyperparameters of deep neural network models.

Model	Topology
DNN	Dense1_units = 100; Dropout = 0.2; Activation = relu; Dense2_units = 50; Dropout = 0.2; Activation = relu; dense_final_units: 6; Activation = softmax; Optimizer: Adam; Learning rate: 0.001; batch size = 128, epoch = 40
1D-CNN	Conv1D: (num_filters = 15, filter_size:2); Activation = relu; Maxpooling1D; Conv1D: (num_filters = 15, filter_size = 2); Activation = relu; Maxpooling1D; Flatten; Dense units = 100; Activation: relu; dropout = 0.05; dense_final_units = 6; Activation = Softmax; optimizer = Adam, epoch = 15
SVM	kernel = rbf; gamma = 1 ; max_iter = 500; decision_function_shape = ovo
XGBOOST	XGBClassifier()
CNN-LSTM	Conv1D: (num_filters = 15, filter_size = 2); Activation: relu; maxpooling1D; LSTM_units = 100; Dense_units = 100; Activation = relu; Dropout = 0.05; Dense_units = 50; Activation = relu; Dropout = 0.05; Dense_final_units = 6; Activation = softmax; Optimizer = adam

4.5. Evaluation on Android Applications Dataset-3

In this experiment, we identify malicious Android applications (also known as app.) using machine learning and deep learning techniques. Here, we use system calls as a feature for representing each application. First, we create an Android virtual device and install applications to be inspected. A total of 2000 malware applications are randomly chosen from Drebin dataset [37], and 2000 legitimate applications are downloaded from the Google Playstore. While running applications, system calls are recorded using strace utility, during this event we employ Android Monkey (a utility in Android SDK for fuzz testing application) to simulate the collection of events (e.g., changing the location, battery charging status, sending SMS, dialling to a number, swipes, clicking on widgets of an app. etc.). In particular, in this work we execute an application with 1500 random events for one minute, however, the analysis could also be performed with varying events.

Relevant system calls are selected using the mRMR feature selection approach, and further each app. is represented using a numerical vector employing Term Frequency Inverse Document Frequency (TF-IDF). The performance of machine learning classifiers on the sequence of system call (two calls considered in sliding window fashion) is shown in Table 4. It was observed that distinguishing feature vectors were obtained by considering two consecutive system calls. Some examples of system call sequence are shown in Figure 10.

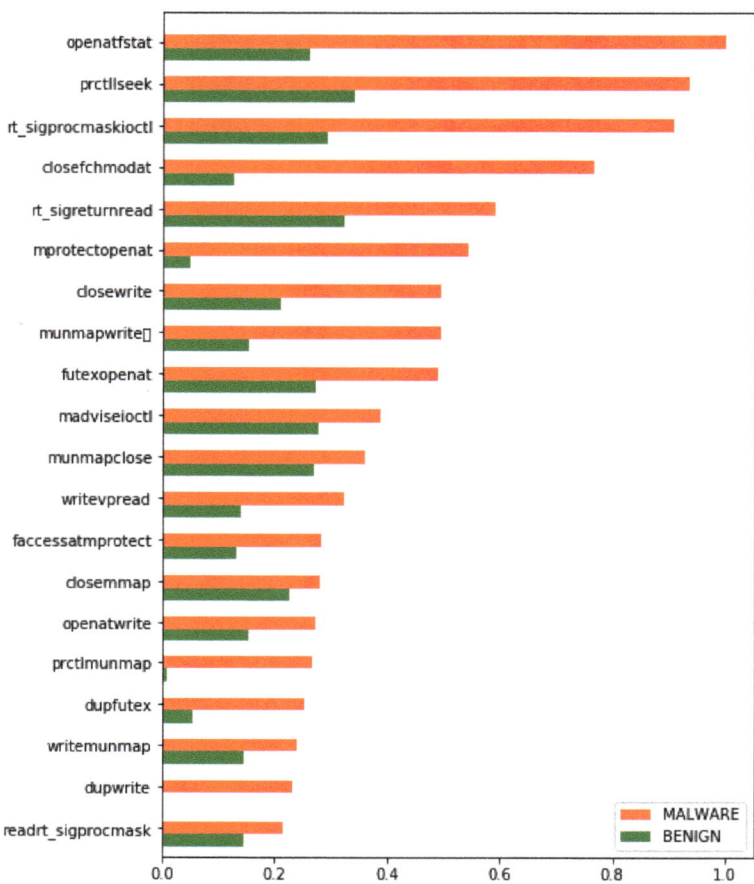

Figure 10. System call sequence.

We considered 40% of top system calls from the list of unique calls extracted from entire training set.

Table 4. Performance matrix of machine learning classifiers on system call sequence.

Classifier	Accuracy	F1-Measure	TPR	AUC
Random Forest	97.80%	97.8%	97.7%	0.998
AdaBoostM1	97.37%	96.4%	96.8%	0.97
J48	95.12%	95.1%	96.0%	0.965
SVM	95.02%	95.3%	95.1%	0.963
XGBoost	99.82%	99.82%	99.71%	0.998

From Table 4, we can visualize the best outcome for the XGBoost classifier. However, this result is obtained with an extra effort i.e., feature engineering which is a critical task in the machine learning pipeline. To eliminate the task of feature engineering, we make use of deep neural network architecture, which is a collection of layers, with each layer consisting of several neurons. A neuron acts as a processing unit that collects multiple inputs, multiplies weight, and finally applies the activation function. We use a deep neural network with an input layer consisting of 500 neurons and the second layer contains

250 neurons. In all layers, we use the Rectified Linear Unit (ReLU) activation function. The sigmoid activation function was used in the output layer since malware identification is a binary classification problem. For faster convergence and to avoid overfitting, the Adam optimizer and cross-entropy loss function are utilized. Table 5 is the results obtained at varying values of dropout, the best results are obtained with a dropout rate of 0.1.

Table 5. Performance of multi-level perceptron on varying the dropout rate.

Dropout	Accuracy (%)	F1-Score (%)	Precision (%)	Recall (%)
0.1	98.48	98.48	98.48	98.48
0.2	97.92	97.92	97.97	97.87
0.3	98.38	98.38	98.48	98.28
0.4	98.13	98.14	97.41	98.89
0.5	98.03	98.04	97.50	98.58
0.6	97.97	97.96	98.57	97.37

4.6. Evaluation on Synthetic Samples Dataset-4

Malware constructors generate variants from the base virus by inserting equivalent instructions, reordering, and subroutine permutations as code obfuscation techniques. The segments mutate from one generation to another where mutant code is transformed by the metamorphic engine to evade AntiVirus (AV) signature detection. This motivates the use of machine learning techniques to explore metamorphism among variants and within different families among synthetic samples, and to understand the extent of obfuscation induced by the virus kits. Malware data set comprising of 800 NGVCK viruses were used. Prior studies in [57] reported that the NGVCK samples could easily bypass strong statistical detectors based on HMM by using the opcode sequence. Likewise, 1200 benign executables were downloaded from different sources, which include games, web browsers, media players, and executables of system 32 from a fresh installation of the Windows XP operating system. As in previous experiments, we scan all benign with VirusTotal to assure that none of the benign samples is infected. The complete data set was divided such that 80% of samples are used for training and the remaining 20% are used as a test set. In this experiment, executables based on API calls were analyzed.

We extracted unique opcode bigrams from the training set and found 733 of them. Prominent opcodes are filtered out using the mRMR approach. We also studied the impact of varying feature lengths beginning with 50 bigram opcode until 250 bigrams are included. The feature space is extended in increments of 50 opcodes at a time. We found that an increase in bigrams had a marginal influence on the classifier performance. As we progressively extend the feature vector, the informative attributes begin to appear, which eventually improves the results. However, if we further increase the features beyond a certain limit there is a drop in accuracy, primarily due to the addition of noise. We developed a classification model using different algorithms such as J48, AdaBoostM1 with J48 as a base classifier, and Random Forest. Table 6 compares the best outcome of classifiers attained at a feature length of 150 bigrams.

Table 6. Performance metrics of machine learning models.

Classifier	Accuracy (%)	F-Measure (%)	Precision (%)	Recall (%)
J48	99.5	99.4	99.5	99.3
AdaboostMI(J48)	99.5	99.4	99.5	99.3
Random Forest	99.7	99.6	99.5	99.7

To understand the extent of metamorphism in virus generation kits, 677 viruses were created using different infection mechanisms to form malware families. In particular, we generated using virus kits (NGVCK, MPCGEN, G2, and PSMPC) and also downloaded real malware samples downloaded from VX Heavens. Data set description is given in Table 7.

Table 7. Data set description with samples, number of families, and number of variants of each family.

Constructors	Number of Families	Number of Variants
NGVCK	5	21
G2	5	21
PSMPC	5	21
MPCGEN	5	21
Real Malware Samples	11	5–77

Mnemonics are extracted from each malware sample and aligned using the global and local sequence alignment method. Sequence alignment places one opcode sequence over another to determine if sequences are identical. In the process of alignment, two opcode sequence gaps may be inserted. We have adopted a simple scoring scheme where a match is assigned a value of +1, and every mismatch and gap score is assumed as −1. A similarity matrix is constructed using pairwise alignment of malware samples within the family. We record minimum, average, and highest similarity distance for all malware samples. Likewise, the similarity distance of base malware across malware families is computed.

Two families are said to overlap if the similarity distance computed for base malware samples $Base_i$ and $Base_j$ is within the range of minimum and average similarity distance determined for families i or j. This means the greater the distance of a sample from the base malware, the lesser the similarity. Conversely, a high score depicts a higher similarity between any two samples. Table 8 depict a segment of pairwise alignment of two samples generated using the NGVCK constructor. Each row preceded with a hash symbol represents a gap and an asterisk designate a mismatch of an opcode for any two malware samples.

The local alignment technique is employed to identify a common code among obfuscated samples as the code varies in the subsequent generation to identify conserved code regions. We found variants generated from MPCGEN are similar to G2 and PSMPC. In Figure 11, MPCGEN-F1 and MPCGEN-F3 have high similarities with a base malware of G2 and PSMPC (G2-F1, G2-F3, PSMPC-F1, and PSMPC-F3).

Table 8. Pairwise alignment of two samples generated using the NGVCK constructor. The sequence shows match, mismatch, and gaps inserted for aligning the samples.

Sample 1	Sample 2
push	push
retn	retn
# -	mov
# -	sub
and	and
lea	lea
mov	mov
mov	mov
# -	popa
# -	sub
jmp	jmp
* inc	mov
* shr	and
* ror	mov
* cmp	dec

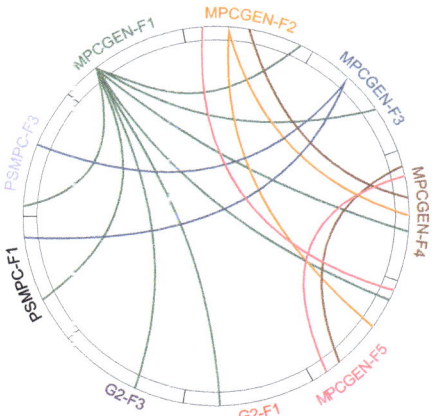

Figure 11. Overlapping MPCGEN malware families.

To examine obfuscation techniques using malware constructors, we calculated alignments of sequences and recorded mismatch among mnemonics. There was a visible instruction replacement for NGVCK samples in comparison to other synthetic generators. In Table 9, prominent mismatch opcodes are shown for four generators as the rest has shown a similar trend. mov, push, lea, pop, and jmp are primarily used as replacement instructions.

Table 9. Replacement of opcodes for malware generator (NGVCK, G2, PSMPC, and MPCGEN). For all generators, mov, push, pop, and jump instructions are replaced.

NGVCK	G2	PSMPC	MPCGEN
add mov	int call	jnz loop	mov pop
push mov	mov pop	-	cmp mov
mov pop	lea mov	-	int mov
call mov	xor cwd	-	mov lea
mov sub	mov movsb	-	jmp int
push add	rep movsb	-	call add
mov xor	xor mov	-	add movsw
and mov	cwd mov	-	lea jmp
mov jz	int inc	-	movsw mov
mov cmp	movsb movsw	-	push pop

To ascertain overlap among real malware samples of VX-Heavens and the obfuscated families, we studied the overlapping of the opcode sequence of real malware samples with synthetic ones. Initially, we determine base malware alignment (a sample that is closer to all samples in a family). Figure 12 shows the overlap of Win32.Agent with NGVCK indicating real samples that also use code modification similar to synthetic constructors. Win32.Bot and Win32.Downloader overlap Win32.Autorun, Win32.Downloader, Win32.Mydoom, and Win32.Xorer families indicating that worm families preserve the common base code to differ in syntactic structure due to obfuscation or an extension of malevolence.

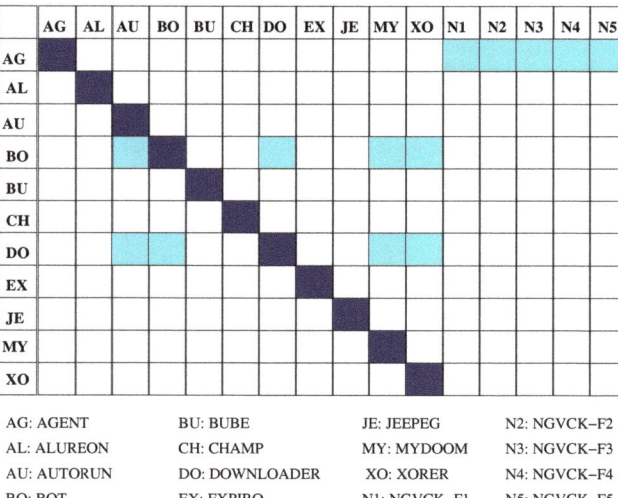

	AG	AL	AU	BO	BU	CH	DO	EX	JE	MY	XO	N1	N2	N3	N4	N5
AG	■											■	■	■	■	■
AL		■														
AU			■													
BO				■			■			■						
BU					■											
CH						■										
DO				■			■									
EX								■								
JE									■							
MY										■						
XO											■					

AG: AGENT	BU: BUBE	JE: JEEPEG	N2: NGVCK-F2
AL: ALUREON	CH: CHAMP	MY: MYDOOM	N3: NGVCK-F3
AU: AUTORUN	DO: DOWNLOADER	XO: XORER	N4: NGVCK-F4
BO: BOT	EX: EXPIRO	N1: NGVCK-F1	N5: NGVCK-F5

Figure 12. Overlapping families of real malware with other families.

5. Conclusions and Future Work

In this paper, we address the detection of malicious files using diverse datasets comprising of real and synthetic malware samples. The solution employs a collection of machine learning and deep learning approaches. Machine learning models were trained on prominent features derived using mRMR and ANOVA. Our results show that the Random Forest classifier attained better results, comparing all other machine learning algorithms used in this work. We also conclude that models trained on static features did not attain good results due to the sparse vectors. We demonstrate the efficacy of APIs and system call sequence in identifying samples. Moreover, to improve accuracy, we implemented our solution using distinct deep learning methods, and demonstrate fine-tuned a deep neural network that resulted in an F1-score of 99.1% and 98.48% on Dataset-2 and Dataset-3, respectively. Finally, we performed an exhaustive analysis of code obfuscation on variants generated using NGVCK and other virus kits. We found that NGVCK samples are appropriately detected by using a simple feature, such as opcode bigram. We also demonstrated that there exists inter-constructor overlaps especially amongst G2, MPCGEN, and PSMPC indicating the use of a generic code for infection. Our results also show that malware constructors employ naive obfuscation techniques, particularly they utilize junk instructions, a replacement of equivalent instructions involving mov, push, pop, jump, and lea.

In future, we would like to analyze malware and benign samples using an ensemble of features and classify unseen samples using ensembles of classifiers employing majority voting. We would also like to experiment on multi-class classification, i.e., labeling malware to its respective family. Moreover, we plan to investigate the efficacy of the machine learning and deep learning models on evasive samples generated through feature manipulations, and propose a countermeasure against adversarial attacks.

Author Contributions: Conceptualization, M.A., A.J., S.A., P.V., F.M. (Francesco Mercaldo), F.M. (Fabio Martinelli) and A.S.; methodology, M.A., A.J., S.A., P.V., F.M. (Francesco Mercaldo), F.M. (Fabio Martinelli) and A.S.; software, M.A., A.J., S.A., P.V., F.M. (Francesco Mercaldo), F.M. (Fabio Martinelli) and A.S.; validation, M.A., A.J., S.A., P.V., F.M. (Francesco Mercaldo), F.M. (Fabio Martinelli) and A.S. All authors have read and agreed to the published version of the manuscript.

Funding: This research received no external funding.

Acknowledgments: This work has been partially supported by MIUR-SecureOpenNets, EU SPARTA CyberSANE and E-CORRIDOR projects.

Conflicts of Interest: The authors declare no conflict of interest.

Abbreviations

The following abbreviations are used in this manuscript:

List of Abbreviations

1D-CNN	One-Dimensional Convolutional Neural Network
ANOVA	Analysis of Variance
API	Application Programming Interface
AV	AntiVirus
CART	Classification and Regression Tree
CNN	Convolutional Neural Network
CNN-LSTM	CNN Long Short-Term Memory Network
DNN	Deep Neural Network
FN	False Negative
FP	False Positive
FPR	False Positive Rate
FVT	Frequency Vector Table
G2	Second Generation Virus Generator
HMMs	Hidden Markov Models
k-NN	K-Nearest Neighbors
MID	Mutual Information Difference
MIQ	Mutual Information Quotient
MPCGEN	Mass Produced Code Generation Kit
mRMR	Minimum Redundancy and Maximum Relevance
NGVCK	Next Generation Virus Construction Kit
OS	Operating System
PE	Portable Executables
PSMPC	Phalcon/Skin Mass–Produced Code generator
QEMU	Quick EMUlator
ReLU	Rectified Linear Activation Function
RF	Random Forest
SMS	Short Message Service
SVM	Support Vector Machine
syscall	System Call
TF-IDF	Term Frequency Inverse Document Frequency
TPR	True Positive Rate
R	Recall
RDTSC	Read Time Stamp Counter
P	Precision
TP	True Positive
TN	True Negative
FP	False Positive
FN	False Negative
VCL32	Virus Creation Lab for Win32
XGBoost	eXtreme Gradient Boosting

List of Mathematical Symbols

Σ	Summation notation
$\log$	Logarithm
x, y	Features or Attributes
$P(x)$	Probability distribution of x
$I(x, y)$	Mutual information between feature x and y
$P(x_i, y_j)$	Joint probabilistic distribution of feature x and y
$P(x_i), P(y_j)$	Marginal probabilities of x and y
i	Level or state of feature x

j	Level or state of feature y
S	Set obtained from cross product of set of states of x and y
$V(x)$	Relevance value of an attribute x
h	Target variable or class
$I(h, x)$	Mutual information between class h and feature x.
$W(x)$	Redundancy value of feature x
N	Total number of attributes
$I(y_j, x)$	Mutual information of features y_j and x respectively
$MID(x)$	Difference between the relevance value $V(x)$ and the redundancy value $W(x)$
$MIQ(x)$	Obtained by dividing the relevance value with the redundancy Value
SS_B^p	Sum of squares of feature p belonging to group B
SS_W^p	Sum of squares of feature p belonging to group W
SS_T^p	Total sum of squares of feature p
k	Number of classes (malware/benign)
l	Number of states of feature p
μ_p	Mean of frequencies of feature p
X_{ij}	Feature at class I and state j
μ_i^p	Mean of frequencies of feature p in ith discretization state
DF_W^p	Degree of freedom of feature p within the group W
$(k*l)^p$	Number of observations of feature p
l^p	Number of samples of feature p
DF_B^p	Degree of freedom of feature p between the group B
$F(DF_B^p, DF_W^p)$	F-score
X	A tuple or feature represented by an n-dimensional attribute vector

References

1. Saraiva, D.A.; Leithardt, V.R.Q.; de Paula, D.; Sales Mendes, A.; González, G.V.; Crocker, P. Prisec: Comparison of symmetric key algorithms for iot devices. *Sensors* **2019**, *19*, 4312. [CrossRef] [PubMed]
2. Bulazel, A.; Yener, B. A survey on automated dynamic malware analysis evasion and counter-evasion: Pc, mobile, and web. In Proceedings of the 1st Reversing and Offensive-Oriented Trends Symposium, Vienna, Austria, 16–17 November 2017; pp. 1–21. [CrossRef]
3. Mandiant. Available online: https://www.fireeye.com/mandiant.html (accessed on 6 July 2021).
4. Dinaburg, A.; Royal, P.; Sharif, M.; Lee, W. Ether: Malware analysis via hardware virtualization extensions. In Proceedings of the 15th ACM Conference on Computer and Communications Security, Alexandria, VA, USA, 27–31 October 2008; ACM: New York, NY, USA, 2008; pp. 51–62. [CrossRef]
5. Alaeiyan, M.; Parsa, S.; Conti, M. Analysis and classification of context-based malware behavior. *Comput. Commun.* **2019**, *136*, 76–90. [CrossRef]
6. Intel. Available online: http://www.intel.com/ (accessed on 6 July 2021).
7. Virus Total. Available online: https://www.virustotal.com/en/statistics/ (accessed on 6 July 2021).
8. Christodorescu, M.; Jha, S.; Kruegel, C. Mining specifications of malicious behavior. In Proceedings of the the 6th Joint Meeting of the European Software Engineering Conference and the ACM SIGSOFT Symposium on the Foundations of Software Engineering, Dubrovnik, Croatia, 3–7 September 2007; ACM: New York, NY, USA, 2007; pp. 5–14. [CrossRef]
9. Alzaylaee, M.K.; Yerima, S.Y.; Sezer, S. DL-Droid: Deep learning based android malware detection using real devices. *Comput. Secur.* **2020**, *89*, 101663. [CrossRef]
10. Canfora, G.; Medvet, E.; Mercaldo, F.; Visaggio, C.A. Detecting Android malware using sequences of system calls. In Proceedings of the 3rd International Workshop on Software Development Lifecycle for Mobile, Bergamo, Italy, 31 August 2015; pp. 13–20. [CrossRef]
11. Wang, W.; Li, Y.; Wang, X.; Liu, J.; Zhang, X. Detecting Android malicious apps and categorizing benign apps with ensemble of classifiers. *Future Gener. Comput. Syst.* **2018**, *78*, 987–994. [CrossRef]
12. Nataraj, L.; Karthikeyan, S.; Jacob, G.; Manjunath, B. Malware images: Visualization and automatic classification. In Proceedings of the 8th International Symposium on Visualization for Cyber Security, Pittsburgh, PA, USA, 20 July 2011; pp. 1–7. [CrossRef]
13. Ni, S.; Qian, Q.; Zhang, R. Malware identification using visualization images and deep learning. *Comput. Secur.* **2018**, *77*, 871–885. [CrossRef]
14. Yuan, Z.; Lu, Y.; Xue, Y. Droiddetector: Android malware characterization and detection using deep learning. *Tsinghua Sci. Technol.* **2016**, *21*, 114–123. [CrossRef]
15. McLaughlin, N.; Martinez del Rincon, J.; Kang, B.; Yerima, S.; Miller, P.; Sezer, S.; Safaei, Y.; Trickel, E.; Zhao, Z.; Doupé, A.; et al. Deep android malware detection. In Proceedings of the Seventh ACM on Conference on Data and Application Security and Privacy, Scottsdale, AZ, USA, 22–24 March 2017; pp. 301–308. [CrossRef]

16. Karbab, E.B.; Debbabi, M.; Derhab, A.; Mouheb, D. Android malware detection using deep learning on API method sequences. *arXiv* **2017**, arXiv:1712.08996.
17. Iadarola, G.; Martinelli, F.; Mercaldo, F.; Santone, A. Towards an Interpretable Deep Learning Model for Mobile Malware Detection and Family Identification. *Comput. Secur.* **2021**, *105*, 102198. [CrossRef]
18. Zhang, B.; Yin, J.; Hao, J. Using Fuzzy Pattern Recognition to Detect Unknown Malicious Executables Code. In *Fuzzy Systems and Knowledge Discovery*; Lecture Notes in Computer Science; Springer: Berlin/Heidelberg, Germany, 2005; Volume 3613, pp. 629–634. [CrossRef]
19. Sun, H.M.; Lin, Y.H.; Wu, M.F. API Monitoring System for Defeating Worms and Exploits in MS-Windows System. In *Information Security and Privacy*; Springer: Berlin/Heidelberg, Germany, 2006; Volume 4058, pp. 159–170. [CrossRef]
20. Bergeron, J.; Debbabi, M.; Erhioui, M.M.; Ktari, B. Static Analysis of Binary Code to Isolate Malicious Behaviors. In Proceedings of the 8th Workshop on Enabling Technologies on Infrastructure for Collaborative Enterprises, Palo Alto, CA, USA, 16–18 June 1999; IEEE Computer Society: Washington, DC, USA, 1999; pp. 184–189. [CrossRef]
21. Zhang, Q.; Reeves, D.S. MetaAware: Identifying Metamorphic Malware. In Proceedings of the Twenty-Third Annual Computer Security Applications Conference (ACSAC 2007), Miami Beach, FL, USA, 10–14 December 2007; pp. 411–420. [CrossRef]
22. Sathyanarayan, V.S.; Kohli, P.; Bruhadeshwar, B. Signature Generation and Detection of Malware Families. In Proceedings of the 13th Australasian Conference on Information Security and Privacy, Wollongong, Australia, 7–9 July 2008; Springer: Berlin/Heidelberg, Germany, 2008; pp. 336–349. [CrossRef]
23. Rabek, J.C.; Khazan, R.I.; Lewandowski, S.M.; Cunningham, R.K. Detection of injected, dynamically generated, and obfuscated malicious code. In Proceedings of the 2003 ACM Workshop on Rapid Malcode, Washington, DC, USA, 27 October 2003; ACM: New York, NY, USA, 2003; pp. 76–82. [CrossRef]
24. Damodaran, A.; Di Troia, F.; Visaggio, C.A.; Austin, T.H.; Stamp, M. A comparison of static, dynamic, and hybrid analysis for malware detection. *J. Comput. Virol. Hacking Tech.* **2017**, *13*, 1–12. [CrossRef]
25. Wong, W.; Stamp, M. Hunting for metamorphic engines. *J. Comput. Virol.* **2006**, *2*, 211–229. [CrossRef]
26. Nair, V.P.; Jain, H.; Golecha, Y.K.; Gaur, M.S.; Laxmi, V. MEDUSA: MEtamorphic malware dynamic analysis using signature from API. In Proceedings of the 3rd International Conference on Security of Information and Networks, Taganrog, Russia, 7–11 September 2010; ACM: New York, NY, USA, 2010; pp. 263–269. [CrossRef]
27. Suarez-Tangil, G.; Stringhini, G. Eight years of rider measurement in the android malware ecosystem: Evolution and lessons learned. *arXiv* **2018**, arXiv:1801.08115.
28. Zhou, Y.; Jiang, X. Dissecting Android Malware: Characterization and Evolution. In Proceedings of the 33rd IEEE Symposium on Security and Privacy (Oakland 2012), San Francisco, CA, USA, 21–23 May 2012. [CrossRef]
29. Arp, D.; Spreitzenbarth, M.; Huebner, M.; Gascon, H.; Rieck, K. Drebin: Efficient and Explainable Detection of Android Malware in Your Pocket. In Proceedings of the 21st Annual Network and Distributed System Security Symposium (NDSS), San Diego, CA, USA, 23–26 February 2014.
30. Yan, L.K.; Yin, H. DroidScope: Seamlessly Reconstructing the OS and Dalvik Semantic Views for Dynamic Android Malware Analysis. In Proceedings of the 21st USENIX Conference on Security Symposium, Bellevue, WA, USA, 8–10 August 2012; USENIX Association: Berkeley, CA, USA, 2012; p. 29. ISBN 978-931971-95-9.
31. Canfora, G.; Mercaldo, F.; Visaggio, C.A. A classifier of Malicious Android Applications. In Proceedings of the 2nd International Workshop on Security of Mobile Applications, in Conjunction with the International Conference on Availability, Reliability and Security, Regensburg, Germany, 2–6 September 2013. [CrossRef]
32. Su, D.; Liu, J.; Wang, W.; Wang, X.; Du, X.; Guizani, M. Discovering communities of malapps on Android-based mobile cyber-physical systems. *Ad Hoc Netw.* **2018**, *80*, 104–115. [CrossRef]
33. Liu, X.; Liu, J.; Zhu, S.; Wang, W.; Zhang, X. Privacy risk analysis and mitigation of analytics libraries in the android ecosystem. *IEEE Trans. Mob. Comput.* **2019**, *19*, 1184–1199. [CrossRef]
34. Casolare, R.; Martinelli, F.; Mercaldo, F.; Santone, A. Detecting Colluding Inter-App Communication in Mobile Environment. *Appl. Sci.* **2020**, *10*, 8351. [CrossRef]
35. VX Heavens. Available online: https://vx-underground.org/archive/VxHeaven/index.html (accessed on 6 July 2021).
36. Virusshare. Available online: https://virusshare.com/ (accessed on 6 July 2021).
37. Drebin. Available online: https://www.sec.cs.tu-bs.de/~danarp/drebin/download.html (accessed on 6 July 2021).
38. Alkhateeb, E.; Stamp, M. A Dynamic Heuristic Method for Detecting Packed Malware Using Naive Bayes. In Proceedings of the 2019 International Conference on Electrical and Computing Technologies and Applications (ICECTA), Ras Al Khaimah, United Arab Emirates, 19–21 November 2019; pp. 1–6. [CrossRef]
39. Xen Project. Available online: http://www.xen.org (accessed on 6 July 2021).
40. Windows API Library. Available online: https://docs.microsoft.com/en-us/windows/win32/apiindex/windows-api-list (accessed on 6 July 2021).
41. Vnc Viewer. Available online: https://www.realvnc.com/en/connect/ (accessed on 6 July 2021).
42. You, I.; Yim, K. Malware obfuscation techniques: A brief survey. In Proceedings of the 2010 International Conference on Broadband, Wireless Computing, Communication and Applications, Fukuoka, Japan, 4–6 November 2010; pp. 297–300. [CrossRef]
43. Borello, J.M.; Mé, L. Code obfuscation techniques for metamorphic viruses. *J. Comput. Virol.* **2008**, *4*, 211–220. [CrossRef]

44. ObjDump. Available online: https://web.mit.edu/gnu/doc/html/binutils_5.html (accessed on 6 July 2021).
45. Witten, I.; Frank, E. *Practical Machine Learning Tools and Techniques with Java Implementation*; Morgan Kaufmann: Burlington, MA, USA, 1999; ISBN 1-55860-552-5.
46. Factorial Anova. Available online: http://en.wikipedia.org/wiki/Analysis_of_variance (accessed on 6 July 2021).
47. Ding, C.; Peng, H. Minimum Redundancy Feature Selection from Microarray Gene Expression Data. In Proceedings of the IEEE Computer Society Conference on Bioinformatics, Stanford, CA, USA, 11–14 August 2003; p. 523. [CrossRef]
48. Cortes, C.; Vapnik, V. Support-vector networks. *Mach. Learn.* **1995**, *20*, 273–297. [CrossRef]
49. Rish, I. An empirical study of the naive Bayes classifier. In Proceedings of the IJCAI 2001 Workshop on Empirical Methods in Artificial Intelligence, Seattle, WA, USA, 4 August 2001; Volume 3, pp. 41–46.
50. Quinlan, J.R. *C4.5: Programs for Machine Learning*; Elsevier: Amsterdam, The Netherlands, 2014; ISBN 1-55860-238-0.
51. Breiman, L. Random forests. *Mach. Learn.* **2001**, *45*, 5–32. [CrossRef]
52. Chen, T.; Guestrin, C. Xgboost: A scalable tree boosting system. In Proceedings of the 22nd Acm Sigkdd International Conference on Knowledge Discovery and Data Mining, San Francisco, CA, USA, 13–17 August 2016; pp. 785–794. [CrossRef]
53. Narudin, F.A.; Feizollah, A.; Anuar, N.B.; Gani, A. Evaluation of machine learning classifiers for mobile malware detection. *Soft Comput.* **2016**, *20*, 343–357. [CrossRef]
54. Singh, J.; Singh, J. A survey on machine learning-based malware detection in executable files. *J. Syst. Archit.* **2021**, *112*, 101861. [CrossRef]
55. Mahdavifar, S.; Ghorbani, A.A. Application of deep learning to cybersecurity: A survey. *Neurocomputing* **2019**, *347*, 149–176. [CrossRef]
56. Raff, E.; Zak, R.; Cox, R.; Sylvester, J.; Yacci, P.; Ward, R.; Tracy, A.; McLean, M.; Nicholas, C. An investigation of byte n-gram features for malware classification. *J. Comput. Virol. Hacking Tech.* **2018**, *14*, 1–20. [CrossRef]
57. Lin, D.; Stamp, M. Hunting for undetectable metamorphic viruses. *J. Comput. Virol.* **2011**, *7*, 201–214. [CrossRef]

Article

A Novel Monte-Carlo Simulation-Based Model for Malware Detection (eRBCM)

Muath Alrammal *, Munir Naveed and Georgios Tsaramirsis

Faculty of Computer Information Science, Abu Dhabi Women College, Higher Colleges of Technology, Abu Dhabi 41012, United Arab Emirates; mnaveed@hct.ac.ae (M.N.); gtsaramirsis@hct.ac.ae (G.T.)
* Correspondence: malrammal@hct.ac.ae

Abstract: The use of innovative and sophisticated malware definitions poses a serious threat to computer-based information systems. Such malware is adaptive to the existing security solutions and often works without detection. Once malware completes its malicious activity, it self-destructs and leaves no obvious signature for detection and forensic purposes. The detection of such sophisticated malware is very challenging and a non-trivial task because of the malware's new patterns of exploiting vulnerabilities. Any security solutions require an equal level of sophistication to counter such attacks. In this paper, a novel reinforcement model based on Monte-Carlo simulation called eRBCM is explored to develop a security solution that can detect new and sophisticated network malware definitions. The new model is trained on several kinds of malware and can generalize the malware detection functionality. The model is evaluated using a benchmark set of malware. The results prove that eRBCM can identify a variety of malware with immense accuracy.

Keywords: malware detection; Monte-Carlo simulation; reinforcement learning

1. Introduction

As the Internet has become essential in our life, the number of users who use internet services such as e-commerce and e-banking, has increased rapidly. Unfortunately, this increment is accompanied by an increased number of cyber-criminals who use malware (malicious programs) to achieve their malicious intentions [1].

Cyber-criminals launch new malware/attacks every year that are more sophisticated and harmful than previous years. Malware can adapt to the environment according to the security barriers set in an IT environment. Millions of new definitions are generated daily to exploit the vulnerabilities and compromise commercial information systems [2].

To overcome this severe threat, security companies such as Kaspersky and Symantec have introduced several anti-malware products to protect individuals and companies [2]. These products are for known malware definitions. While such solutions can detect known malware with high accuracy, they often lack the ability to detect unknown malware. Moreover, referencing all the different malware has become a complex task because of the enormous increase in the number of malware programs, making it difficult to find lasting solutions. These limitations have made it necessary to explore intelligent approaches that are flexible and adaptable in detecting unknown malware.

Most of the new intelligent approaches to malware detection are trained using the selective features of known malware that can represent malware in its best form. These representations are then used as training instances for a suitable machine-learning algorithm that generalizes or maps such features-based malware detection mechanisms [3–13]. This work extends a previously explored approach called RBCM, which is also based on reinforcement learning [3]. The RBCM extension is called eRBCM, and merges the most beneficial features of Monte-Carlo-based real-time learning (MOCART) [4] and random forest [5–7] to make it more scalable for higher-order training datasets.

The rest of this paper is organized as follows: Section 2 presents the various approaches adopted to detect and analyze network malware; Section 3 describes our motivations and contributions; Section 4 provides a short introduction to MOCART; Section 5 illustrates the enhancements to our previous approach (RBCM) [14], made to avoid converging to local minima in the search spaces with a narrow range of values in an observation dataset; Section 6 shows the experimental set-up and compares the performance of eRBCM with its state-of-the-art rivals; and Section 7 presents our conclusions and future work.

2. Related Work

According to the malware detection taxonomy outlined by [8], machine-learning approaches can be classified based on three major dimensions: malware targets, malware features, and the AI model used to generalize malware detection. This section focuses on the third dimension, the machine-learning algorithm, since our study evaluates algorithm performance in the malware detection task. Machine-learning algorithms are scalable to generalize non-linear problem spaces, which is the main motivation for exploring such approaches to optimize malware detection.

Different malware detection approaches in the literature have adopted different machine-learning techniques, such as random forest (RF) [5–7], neural network [9–11], decision tree [12,13], naïve Bayes [14,15], KNN and SVM [15], ARIMA [16], and reinforcement learning [17,18].

The RF machine-learning technique has been applied in several malware classification problems in the literature [5–7] because of its competitive performance compared to other algorithms. In an original approach proposed in [19], where the malware features were modelled as grayscale images, a comparison between three machine learning techniques revealed that RF outperformed the naïve Bayes and KNN algorithms.

RF was explored in [6] to generalize the malware detection and classification. The authors presented a machine-learning technique called AMICO [6], which was trained using the network-traffic-based selection parameters. The main purpose was to evaluate the payload information in network traffic. Parameters such as IP address, source URL, target URL, file contents, etc. were analyzed to identify the malware patterns. In the sandbox environment, a download reconstruction module was used to generate the network traffic in real-time. The traffic was based on executable files and the malware detection technique was evaluated using real-time generated data. The training data constructed to generalize the AI model was based on both malware-based traffic data and normal traffic.

To distinguish between malware and benign files, Vadrevu et al. [7] opted for a supervised learning approach based on a RF algorithm, where the training set of labelled malicious instances was evaluated over a period of one to two months. The simulated data contained a fair distribution of both kinds of samples. The model trained on this data was tested using an academic network, and the test results showed that AMICO could detect 90% of the malicious content that travelled over the network during the testing phase.

A classifier based on a decision tree was used in [7] to detect malicious contents. The "Malware Target Recognition (MaTR)" model is a hybrid of a decision tree classifier and is optimized by using a sophisticated heuristic-based feature search to keep the rules exploration-focused towards the promising area of the search space. In their work, the heuristics are built using the structural information of malicious contents and structural anomalies. Examples of malicious content structure include file path, attributes, and size, while examples of structural anomalies include entry point and section names. The classifier was trained using a benchmark dataset called VX-Heaven. The heuristics were built in the pre-processing stages and remained part of the training instances to extract quality rules. The classifier was tested on malicious contents that were not used during the training phase. The test data showed an accuracy of 99% for the decision-tree-based classifier's malware detection.

Neural-network-based approaches in malware detection were introduced in [9–11], while a recurrent neural network (RNN)-based model was explored by Andrade et al. [9].

The model was trained using a benchmark dataset that is publicly available for exploring new security solutions. Their neural network model creates new connections among the neurons based on cycles to increase the memory-based connectivity. The model also balances the trade-off between the long-term and short-term memory approaches. Short-term memory emphasizes the exploration of solution space, while long-term memory exploits the already known best regions of the solution space. The experimental results showed that RNN detected malicious content with 67% accuracy.

There are several approaches for app malware detection. Approaches including EspyDroid [20], AndroShield [21], Droidcat [22], and RevealDroid [23] are used as solutions for obfuscation camouflage techniques such as junk code insertion, package renaming, and altering control-flow [24–26].

Other approaches for app malware detection, such as API-Graph [27], DroidEvolver [28], and DroidSpan [29], are oriented towards solving the problem of sustainability (performance over time). However, it is unclear how these approaches address this problem in the case of a network attack or malware.

This paper focuses mainly on exploring a machine-learning model that can generalize the patterns of a variety of malware.

3. Motivations and Contributions

Our motivations are summarized below.

Because of the enormous increase in new malware samples, traditional approaches are not scalable to the sophistication of new attacks and lack the capability to detect and analyze these attacks. Intelligent, self-adaptive approaches for efficient network malware detection and analysis are required [2].

There is a need for an approach that can be easily scaled for large and high-dimensional malware datasets to avoid extensive training episodes. This is essential in order to generalize the characteristics of different kinds of malware. The security solution can be trained on different datasets without changes in the learning structures.

Our contributions are summarized as follows.

We improve RBCM [3] to avoid being trapped in local minima. The current version combines the best features of MOCART [4] and RF [5–7]. Monte-Carlo simulations are optimized to dynamically select the region and scale of samples used by the learning model. The dynamic sampling technique is used to enhance the performance of the RBCM learning model, which selects a sampling region of lower error and fixed size. This drawback decreases RBCM's performance in cases where the sample space is limited or there are large areas of low-quality samples that reduce the error with respect to the current surroundings, but the model does not learn new knowledge.

We test eRBCM using the three datasets: Microsoft Malware [30], ARP attack, and ICMP attack [31]. Furthermore, we provide a comparison of eRBCM with four state-of-the-art, best-performing prediction algorithms.

4. Monte-Carlo-Based Real-Time Learning (MOCART)

MOCART [4] is a Monte-Carlo (MC)-simulation-based machine-learning algorithm that applies the Monte-Carlo tree search to obtain estimates from one node of the solution space to another to reach the goal node. The MC simulations explore the solutions using a sample space and build a learning structure. In MOCART, MC simulations build a value function that can predict the outcomes for each action in an uncertain or unknown environment. The simulations use a model of system which can predict outcomes for a deterministic or nondeterministic problem space. As a result of these characteristics, MOCART has been used in several domains, especially nondeterministic domains. Because of these capabilities, MOCART is particularly suitable for malware detection, as the behavior of a sophisticated piece of malware can be non-deterministic, and it might behave differently at the same state in a problem space. This is particularly true for a new set of malwares that are sensitive to sandboxing and Trojans. However, MOCART underperforms in domains where the number of possible samples generated during simulations are limited or if the simulation model is biased towards more exploitation than exploration.

5. Reinforcement Learning Model RBCM

To generalize the pattern recognition of various malware attacks, an updated version of reinforcement learning called *e*RBCM is explored in this work. *e*RBCM combines the best features of MOCART and RF.

The sampling techniques are modified in RBCM to keep finding new samples until the error rate remains below a threshold θ.

RBCM suffers from local minima when space dimensions are of a small scale or data has fewer variations with respect to class labels [4]. *e*RBCM increases the number of samples in the simulation model if class labels are not equally distributed. The generative model of *e*RBCM is shown in Figure 1 for a sample S, simulation length d, and extension n.

Procedure Simulate (S, d, n, e)
1. Generate d samples at S
2. Evaluate Learning Structures on d samples
3. Measure *rmse* for d samples
4. If *rmse* < e
5. Generate $d + n$ samples
6. Update Learning Structure
7. Evaluate Learning Structure on $d + n$ sample
8. $e = rmse$

Figure 1. Generative model for *e*RBCM.

The generative model of *e*RBCM extends the simulation length by n, as shown in step 6 of Figure 1. The update decision is made by using epsilon e, which depends on the current root mean square error (RMSE) of the learning structure (the learning structure is a Q function). This is a validation RMSE of the Q function on unseen data. The decision parameter is dynamic and keeps reducing itself depending on the RMSE, which makes the sample exploration self-adaptive and keeps the trade-off between exploration and exploitation in balance. Figure 2 explains the sampling process with respect to the depth of search for samples in the direction of solutions.

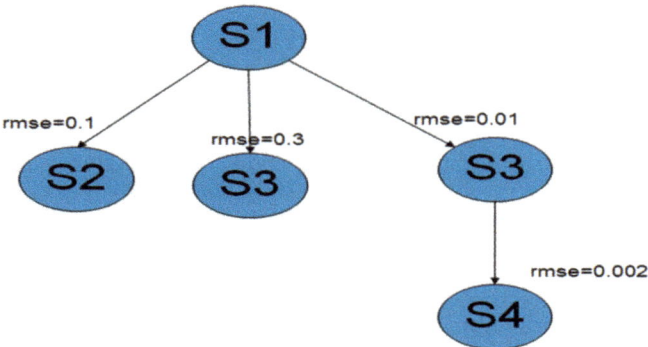

Figure 2. Selection process of deeper search.

The search space at S1 was simulated with three neighboring states and only S3 was extended as it met the criteria for decision making. The state S3 produced the smallest RMSE of its siblings and a more reliable and stronger heuristic to select the direction to explore deeper into. This also assisted the Q function to be updated with the weight values that reduce the error of the network. The epsilon value was updated on each extension of the search process; for example, epsilon at S4 will be 0.002. If no neighbor of S4 produces a lower RMSE than S4, the search will stop at this state of the space. This process is also intuitive to bring the search out of local minima, as the RMSE will never be reduced below the best local minimum and the generative model will explore more spaces.

Due to dynamic changes in epsilon in the generative model, the model can learn biased strategies to explore the space rather than exploiting the best results. However, because of a fixed number of extensions, the search policy is kept balanced at the exploitation of the best and the search of new states in the search process. The adaptive use of epsilon also introduces the benefit of avoiding the visit for the same sample more than once. It reduces the time of searching for the best solution in the space and gives eRBCM the advantage of quicker convergence than CNN.

6. Reinforcement Learning Model Experimentation

6.1. Experimentation

All experiments were performed using Windows 10 Enterprise with 16 GB RAM and dual Intel Core (TM) i7-4702MQ CPUs, each of 2.20 GHz speed. The benchmark malware files were analyzed using different programs for deep visibility of attack data. The tools used were Wireshark and Network Miner. All tools were run in a special operating system called Security Onion. The attack files were further processed to generate a training dataset. The benchmark malware datasets analyzed were: Microsoft Malware dataset [30], ARP attack dataset [31], and ICMP attack dataset [31]

6.1.1. Microsoft Malware Data

The dataset in [30] is organized with respect to machines and has several input features (e.g., 'machineidentifier' and 'hasdetected') which are malware detected on the machine. This column is used as the actual output for training the machine-learning algorithms The dataset contains the system details for each observation, including default browser, current OS version, firewall, processor, primary disk type, volume capacity, total physical RAM, casing details, and gaming systems. This dataset is used for training machine-learning algorithms to detect malware on end systems running Windows OS.

6.1.2. ARP Attack Dataset

The ARP dataset [31] is taken from the Contagion malware dataset. ARP attacks exploit the vulnerabilities related to Address Resolution Protocol. ARP vulnerabilities can lead to attacks such as ARP spoofing. These types of attacks require careful analysis of the network characteristics for detection. The dataset for this malware is given in pcap files, which contain the network characteristics of the malware attack. Wireshark is used to extract the pattern of the malware. The data in pcap files is exported to csv which is then used as a training dataset.

6.1.3. ICMP Attack Dataset

The ICMP malware dataset [31] is also in the form of pcap files or network data of ICMP-related attacks (IMCP smurf or ping of death, etc.). These malwares exploit the vulnerabilities of network traffic based on ICMP messages or echo messages. Such messages can penetrate a network without being flagged because most of the security solutions are used to filter TCP/UDP based messages. The pattern of such attacks is extracted using Wireshark and exported to a csv file which is then used to train machine-learning algorithms.

eRBCM was trained using the benchmark datasets. The model testing also included malware definitions not used in the training. The malware categories of ICMP and ARP include several patterns (definitions) of network malware that are part of the benchmark dataset [28]. These models were trained using 200 malwares in both categories. The testing of the eRBCM to measure its performance was conducted on 150 malwares that were different from the 200 used in the training of eRBCM. The eRBCM performance was compared with the following state-of-the-art machine-learning techniques:

1. J48.
2. Convolutional neural network (CNN).
3. Feedforward neural network (FNN).
4. Random forest (RF).

The model performance was measured by applying the correlation coefficient (CC), RMSE, and accuracy. Higher correlation coefficient and accuracy values indicate a better performance, while a model with a lower RMSE is considered superior to those with a higher RMSE.

6.2. Results

The results of each model were averaged over ten runs, with the averages shown in Figure 3. The results show that RF established better correlation-based rules and had a superior performance than other models with respect to the CC. RF extracts the best possible rules as it is an ensemble model of a decision tree and identifies the best tree structure.

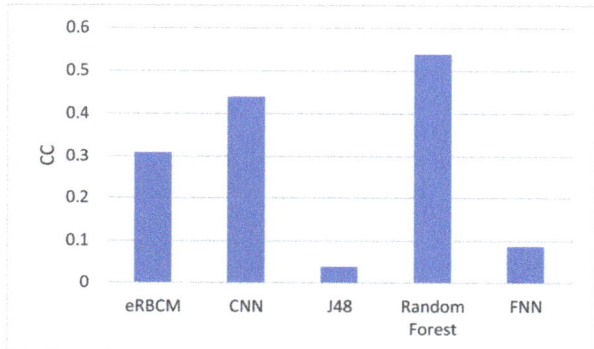

Figure 3. The correlation coefficient (CC) results of each model.

Figure 4 shows each model's accuracy. The accuracy profile indicates that eRBCM's performance was better than its competitors. Because of variations in sample size in each run, the error-rate fluctuated greatly in each episode of testing. The application of convolutional neural network to extract the attack behaviors of the different malware was a promising strategy. The convolutional neural network took several training episodes to converge as compared to eRBCM.

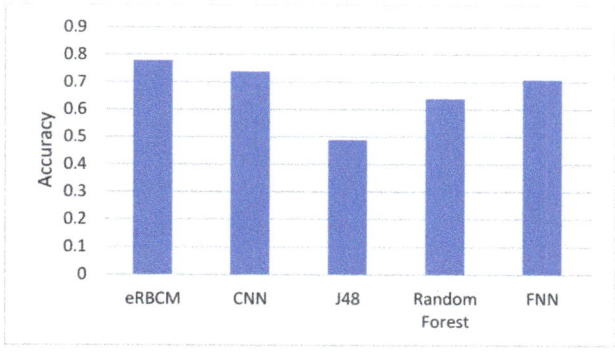

Figure 4. The accuracy results of each model.

While random forest had a higher CC than other models, its performance lacked consistency in relation to accuracy due to the complex nature of malware patterns. When comparing RF and J48, RF performed better with respect to CC and accuracy because it is an ensemble model.

The performances of CNN and FNN were comparable in terms of accuracy, indicating the capability of neural network structures to generalize malware patterns. However, FNN identified fewer similar rules and produced low correlation-based outcomes.

The main success of eRBCM in terms of performance is its self-adaptability to explore and then balance the trade-off between exploration and exploitation. eRBCM can guide its search towards the promising area of a solution space due to epsilon. The generative model explores more on the lower sides of RMSE as compared to regions of higher RMSE.

Figure 5 displays the RMSE results of each machine-learning technique. The results show that eRBCM produced a lower RMSE than most of its rivals. eRBCM performed better than its predecessor, RBCM, and had a consistently better performance than other models because of its adaptive approach in simulations to keep the sample size and space suitable for model learning. The samples were selected based on the quality of the search for a solution during Monte-Carlo simulations.

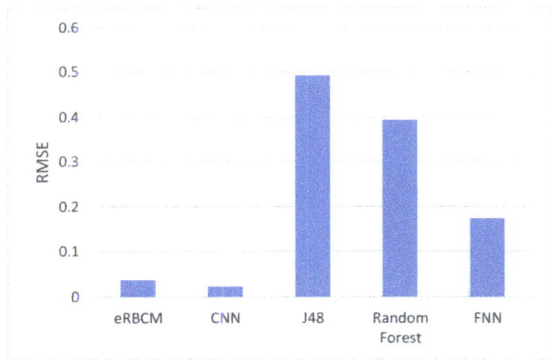

Figure 5. The RMSE performance of each model.

6.3. Look-Ahead Search of eRBCM

Figure 6 provides a deep insight into the sample selection mechanism of eRBCM. The accuracy of each solution search in a simulation depends on a specific number of samples from the search space. The sample selection mechanism is non-linear and requires adaptation to each problem set given to the simulation model.

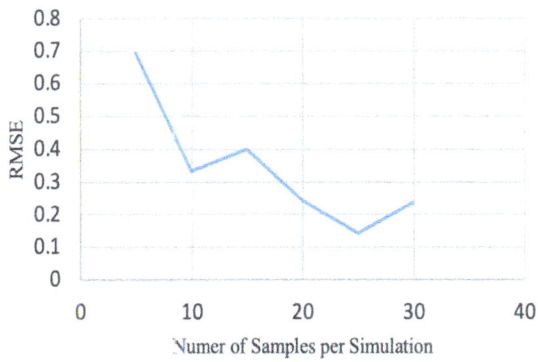

Figure 6. The dependence of RMSE on the number of samples per simulation.

eRBCM's selection mechanism depends on a threshold based on the error rate. It selects a threshold value that minimizes the RMSE. This is the main reason for the successful generalization of attack patterns by eRBCM. The self-adaptivity of epsilon enables eRBCM to explore the larger but focused area of search space compared to RBCM and CNN. eRBCM converges faster than its rival because of the self-tuning of epsilon.

The look-ahead search of the generative model also benefits eRBCM in terms of searching high-quality regions with a smaller number of iterations. The regions of lower RMSE are explored in more depth compared with the regions of higher RSME. This can lead to local minima, but due to the dynamic value of epsilon, the generative model departs such regions in few iterations.

Figure 7 shows the results with respect to RMSE for look-ahead search self-adaptability. The results show that the extended search produced quality solutions with low RMSE. The enhanced performance in the look-ahead search during simulations is explained by the guided exploration of the generative model in the simulation. The higher the n-value of the simulation model (as given in Figure 1), the more eRBCM explores more promising states of the solution space.

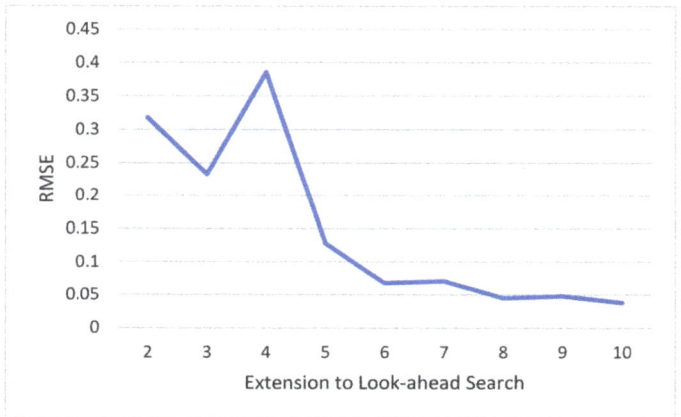

Figure 7. The dependence of RMSE on the number of states extended in a simulation.

With shallow searches in the region of quality solutions, eRBCM remains biased towards the exploitation of the best solutions found and it converges to suboptimal solutions as shown in Figure 7. With extensions to look-ahead search, the deep search provides an optimal balance of exploration and exploitation of the current best-found solutions. It also explains the phenomena shown in Figure 2 relating to the look-ahead search of the generative model of eRBCM. At a deeper search, the eRBCM generative model mirrors the natural selection mechanism of evolutionary techniques. It provides a new solution as a mutation of the existing best solution, as shown in Figure 2. At state S3, for example, the generative model generates a new state S4 which is a mutation of S3.

7. Conclusions

In this paper, we presented a new approach called eRBCM to detect malware. The new model was designed using the reinforcement learning approach, which utilizes the strength of Monte-Carlo simulations and builds a strong machine-learning model to detect complex malware patterns. It combines the most beneficial elements of MOCART's reinforcement learning and RF's exploration capabilities. A large number of experiments were conducted using different malware benchmarks, including ARP attack, ICMP attack, and Microsoft Malware. eRBCM was consistently better than its competitors in terms of learning the new malware patterns and detecting unknown malware. This was mainly explained by *e*RBCM's self-adaptability to exploration and intelligent tuning of the balance for the trade-off between exploration and exploitation.

For future work, we plan to test our approach with various attacks to measure its scalability and accuracy. Furthermore, *e*RBCM will be explored for mobile malware using benchmark datasets. The mobile malware will be analyzed using sophisticated forensics tools, identifying key patterns via an innovative pre-processing stage. The malware will be scanned and categorized based on its malicious agenda. In each category, the common parameters will be explored using clustering, with these clusters used to generate a training dataset.

Author Contributions: Conceptualization, M.A.; methodology, M.N.; software, G.T.; validation, G.T.; formal analysis, M.A., M.N. and G.T. All authors have read and agreed to the published version of the manuscript.

Funding: This research was funded by Higher Colleges of Technology, grant number 1394.

Acknowledgments: The authors would like to thank, Higher Colleges of Technology for their support to this project.

Conflicts of Interest: The authors declare no conflict of interest.

References

1. Ye, Y.; Li, T.; Adjeroh, D.; Iyengar, S.S. A Survey on Malware Detection Using Data Mining Techniques. *ACM Comput. Surv.* **2017**, *50*, 1–40. [CrossRef]
2. Babaagba, K.O.; Adesanya, S.O. A Study on the Effect of Feature Selection on Malware Analysis using Machine Learning. In Proceedings of the 2019 8th International Conference on Educational and Information Technology, Cambridge, UK, 2–4 March 2019; pp. 51–55. [CrossRef]
3. Naveed, M.; Alrammal, M. Reinforcement learning model for classification of Youtube movie. *J. Eng. Appl. Sci.* **2017**, *12*, 8746–8750. [CrossRef]
4. Naveed, M.; Crampton, A.; Kitchin, D.; McCluskey, L. Real-Time Path Planning using a Simulation-Based Markov Decision Process. In *Research and Development in Intelligent Systems XXVIII*; Springer: London, UK, 2011; pp. 35–48.
5. Kwon, B.J.; Mondal, J.; Jang, J.; Bilge, L.; Dumitraş, T. The Dropper Effect. In Proceedings of the 22nd ACM SIGSAC Conference on Computer and Communications Security, Denver, CO, USA, 12–16 October 2015; pp. 1118–1129. [CrossRef]
6. Mao, W.; Cai, Z.; Towsley, D.; Guan, X. Probabilistic Inference on Integrity for Access Behavior Based Malware Detection. In Proceedings of the 18th International Symposium on Research in Attacks, Intrusions, and Defenses—RAID 2015, Kyoto, Japan, 2 November 2015; pp. 155–176.
7. Vadrevu, P.; Rahbarinia, B.; Perdisci, R.; Li, K.; Antonakakis, M. Measuring and Detecting Malware Downloads in Live Network Traffic. In Proceedings of the European Symposium on Research in Computer Security—ESORICS 2013, Egham, UK, 9–13 September 2013; pp. 556–573.
8. Ucci, D.; Aniello, L.; Baldoni, R. Survey of machine learning techniques for malware analysis. *Comput. Secur.* **2018**, *81*, 123–147. [CrossRef]
9. Andrade, E.D.O.; Viterbo, J.; Vasconcelos, C.N.; Guérin, J.; Bernardini, F.C. A Model Based on LSTM Neural Networks to Identify Five Different Types of Malware. *Procedia Comput. Sci.* **2019**, *159*, 182–191. [CrossRef]
10. Saxe, J.; Berlin, K. Deep neural network based malware detection using two dimensional binary program features. In Proceedings of the 10th International Conference on Malicious and Unwanted Software (MALWARE), Fajardo, PR, USA, 20–22 October 2015; pp. 11–20. [CrossRef]
11. Firdausi, I.; Lim, C.; Erwin, A.; Nugroho, A.S. Analysis of Machine learning Techniques Used in Behavior-Based Malware Detection. In Proceedings of the 2010 Second International Conference on Advances in Computing, Control, and Telecommunication Technologies, Jakarta, Indonesia, 2–3 December 2010; IEEE: Piscataway, NJ, USA, 2010; pp. 201–203. [CrossRef]
12. Dube, T.; Raines, R.; Peterson, G.; Bauer, K.; Grimaila, M.; Rogers, S. Malware target recognition via static heuristics. *Comput. Secur.* **2012**, *31*, 137–147. [CrossRef]
13. Mirza, Q.K.A.; Awan, I.; Younas, M. CloudIntell: An intelligent malware detection system. *Future Gener. Comput. Syst.* **2018**, *86*, 1042–1053. [CrossRef]
14. Menahem, E.; Shabtai, A.; Rokach, L.; Elovici, Y. Improving malware detection by applying multi-inducer ensemble. *Comput. Stat. Data Anal.* **2009**, *53*, 1483–1494. [CrossRef]
15. Santos, I.; Devesa, J.; Brezo, F.; Nieves, J.; Bringas, P.G. OPEM: A Static-Dynamic Approach for Machine-Learning-Based Malware Detection. In Proceedings of the International Joint Conference CISIS'12-ICEUTE'12-SOCO'12 Special Sessions 2013, Ostrava, Czech Republic, 24 August 2012; pp. 271–280.

16. Almshari, M.; Tsaramirsis, G.; Khadidos, A.O.; Buhari, S.M.; Khan, F.Q.; Khadidos, A.O. Detection of Potentially Compromised Computer Nodes and Clusters Connected on a Smart Grid, Using Power Consumption Data. *Sensors* **2020**, *20*, 5075. [CrossRef] [PubMed]
17. Alrammal, M.; Naveed, M. Monte-Carlo Based Reinforcement Learning (MCRL). *Int. J. Mach. Learn. Comput.* **2020**, *10*, 227–232. [CrossRef]
18. Naveed, M.; Alrammal, M.; Bensefia, A. HGM: A Novel Monte-Carlo Simulations based Model for Malware Detection. *IOP Conf. Ser. Mater. Sci. Eng.* **2020**, *946*, 012003. [CrossRef]
19. Karanja, E.; Masupe, S.; Jeffrey, M.G. Analysis of internet of things malware using image texture features and machine learning techniques. *Internet Things* **2019**, *9*, 100153. [CrossRef]
20. Gajrani, J.; Agarwal, U.; Laxmi, V.; Bezawada, B.; Gaur, M.S.; Tripathi, M.; Zemmari, A. EspyDroid[+]: Precise reflection analysis of android apps. *Comput. Secur.* **2019**, *90*, 101688. [CrossRef]
21. Amin, A.; Eldessouki, A.; Magdy, M.T.; Abdeen, N.; Hindy, H.; Hegazy, I. AndroShield: Automated Android Applications Vulnerability Detection, a Hybrid Static and Dynamic Analysis Approach. *Information* **2019**, *10*, 326. [CrossRef]
22. Cai, H.; Meng, N.; Ryder, B.; Yao, D. DroidCat: Effective Android Malware Detection and Categorization via App-Level Profiling. *IEEE Trans. Inf. Forensics Secur.* **2018**, *14*, 1455–1470. [CrossRef]
23. Garcia, J.; Hammad, M.; Malek, S. Lightweight, Obfuscation-Resilient Detection and Family Identification of Android Malware. *ACM Trans. Softw. Eng. Methodol.* **2018**, *26*, 1–29. [CrossRef]
24. Zheng, M.; Lee, P.P.C.; Lui, J.C.S. ADAM: An Automatic and Extensible Platform to Stress Test Android Anti-Virus Systems. In *Detection of Intrusions and Malware, and Vulnerability Assessment, Proceedings of the 9th International Conference, DIMVA 2012, Heraklion, Crete, Greece, 26–27 July 2012*; Lecture Notes in Computer Science; Flegel, U., Markatos, E., Robertson, W., Eds.; Springer: Berlin/Heidelberg, Germany, 2013; pp. 82–101. [CrossRef]
25. Schrittwieser, S.; Katzenbeisser, S.; Kieseberg, P.; Huber, M.; Leithner, M.; Mulazzani, M.; Weippl, E. Covert Computation—Hiding code in code through compile-time obfuscation. *Comput. Secur.* **2014**, *42*, 13–26. [CrossRef]
26. Wei, F.; Li, Y.; Roy, S.; Ou, X.; Zhou, W. Deep Ground Truth Analysis of Current Android Malware. In *Detection of Intrusions and Malware, and Vulnerability Assessment*; Lecture Notes in Computer Science Book Series; Springer: Cham, Switzerland, 2017; Volume 10327, pp. 252–276. [CrossRef]
27. Zhang, X.; Zhang, Y.; Zhong, M.; Ding, D.; Cao, Y.; Zhang, Y.; Zhang, M.; Yang, M. Enhancing State-of-the-art Classifiers with API Semantics to Detect Evolved Android Malware. In Proceedings of the 2020 ACM SIGSAC Conference on Computer and Communications Security, Virtual Event, 9–13 November 2020; pp. 757–770. [CrossRef]
28. Xu, K.; Li, Y.; Deng, R.; Chen, K.; Xu, J. DroidEvolver: Self-Evolving Android Malware Detection System. In Proceedings of the 2019 IEEE European Symposium on Security and Privacy (EuroS P), Stockholm, Sweden, 17–19 June 2019; pp. 47–62. [CrossRef]
29. Cai, H. Assessing and Improving Malware Detection Sustainability through App Evolution Studies. *ACM Trans. Softw. Eng. Methodol.* **2020**, *29*, 1–28. [CrossRef]
30. Ronen, R.; Radu, M.; Feuerstein, C.; Yom-Tov, E.; Ahmadi, M. Microsoft Malware Classification Challenge. 2018. Available online: https://www.researchgate.net/publication/323470001 (accessed on 22 November 2021).
31. Parkour, M. Contagio Malware Dump. Available online: https://contagiodump.blogspot.com/ (accessed on 26 February 2020).

Article

Detecting Browser Drive-By Exploits in Images Using Deep Learning

Patricia Iglesias *, Miguel-Angel Sicilia * and Elena García-Barriocanal

Computer Science Department, University of Alcala, 28805 Madrid, Spain
* Correspondence: mariapatricia.iglesi@uah.es (P.I.); msicilia@uah.es (M.-A.S.); elena.garciab@uah.es (E.G.-E.)

Abstract: Steganography is the set of techniques aiming to hide information in messages as images. Recently, stenographic techniques have been combined with polyglot attacks to deliver exploits in Web browsers. Machine learning approaches have been proposed in previous works as a solution for detecting stenography in images, but the specifics of hiding exploit code have not been systematically addressed to date. This paper proposes the use of deep learning methods for such detection, accounting for the specifics of the situation in which the images and the malicious content are delivered using Spatial and Frequency Domain Steganography algorithms. The methods were evaluated by using benchmark image databases with collections of JavaScript exploits, for different density levels and steganographic techniques in images. A convolutional neural network was built to classify the infected images with a validation accuracy around 98.61% and a validation AUC score of 99.75%.

Keywords: steganography; steganalysis; polyglots; neural networks; deep learning

1. Introduction

Steganography is a set of techniques designed to hide information or objects by embedding them in another object called a host, so that they go unnoticed. Stenography has been applied since ancient Greece but it has recently grown in importance as an area of study in communications and computer security [1] due to its application by illegal or malicious organisations to evade security measures and extract information by hiding malicious code in digital objects such as video, images, audio or documents, among other uses. The fight against stenography is the discipline of steganalysis, which aims to detect the existence of hidden information in the host.

Depending on the steganography technique, the detection can be easier or more difficult. One of the most common techniques is the addition of a signature in the stego file, but other techniques such as adding the data after the host EOF are also very common. More sophisticated techniques distribute stego in different ways exploiting different characteristics of the images. This part is reviewed in Section 2.

Many of the current widely used tools are capable of identifying EOF steganography or signature [2], but the implementations are not as effective in detecting more sophisticated techniques and they are able to avoid the security controls.

Steganography techniques can be applied to introduce malicious code based on polyglot techniques embedded in a stego image. A polyglot is an image and JavaScript code at the same time. If in the web page it is invoked as the next block of code, an image is displayed:

`<img src=''polyglot_stego_image.jpg''/>`

However, if in the web page it is invoked as the next block of code, a JavaScript code embedded in the image is executed:

`<script src=''polyglot_stego_image.jpg''> </script>`

During the recent COVID-19 pandemic, one of the main attack categories reported by the European Union Agency for Cybersecurity (ENISA) was the delivery of malware using undetected and sophisticated mechanisms [3]. One such sophisticated type of attack is the binomial steganography-polyglot, which has been exploited for real attacks, as it is currently undetectable by standard security measures [4]. The malicious code can be executed using "polyglot" techniques, which consist of embedding the code in such a way that it is executable when is read by the web browser. In this type of attack, it is important to detect the stego image before it is executed by the browser, which requires some kind of detection model.

The attacker shows other examples of applications (https://www.bleepingcomputer.com/tag/steganography/ (accessed on 1 November 2022)), for instance, Zeus malware to set up a man-in-the-middle attack (https://www.silicon.co.uk/security/virus/zeus-banking-trojan-205640 (accessed on 1 November 2022)), Lockibot malware family to download the malicious malware as second step embedded in an image or in September of 2022, the latest Window logo cyber espionage attack (https://www.cybertalk.org/2022/09/30/hackers-hide-malware-in-windows-logo/ (accessed on 1 November 2022)) for Middle East countries.

This paper proposes a new approach to detect polyglots. It is based on the early detection of the stego image created with LSB (Least Significant Bit) steganography, LSB with Fermat and Fibonacci generators and F5 using deep learning (DL) techniques. Specifically, a convolutional neural network (CNN) is used to classify the infected images and the clean images. The main advantages of the work reported here over other approaches are (1) that the images are only resized in the pre-processing part, trying to keep the images as close as possible to the original and reducing the time processing and resources consumption; (2) a very good performance of the algorithm has been obtained in the detection of different LSB steganographies (LSB, LSB set with generator function (Fibonacci, Fermat), LSB in the description) and F5 [5]; (3) a higher quality of steganography images is able to be detected with lower relation of bits per pixel (BPP).

The rest of the paper is structured as follows: Section 2 briefly reviews the background of steganography and steganalysis techniques, including the application in polyglot attacks based on steganography. Section 3 proposes a method to detect polyglots and a description of the different setups and experiments that were performed in order to design the algorithm. Section 4 exposes the results and analysis of the different experiments. Finally, the Section 5 provides the conclusion, including remarks and outlook.

2. Background

In this section, a brief report on the the state of the art is presented, including relevant steganography techniques and the corresponding detection approaches, with a focus on the use of least significant bit stenography (LSB) and F5 for embedding polyglots. Finally, previous works applying a deep learning approach to steganalysis are described.

2.1. Stenography

As previously stated, steganography is the art of hiding information in a host in such a way that it is not detectable [6]. There are different types of steganography depending on the object where the message is hidden, e.g., text, image, audio, video and network or protocol stenography. This paper focuses on the image stenography.

Several groups of algorithms can be applied to embed data into images [7]. Some examples of groups of techniques are:

- Based on spatial domain. They are based on the statistics of the image and create a hidden channel using a replacement method. It can be implemented in a sequential way, e.g., using the least significant bits (LSB) or in a random sequence, for instance, by using the least significant bits (LSB) with Fermat or Fibonacci formulas generator (https://stegano.readthedocs.io/en/latest/software.html#the-command-stegano-lsb-set (accessed on 15 July 2021)).

- Based on the frequency domain. It spreads the data over the frequency domain of the signal. Almost all robust methods of steganography are based in the Frequency Domain. Some examples are F5 algorithm (a Discrete Cosine Transform (DCT)), OutGuess (https://www.rbcafe.es/software/outguess/ (accessed on 1 August 2022)), YASS (https://github.com/logasja/yass-js (accessed on 1 August 2022)), etc. There are more robust methods than LSB, although they have the limitation of the number of least significant bits of an image.
- Based on spread spectrum image steganography (SSIS). They are based on modulating a narrow band above the carrier.
- Based on machine learning algorithms [8].
- Manually inserting the code in the image randomly, etc.

2.2. Steganalysis: Frameworks and Techniques

Steganalysis is the process of detecting steganography by observing variations at different levels between the cover object and the final stego file. The aim of steganalysis is to identify suspicious information flows and to determine whether or not they have encoded hidden messages [9,10].

The steganalysis techniques depend on what information is available, whether it is just the stego, both the stego and the cover file, the stego and the message, or the stego and the steganography technique used. The less information available, the more difficult the steganalysis becomes. There are frameworks, such as the one proposed by Xiang-Yang et al. [11], for blind steganalysis.

In addition to the general framework, many authors suggest different taxonomies of steganalysis techniques as Nissar and Mir's one [12] or the Karampidis et al. taxonomy [13], both of them well-known and commonly accepted. Yet neither Karampidis et al. nor Nissar and Mir include the latest techniques as machine learning or deep learning as a technique to approach steganalysis, and there are only a few previous studies using these approaches to address the specificity of polyglots, which are described below.

2.3. Polyglot Attacks with Steganography

Steganography attacks are based in the broad use of multimedia files and the difficulties of the traditional security tools to detect stegos in the files. Some security approaches detect the strange behaviour, not the steganography infection, so a system could be infected for a long time without notice. The infection begins just when the user downloads the file where a polyglot is hidden. Typically, these polyglots are sent by email by phishing [4].

Polyglots are able to execute the code, e.g., in a power shell. The usual behaviour of the attack is based on botnets that works as command and control (C&C). This means that there is hidden some malicious code (downloader) which is "zombie" until the control botnet contacts to it and sends the instructions.

According to Kaspersky ICS CERT, steganography is mainly used in industry target attacks and in different areas [4]. Some of the most famous pieces of malware using steganography for espionage were Loki Bot (https://www.zdnet.com/article/lokibot-information-stealer-now-hides-malware-in-image-files/ (accessed on 1 October 2022)) and ZeroT (https://attack.mitre.org/software/S0230/ (accessed on 1 October 2022)). Other examples of bank trojans that use steganography in their attacks, some of them from the Bebloh family [14], e.g., Shiotob or URLZone, or the Ursnif family [15], e.g., Gozi or ISFB. Additionally, during the COVID-19 pandemic, there were directed attacks with polyglots related to making information about the vaccines be unnoticed.

2.4. Previous Work in Steganalysis and Deep Learning

In steganalysis, approaches outside the deep learning fields have been based nowadays in the computation of rich models followed by ensemble classifiers, as [16] or [17] propose.

Regarding deep learning, the first references using neural networks date back to 2005 [18,19]. These works proposed a feed-forward network used as classifier to detect if there is presence of steganography or not. However, the first approaches in CNN, based in Local Contrast Normalization or Local Response Normalization, appear in 2015 [20]. The performance of this CNN for stegoanalysis was not as good as that of traditional approaches, but also in 2015, the first batch normalisation-based CNN approach emerged with similar performance to the other existing ones [21].

Since CNN requires large amounts of memory and time for the training, the steganography and AI communities worked on approaches to reduce them, resulting in approaches based in transfer learning [22], as the one proposed by [21]. However, these early approaches do not have good accuracy.

From 2015 to 2016, efforts focused on spatial steganalysis [23,24], and in 2017, work was reoriented to JPEG steganalysis [23,25]. Then came the GAN model [26,27], but it does not seem to be very successful since the accuracy was low in comparison to the traditional works. The GAN approach consisted of generating JPEG-infected images to train a CNN and generate a model that could subsequently classify between the presence of steganography or lack thereof. As it will be described in next sections, steganography is very sensitive to any modification in the images, and, as the GAN approach added additional noise in the images, the results were not as good as expected.

SRNET [28] is also a CNN for classification of images that tries to add information to help to detect singularities due to the structure of the network and get rid of the Relu. The use of Relu and a softmax function for classification, as in the model proposed in this paper, provides better results for detecting stego images, as we review in Section 4 Results and discussion.

In 2021, another CNN approach was published [29]. It uses a pre-processing stage, feature extraction, separable convolutions and a classification module. In addition to pre-processing, this approach implements a HPF (High-Pass Filter) and 30 filters consisting of padding and strides. The authors of [30] performed a study of a filter subset selection method for steganalysis CNN. This study shows that the application of redundant filters produces over-fitting, introduces noise and exploits the performance of the steganalysis CNN models. These assertions have been taken as the basis for the approach proposed in this paper, so the work reported here goes further by minimising pre-processing, only using a single rescaling step.

The metric Bits per Pixels measures the quality of the steganography implementing the relation between the number of secret bits embedded and the number of bits of the original image. This metric is used for [29] to establish the performance of the steganalysis method, being able to detect infected images with a 0.2 bpp and getting an accuracy of 80.3%. In the final model proposed in this paper, the model is able to detect images with a 0.0027 bpp and a validation accuracy of 98.61% is obtained, so it is able to detect the infection in higher-quality steganography images (with a smaller proportion of infected data in the host) and better validation accuracy.

This new approach only features resizing in the pre-processing stage, it uses a sigmoid function in the classification, instead of a softmax as [29] did, and it uses a drop-out for adding more flexibility to the model. The approach also uses Adam optimization and drop-out in order to improve the generalization of the model, as suggested in [31].

3. Proposal for a Steganalysis Approach to Polyglot Detection

3.1. Description of the Approach

The steganalysis approach reported in this paper is based on the use of convolutional neural networks (CNN) for image classification. CNNs are selected due to the performance of this type of Neural Networks (NN) in the classification of images [32,33] and the ability to learn the different dimensions of the image to distinguish the nuances of the infected images in comparison to the clean images as used for face recognition in [34].

As CNNs take images (both clean and stego) as input data for training, they are able to learn their features and classify a new image as infected or clean. It learns patterns more difficult to identify by the visual analysis, the analysis of channels or statistical analysis. This is an important feature to take into account in the selection of this technique.

The steps that the authors have performed for the final proposal CNN are described below and shown in Figure 1. A data set has been created by collecting clean images from Coco dataset and ILSVR dataset. On the other hand, stego images were collected from StegoAppDB [35]. As only a stego F5 dataset was found, a stego dataset for polyglots must be created. Identified polyglots in Javascript were then collected. With the polyglots and the cleaned images, stegos were generated with different LSB techniques, sequential and random.

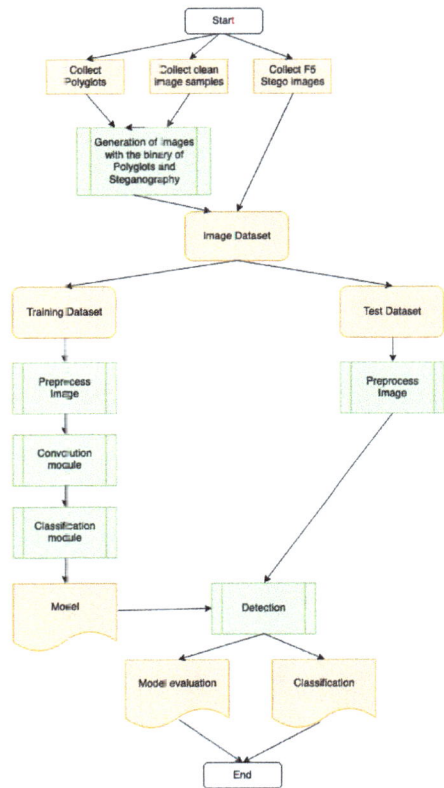

Figure 1. Framework of Blind Steganalysis.

Once that LSB stego dataset, the clean images dataset and the StegoDB App dataset have been created, the clean images and the stegos images are joined and then the images with the metadata of stego or clean label are split into two datasets: a training dataset and a test dataset. The training dataset is used to train the 2D CNN classification model and the test dataset to validate the performance of the produced model. As a result of the process, a model evaluation was obtained in which the validation accuracy was calculated, obtaining the precision in the classification of the stego images.

The model uses a rescaling pre-processing stage very light in comparison to previous tests carried out by other authors and described below. The CNN contains 10 blocks of CNN + Batch Normalisation + Relu that classify into the classes mentioned above, i.e., clean and infected. To obtain good results, a large variety of colour images, objects, different embedded malware/polyglots and different steganography techniques were required, as explained in the following subsections.

The first model designed is based on a CNN having three parts: the preparation module, the convolution module and the classification module. The preparation module was composed of two further sub-modules in where data are re-scaled and data augmentation (rotating) and batch normalization (batch size = 32) are applied. The convolution module was composed of 10 sub-modules (conv2D, batch normalization and activation) and, finally, the classification module used a sigmoid activation on the basis that only two possible values should be handled: the presence of steganography or lack thereof.

The output of the network is the probabilities that an image belongs to the "stego" class and to the "non-stego" class. Due to poor results in validation accuracy (see Figure 2) using this model, a second version of the model was tested, dropping out the pre-processing sub-module and removing the data augmentation based on rotation. The dropping out removes part of information for the training in each iteration making more flexible the model for possible modifications and the elimination of the data augmentation reduces the errors in the training, as there were no real stego images. The second and final model (see Figure 3) obtains very good results for classification, as described in Section 4.

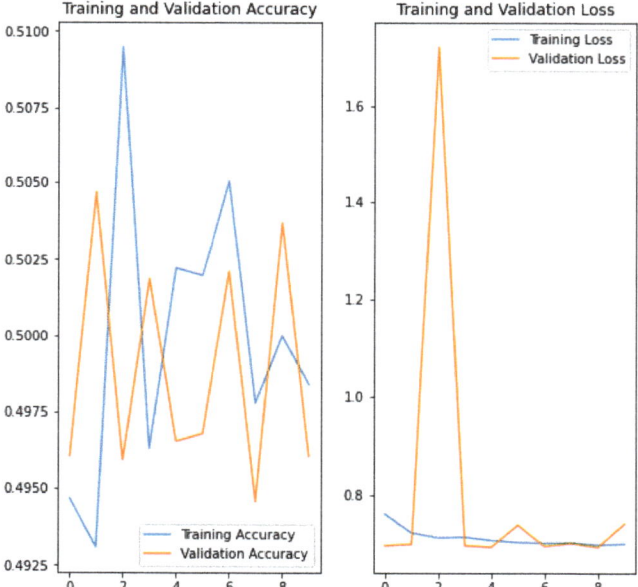

Figure 2. Validation Accuracy and Training Loss of Watermelon & Model 1 & 20 polyglots.

Regarding the dataset to train the model, watermelon images were first used in the classification model. The dataset contained both stego watermelon images generated with a polyglot using the LSB technique and clean watermelon images. Initial results gave an accuracy of 0.9672, which may suggest over-fitting. Although they will be discussed in detail in the following subsection "Experimental setup", a number of issues were identified that could potentially lead to over-fitting, namely:

- The type and variety of objects displayed in the images. COCO were used as sources since the images they contain show different types of objects. The ILSVRC dataset was also used as source of images to increase variety and avoid possible biases.
- The number of polyglots embedded in the images. Several trainings were held varying the number of Javascript polyglots and the number of images infected.

- The characteristics of the images. Training was conducted with greyscale or colour images. Polyglots were embedded before and after colour transformation for different tests.
- The homogeneity of the images (same size, orientation, etc.) Image transformation regarding size and orientation has been performed.

Multiple combinations of these cases were made to obtain higher prediction accuracy and to avoid over-fitting.

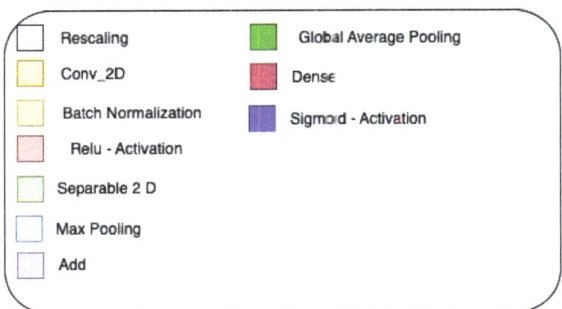

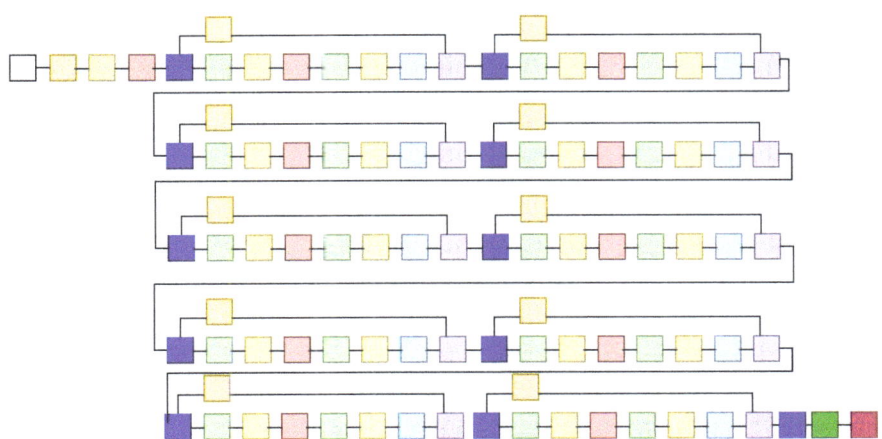

Figure 3. Structure of our proposal of CNN to detect LSB Steganography.

Data enrichment (conversion to gray, rotating and resizing the images) was also applied after the generation of the infected images, but as additional noise was introduced in the identification of the stego images, detection became impossible (accuracy of 0.561) (see Figure 4) and this approach was discarded. After testing all possible combinations in the dataset that could have an impact on accuracy and overfitting, a model with an accuracy of 95.21% (see Figure 5) was obtained. Then, the model was improved with other possible steganography algorithms (random LSB Algorithms (LSB with Fibonacci and Fermat generators and F5), and, after the training, a validation accuracy of 98.61% was obtained (see Figure 6).

The experimental setup is described in Section 3.2.

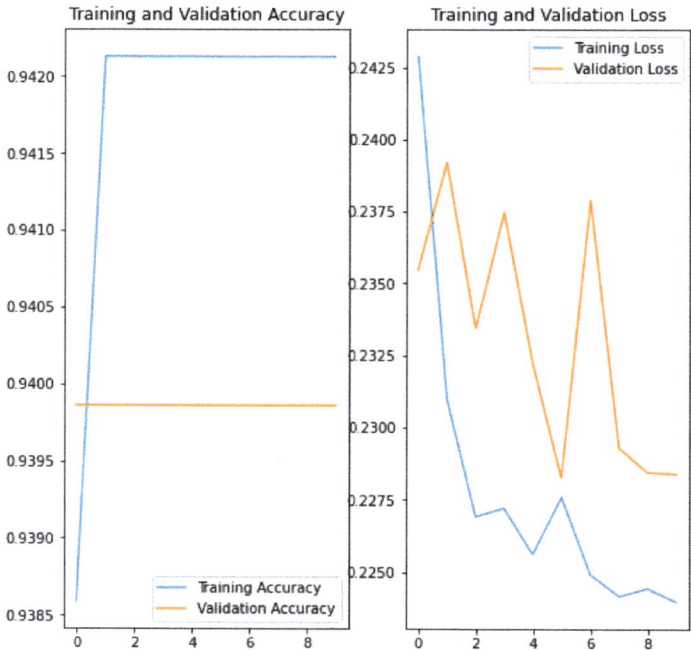

Figure 4. Validation accuracy and Training Loss of Coco Gray & Model 1 & 20 polyglots.

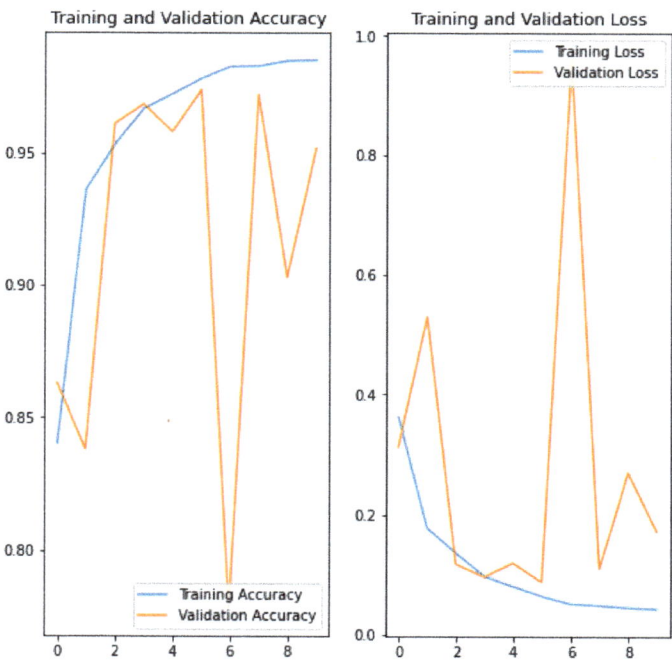

Figure 5. Validation accuracy and Training loss in Coco+ILSVR(205K) Model 2 & 104 Polyglots.

Figure 6. Validation accuracy and Validation AUC ROC in Coco+ILSVR(205K)+F5 & Model 2 & LSB, LSB Random, F5.

3.2. Experimental Setup

In order to obtain the most accurate and least over-fitted model, different datasets were created to train the model. Datasets included both clean images and stego images with known polyglots embedded using LSB stenography. Images were extracted from the COCO dataset [36] or a combination of COCO dataset and ILSVR dataset [37]. The infected datasets were composed of clean images with a different number of hidden polyglot Javascript code of known vulnerabilities [38]. More concretely, the following datasets and procedures for training the model were tested:

1. Watermelon dataset + LSB steganography (v0.1): Dataset of different watermelon images. It contained 1354 clean images and 1946 infected images with 1 polyglot.
2. Watermelon dataset + LSB steganography (v0.2): Dataset of different watermelon images. It contained 1354 clean images and 1946 infected images with 20 polyglots.
3. COCO Dataset + LSB steganography (v1): Using COCO as source from images that contain a variety of items/situations, LSB technique was used to create stego images with polyglots, resulting in a dataset with 37,000 clean images and 3000 infected images.
4. COCO Dataset + LSB steganography + Image modifications (Resizing, Relocation, ...) (v2): Using the the dataset configured in (2), data augmentation and images resizing were performed.
5. COCO Dataset + Gray Conversion + LSB Steganography (v3): Using the clean COCO dataset configured in (2) (40,000 images), 20,280 images images were first converted to greyscale and polyglots were included in 1256 of these greyscale images using the LSB technique.
6. COCO Dataset + LSB Steganography (v4): Using the images from COCO, the number of different polyglots embedded using the LSB technique was increased up to 20 common and known structures in Javascript. The number of infected images

were also increased up to 411,000 images, being 328,000 clean images and 83,000 infected images.
7. COCO Dataset + LSB Steganography (v5): As the previous dataset configuration can suggest overfitting, a new version of the training dataset was designed. Using images from COCO dataset, the number of different polyglots embedded using the LSB technique was increased up to 104 common and known structures in JavaScript. The number of clean images was reduced to 123,460 and the number of infected images to 31,000.
8. COCO Dataset + ILSVR dataset + LSB Steganography (v6): Using images from both COCO dataset and ILSVR dataset [37], the following two datasets were generated, which contained 41.026 clean images and 8.313 embedded images in the first case and 205.130 clean images and 41.026 embedded images in the second. In both cases, stego images were embedded with 104 common structures of polyglots in JavaScript using the LSB technique.
9. COCO Dataset + ILSVR dataset + LSB Steganography + LSB Steganography using Fermat and Fibonacci generation (v7): Based on v4 dataset, 33.347 images infected using Fermat and Fibonacci generator are added. The final dataset is composed of 279.503 images, from them 41.026 LSB infected images and 33.347 images infected using Fibonacci and Fermat generators.
10. COCO Dataset + ILSVR dataset + LSB Steganography + LSB Steganography using Fermat and Fibonacci generation (v7) + F5 [35] (v8) : Based on v5 dataset, 33.347 images infected using Fermat and Fibonacci generator and 621 F5 images are added. The final dataset is composed of 280.124 images, from them 41.026 LSB infected images, 621 F5 infected images and 33.347 images infected using Fibonacci and Fermat generators.

The Python library *Stegano* (https://pypi.org/project/stegano/ (accessed on 1 July 2021)) was used for generation of stegoimages. Images colour variation was implemented using the *Pillow* (https://pypi.org/project/Pillow/ (accessed on 1 July 2021)) library of Python.

Tensorflow and Python were used to generate code of the models, which were based on a convolutional neural network for the classification of images. Implementations were run in multiple environments:

- Local machine;
- Docker Virtual machine based in Tensorflow without GPU;
- Google Colab with no hardware optimizations;
- Google Colab with GPU [39];
- Google Colab with TPU.

Training in the first three environments was discarded due to the resulting long run times or the impossibility to perform the task. The number of images and the use of the neural network required hardware optimisations, and the best-performance models were obtained using Google Colab with TPU.

Regarding the hyperparametry of the models, during the different training processes of both of them, the hyperparameters were never modified in order to be able to compare the results in a robust way.

4. Results and Discussion

The two models generated (with and without pre-processing submodule) were tested. As explained above, the second model (without pre-processing submodule) was generated on the assumption that, although it was not a standard approach, getting rid of image details may condition the classification results.

The results of the validation of both models are in Table 1.

The results confirm that the pre-processing sub-module (model 1) is not needed. When the number of different polyglots that infect images is increased, the model that includes pre-processing seems not to be good (accuracy 56.1% see Figure 4).

Using the second model (without pre-processing) with a wide variety of images and the same increased number of different polyglots, accuracy reaches very good values 95.43%, see Figure 7).

Table 1. Results of Experiments.

Dataset	Model	Number of Polyglots	Type of Stego	Val Accuracy
Watermelon (v0.1)	1	1	LSB	0.9672
Watermelon (v0.2)	1	20	LSB	0.561
Coco RGB (v1)	1	1	LSB	0.9507
Coco RGB (v2)	2	20	LSB	0.9543
Coco Gray (v3)	2	20	LSB	0.9399
Coco RGB (v5)	2	104	LSB	0.9739
Coco RGB (v3+v4)	2	20	LSB + Gray	0.0915
Coco+ILSVR (v6)	2	104	LSB	1
Coco+ILSVR (v7)	2	104	LSB	0.9521
Coco+ILSVR+F5 (v8)	2	NA	LSB, F5	0.9861

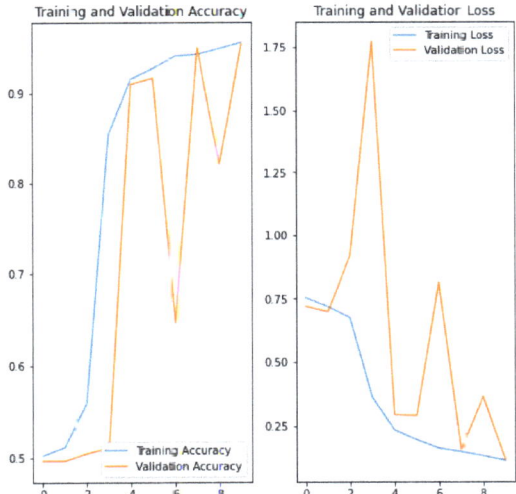

Figure 7. Validation accuracy and Training loss in Coco RGB(v2) Model 2 & 20 Polyglots.

The second model performs very well overall. The results also show that the richer the image, the easier it is to detect embedded polyglots. Thus, the second model more accurately classifies embedded polyglots in colour images than in greyscale images.

It is also noteworthy that if rescaling, greyscale conversion or rotation occurs after LSB stenography, the noise introduced makes classification not possible.

On the other hand, if a large number of different polyglots are used to infect the images, the number of images must be considerably larger to avoid overfitting.

When the second model is trained in a realistic way, i.e., with 280k (approx.) very different images that are provided from different sources and a good number of different polyglots, it can classify with 95.21% accuracy (Figure 5) whether an image is infected or not, with a 0.0027 bpp in the worse relationship, and an error less than 0.06%, which can be considered a good quality indicator of the model and it is supposed to be an advantage over other approaches such as [28,29], that obtain 80.3% of accuracy with a 0.2 bpp in the first case and a 31.3% of error in the second case.

Finally, Model 2 was trained with richer types of steganography methods (LSB + LSB with Fermat and Fibonacci Generators and F5) and it results in a new model with a validation accuracy of 98.61% and a validation AUC score of 99.75% (see Figure 6).

5. Conclusions and Outlook

Convolutional networks have demonstrated their ability to solve image-based tasks such as recognition, classification or segmentation in previous work. In this work, these networks have been used for a stegoanalysis task, namely, for the detection of polyglot payloads in images, which is quite different from the original applications of this type of networks. Results provide evidence of the feasibility of these networks to solve the task and the model provided evidence of better detection results than other previously proposed approaches.

However, results shown here are limited to the detection of stego images using the LSB and F5 stegographic techniques. Future work should expand the range of infected images, including a diversity of stegographic techniques. This is critical in the adversarial environment of malware detection using polyglots, whereby the robustness of models for detecting a diversity of potential variations in the embedded payload represents a significant challenge.

Author Contributions: Conceptualization, M.-A.S., P.I. and E.G.-B.; methodology, M.-A.S., P.I. and E.G.-B.; software, P.I.; validation, M.-A.S., P.I. and E.G.-B.; formal analysis, M.-A.S., P.I. and E.G.-B.; investigation, M.-A.S., P.I. and E.G.-B.; resources, M.-A.S., P.I. and E.G.-B.; data curation, P.I.; writing—original draft preparation, M.-A.S., P.I. and E.G.-B.; writing—review and editing, M.-A.S., P.I. and E.G.-B.; visualization, P.I.; supervision, M.-A.S. and E.G.-B.; project administration, M.-A.S., P.I. and E.G.-B. All authors have read and agreed to the published version of the manuscript.

Funding: This research received no external funding.

Conflicts of Interest: The authors declare no conflict of interest.

References

1. Kadhima, I.J.; Premaratnea, P.; Viala, P.J.; Hallorana, B. Comprehensive survey of image steganography: Techniques, Evaluations, and trends in future research. *Neurocomputing* **2019**, *335*, 299–326. [CrossRef]
2. Muñoz, A. A Simple Steganalysis Tool. Available online: https://stegsecret.sourceforge.net/ (accessed on 1 October 2022).
3. ENISA Threat Landscape 2020: Cyber Attacks Becoming More Sophisticated, Targeted, Widespread and Undetected. Available online: https://www.enisa.europa.eu/news/enisa-news/enisa-threat-landscape-2020 (accessed on 29 November 2021).
4. Steganography in Attacks on Industrial Enterprises. Available online: https://ics-cert.kaspersky.com/reports/2020/06/17/steganography-in-attacks-on-industrial-enterprises (accessed on 29 November 2021).
5. Jiang, C.; Pang, Y.; Xiong, S. A High Capacity Steganographic Method Based on Quantization Table Modification and F5 Algorithm. *Circuits Syst. Signal Process* **2014**, *33*, 1611–1626. [CrossRef]
6. Kour, J.; Verma, D. Steganography Techniques—A Review Paper. *Int. J. Emerg. Res. Manag. Technol.* **2015**, *3*, 132–135.
7. Malik, S.; Mitra, W. Hiding Information—A Survey. *J. Inf. Sci. Comput. Technol.* **2015**, *3*, 232–240.
8. Cho, D.X.; Thuong, D.T.H.; Dung, N.K. A Method of Detecting Storage Based Network Steganography Using Machine Learning. *Procedia Comput. Sci.* **2019**, *154*, 543–548. [CrossRef]
9. Wang, J.; Cheng, M.; Wu, P.; Chen, B. A Survey on Digital Image Steganography. *J. Inf. Hiding Priv. Prot.* **2019**, *1*, 87–93. [CrossRef]
10. Jiao, S.; Zhou, C.; Shi, Y.; Zou, W.; Li, X. Review on Optical Image Hiding and Watermarking Techniques. *Opt. Laser Technol.* **2019**, *109*, 370–380. [CrossRef]
11. Luo, X.Y.; Wang, D.S.; Wang, P.; Liu, F.L. A review on blind detection for image Stenography. *Signal Process.* **2008**, *88*, 2138–2157. [CrossRef]
12. Nissar, A.; Mir, A.H. Classification of Steganalysis Techniques: A Study. *Digit. Signal Process.* **2010**, *20*, 1758–1770. [CrossRef]
13. Karampidis, K.; Kavallieratou, E.; Papadourakis, G. A Review of Image Steganalysis Techniques for Digital Forensics. *J. Inf. Secur. Appl.* **2018**, *40*, 217–235. [CrossRef]
14. Bebloh—A Well-Known Banking Trojan with Noteworthy Innovations. Available online: https://www.gdatasoftware.com/blog/2013/12/23978-bebloh-a-well-known-banking-trojan-with-noteworthy-innovations (accessed on 1 October 2022).
15. Ursnif. Available online: https://attack.mitre.org/software/S0386/ (accessed on 1 October 2022).
16. Tabares-Soto, R.; Arteaga-Arteaga, H.; Mora-Rubio, A.; Bravo-Ortiz, M.A.; Arias-Garzón, D.; Grisales, J.A.A.; Jacome, A.B.; Orozco-Arias, S.; Isaza, G.; Pollan, R.R. Strategy to improve the accuracy of convolutional neural network architectures applied to digital image steganalysis in the spatial domain. *J. Comput. Sci.* **2021**, *7*, e45. [CrossRef] [PubMed]

17. Chaumont, M. Deep Learning in steganography and steganalysis from 2015 to 2018. In *Digital Media Steganography: Principles, Algorithms, Advances*; Hassaballah, M., Ed.; Elsevier: Amsterdam, The Netherlands, 2019.
18. Shi, Y.Q.; Xuan, G.R.; Zou, D.K. Image steganalysis based on moments of characteristics functions using wavelet characteristics functions using wavelet decomposition, prediction-error image, and neural network. In Proceedings of the IEEE International Conference on Multimedia and Expo, Amsterdam, The Netherlands, 6 July 2005; pp. 269–272.
19. Lie, W.N.; Lin, G.S. A Feature-based classification technique for blind steganalysis. *IEEE Trans. Multimed.* **2005**, *7*, 1007–1020.
20. Tan, S.; Li, B. Stacked Convolutional Auto-Encoders for Steganalysis of Digital Images. In Proceedings of the Signal and Information Processing Association Annual Summit and Conference, Siem Reap, Cambodia, 9–12 December 2014.
21. Qian, Y.; Dong, J.; Wang, W.; Tan, T. Deep Learning for Steganalysis via Convolutional Neural Networks. In Proceedings of the Media Watermarking, Security, and Forensics, San Francisco, CA, USA, 8–12 February 2015; Volume 9404.
22. Jin, B.; Cruz, L.; Goncalves, N. Deep Facial Diagnosis: Deep Transfer Learning From Face Recognition to Facial Diagnosis. *IEEE Access* **2020**, *8*, 123649–123661. [CrossRef]
23. Boroum, M.; Chen, M.; Fridich, J. Deep Residual Network for Steganalysis of Digital Images. *IEEE Trans. Inf. Forensics Secur.* **2019**, *14*, 1181–1193. [CrossRef]
24. Li, B.; Wei, W.; Ferreira, A.; Tan, S. ReST-Net: Diverse Activation Modules and Parallel Subnets-Based CNN for Spatial Image Steganalysis. *IEEE Signal Process. Lett.* **2018**, *25*, 650–654. [CrossRef]
25. Xu, G. Deep Convolutional Neural Network to Detect J-UNIWARD. In Proceedings of the 5th ACM Workshop on Information Hiding and Multimedia Security, Philadelphia, PA, USA, 20–22 June 2017; pp. 63–67.
26. Shi, H.; Dong, J.; Wang, W.; Qian, Y.; Zhang, X. SSGAN: Secure Steganography Based on Generative Adversarial Networks. In Lecture Notes in Computer Science, Proceedings of the 18th Pacific-Rim Conference on Multimedia, Harbin, China, 28–29 September 2017; Springer: Cham, Switzerland, 2017; pp. 534–544.
27. Tang, W.; Tan, S.; Li, B.; Huang, J. Automatic Steganographic Distorsion Learning Using Generative Adversarial Networks. *IEEE Signal Process. Lett.* **2017**, *24*, 1547–1551. [CrossRef]
28. Yasrab, R. SRNET: A shallow skip connection based convolutional neural network design for resolving singularities. *J. Comput. Sci. Technol.* **2019**, *34*, 924–938. [CrossRef]
29. Reinel, T.S.; Brayan, A.A.H.; Alej, B.O.M.; Alej, M.R.; Daniel, A.G.; Alej, A.G.J.; Buenaventura, B.-J.A.; Simon, O -A.; Gustavo, I.; Raúl, R.-P. GBRAS-Net: A convolutional neural network architecture for spatial image steganalysis. *IEEE Access* **2021**, *9*, 14340–14350. [CrossRef]
30. Wu, L.; Han, X.; Wen, C.; Li, B. A Steganalysis framework based on CNN using the filter subset selection method. *Multimed. Tools Appl.* **2020**, *79*, 19875–19892. [CrossRef]
31. Zheng, Q.; Yang, M.; Yang, J.; Zhang, Q.; Zhang, X. Improvement of Generalization Ability of Deep CNN via Implicit Regularization in Two-Stage Training Process. *IEEE Access* **2018**, *6*, 15844–15869. [CrossRef]
32. Liu, Y.; Dou, Y.; Qiao, P. Beyond top-N accuracy indicator: A comprehensive evaluation indicator of CNN models in image classification. *IET Comput. Vis.* **2020**, *14*, 407–412. [CrossRef]
33. Zhao, M.; Chang, C.H.; Xie, W.; Xie, Z.; Hu, J. Cloud shape classification system based on multi-channel cnn and improved fdm. *IEEE Access* **2020**, *8*, 44111–44124. [CrossRef]
34. Jin, B.; Cruz, L.; Goncalvez, N. Pseudo RGB-Face Recognition. *IEEE Sens. J.* **2022**, *22*, 21780–21794. JSEN.2022.3197235. [CrossRef]
35. Newman, J.; Lin, L.; Chen, W.; Reinders, S.; Wang, Y.; Wu, M.; Guan, Y. StegoAppDB: A steganography apps forensics image database. *Electron. Imaging* **2019**, *2019*, 536. [CrossRef]
36. Lin, T.Y.; Maire, M.; Belongie, S.; Bourdev, L.; Girshick, R.; Hays, J.; Perona, P.; Ramanan, D.; Zitnick, C.L.; Dollár, P. Microsoft COCO: Common Objects in Context. In Lecture Notes in Computer Science, Proceedings of Computer Vision—ECCV 2014, Zurich, Switzerland, 6–12 September 2014; Fleet, D., Pajdla, T., Schiele, B., Tuytelaars, T., Eds.; Springer: Cham, Switzerland, 2014.
37. Russakovsky, O.; Deng, J.; Su, H.; Krause, J.; Satheesh, S.; Ma, S.; Huang, Z.; Karpathy, A.; Khosla, A.; Bernstein, M.; et al. ImageNet Large Scale Visual Recognition Challenge. *Int. J. Comput. Vis. (IJCV)* **2015**, *115*, 211–252. [CrossRef]
38. A Collection of JavaScript Engine CVEs with PoCs. Available online: https://github.com/tunz/js-vuln-db (accessed on 29 November 2021).
39. Zhao, M.; Jha, A.; Liu, Q.; Millis, B.; Mahadevan-Jansen, A.; Lu, L.; Landman, B.; Tyska, M.J.; Huo, Y. Faster Mean-shift: GPU-accelerated clustering for cosine embedding-based cell segmentation and tracking. *Med. Image Anal.* **2021**, *17*, 102048. [CrossRef]

Disclaimer/Publisher's Note: The statements, opinions and data contained in all publications are solely those of the individual author(s) and contributor(s) and not of MDPI and/or the editor(s). MDPI and/or the editor(s) disclaim responsibility for any injury to people or property resulting from any ideas, methods, instructions or products referred to in the content.

Article

Digital Forensics Classification Based on a Hybrid Neural Network and the Salp Swarm Algorithm

Moutaz Alazab *, Ruba Abu Khurma, Albara Awajan and Mohammad Wedyan

Faculty of Artificial Intelligence, Al-Balqa Applied University, Amman 1705, Jordan; rubaabukhurma82@gmail.com (R.A.K.); a.awajan@bau.edu.jo (A.A.); mwedyan@bau.edu.jo (M.W.)
* Correspondence: m.alazab@bau.edu.jo

Abstract: In recent times, cybercrime has increased significantly and dramatically. This made the need for Digital Forensics (DF) urgent. The main objective of DF is to keep proof in its original state by identifying, collecting, analyzing, and evaluating digital data to rebuild past acts. The proof of cybercrime can be found inside a computer's system files. This paper investigates the viability of Multilayer perceptron (MLP) in DF application. The proposed method relies on analyzing the file system in a computer to determine if it is tampered by a specific computer program. A dataset describes a set of features of file system activities in a given period. These data are used to train the MLP and build a training model for classification purposes. Identifying the optimal set of MLP parameters (weights and biases) is a challenging matter in training MLPs. Using traditional training algorithms causes stagnation in local minima and slow convergence. This paper proposes a Salp Swarm Algorithm (SSA) as a trainer for MLP using an optimized set of MLP parameters. SSA has proved its applicability in different applications and obtained promising optimization results. This motivated us to apply SSA in the context of DF to train MLP as it was never used for this purpose before. The results are validated by comparisons with other meta-heuristic algorithms. The SSAMLP-DF is the best algorithm because it achieves the highest accuracy results, minimum error rate, and best convergence scale.

Keywords: digital forensic; optimization; multilayer perceptron; salp swarm algorithm; connection weights

1. Introduction

Great technological development has led to the use of the Internet in many areas of life [1]. Many companies have taken advantage of the Internet to provide many services using e-commerce without having to submit to market restrictions. This reflected positively in the country's economy by increasing competitiveness and achieving great returns. This has caused a huge positive shift in the number of customers who use the Internet to buy, sell, and transfer large amounts of money [2]. These large sums are tempting for many hackers and scammers to engage in many activities that violate privacy. These put a lot of people and companies at risk through the web and cause huge financial losses [3]. Other violations that may occur online include impersonation, loss of privacy, brand theft, and loss of customers' trust in institutions. Hence, the suitability of the Internet for carrying out business and banking operations is called into question.

DF is a direct result of cybercrime, which is typically applied in computer-related crimes [4–6]. This includes Intellectual property infringement, use of unauthorized privileges to deal with the computer systems, privacy infringement, a security breach of confidential data repositories, carrying out terrorist operations via the Internet, misuse of electronic data, etc. DF is defined as an organized process that uses scientific techniques to collect, document, and analyze electronically stored data. This helps utilize the computer equipment and storage media to provide evidence to detect the abnormal events [7].

Suspicious events and unsafe activities may lead organizations to lose a lot of money, in addition to losing their prestige. Therefore, such events may lead to serious repercussions for both individuals and institutions. According to the statistics shown by the published reports, the number of computer violations has reached more than five million violations so far [8]. Many cybercrimes go unrecorded because their victims do not report them to officials. Victims of cybercrime feel confused and humiliated or believe that the authorities are not taking the necessary measures to punish the attackers. In addition to the lack of competencies and expertise from the workforce, the government needs to mitigate computer crimes [9].

Crime scene cyber investigations include conducting forensic analysis of all types of storage and digital media increasing the volume of data obtained. The value of data depends on the extent to which it is used in decision-making. The process of analyzing digital data during forensic practice is traditionally manual in most cases. Investigators perform reports using some statistical tools to give a picture of the collected data [10]. However, the cost of digital investigations increases with increasing data dimensions. Furthermore, forensic cases increase the data analysis complexity, which leads to the deterioration of the manual process quickly [11]. It is necessary to use more advanced methods with potential beyond the capacity of the conventional manual analysis in dealing with big data. Machine learning (ML) is an efficient and more sophisticated method that facilitates the production of useful knowledge for decision-makers. This is carried out by analyzing data sets from different perspectives and fixing them into meaningful forms [12].

The primary goal of DF analysis is to identify who is responsible for these cyber security crimes. This is performed by the selection, classification, and reordering of the sequence of events of the digital crime. Acts reorganization analyzes DF and prepares a schedule of these cyber events. It determines digital evidence, which is information with true value gained from trusted sources that admit or do not admit to committing cybercrime. Over there, there are many sources from which reliable information can be collected to rebuild cybercrime events, such as web browsers, history files, cookies messages, temporary files, log files, and system files [13].

Some of the valuable sources of information that can aid the DF are the system files and their metadata. These files are normally modified by users and the usual use of computer machines. The reconstruction process for digital events can benefit from the system files affected by cyber-crime. However, these files may be modified by normal programs [14]. For this reason, the reconstruction of digital events and the determination of the timeline sequence of events and activities are essential. This helps to recognize whether the type of application that manipulates the file system is reliable or malicious.

The main problem to be solved in this paper is the classification of files that have been accessed, manipulated, changed, and deleted by application programs. This depends on using some features that are represented by footprints. Identifying the affected files by specious acts facilitates the process of event reconstruction in the file system.

This paper presents one of the Neural Network techniques for rebuilding the acts of a digital crime. Multi-layer perceptron (MLP) is one of the artificial neural networks (ANNs) that imitates the neural human system [15]. It has been used in many applications as a supervised classifier. The main advantage of MLP is that it can learn and tackle many complex problems with promising results in science and engineering. It can be adopted efficiently for either supervised or unsupervised learning. In addition, it has a large ability to tackle parallelism, fault tolerance, and robustness to noise. MLP has a great capacity to generalize as well.

The most common problems of gradient-based MLP are stagnation in local minima, a tendency to the starting values of its parameters, and slow convergence. Due to the local minima problem, several studies in the literature proposed different approaches to train MLP. Swarm-based algorithms have been widely used to train MLPs. These algorithms simulate the natural survival of animals such as the Gradient-Based Optimizer (GBO) [16], Slime mould algorithm (SMA) [17], Heap-based optimizer (HBO) [18], and Harris hawks optimization (HHO) [19]. However, the local minima problem still exists. Furthermore, based on the no-free-lunch theorem (NFL), there is no preference among swarm algorithms. The superiority of one swarm algorithm over another does not imply superiority in all applications. This means that there is no guarantee that it will perform similarly when applied to other optimization problems. These reasons motivated us to propose a new method to train MLPs using the SSA algorithm in the context of DF.

SSA [20] is one effective meta-heuristic algorithm that belongs to a swarm-based family. It has a set of characteristics that motivated us to select it. First, it has a single parameter that decreases in an adaptive manner relative to the increasing iterations. Second, it performs an extensive exploration in the initial iterations then it adaptively switches to exploit the most promising areas of the search space. Third, SSA preserves the best-found solution so that it never loses the optimal solution. Fourth, follower salps change their locations adaptively following other members of the population, so it has the power to alleviate the local minima problem.

SSA has been implemented to solve different optimization problems. In [21], Yang proposed a memetic SSA (MSSA) with multiple independent salp chains. He aimed to make more exploration and exploitation. Another improvement was using a population-based regroup operation for coordination between different salp chains. The MSSA is applied for an efficient maximum power point tracking (MPPT) of PV systems. The results showed the output energy generated by MSSA in Spring in Hong Kong is 118.57%, which is greater than the other algorithms. In [22], the author integrated the SSA with the Ant Lion Optimization algorithm (ALO) for intrusion detection in IoT networks. The true positive rates reached 99.9% and with a minimal delay. In [20], the authors proposed a new phishing detection system using SSA. They aimed to increase the classification performance and reduce the number of features of the phishing system. Different transfer function (TF) families: S-TFs, V-TFs, X-TFs, U-TFs and Z-TFs were used to generate a binary version of the SSA. The results showed that BSSA with X-TFs obtained the best results. In [23], the target of the proposed approach was to reduce the number of symmetrical features and obtain a high accuracy after implementing three algorithms: particle swarm optimization (PSO), the genetic algorithm (GA), and SSA. The used datasets are the Koirala dataset. The proposed COVID-19 fake news detection model obtained a high accuracy (75.4%) and reduced the number of features to 303. In [24], the authors introduced four optimizations algorithms integrated with an MLP neural network, namely artificial bee colony (AEC-MLP), grasshopper optimization algorithm (GOA-MLP), shuffled frog leaping algorithm (SFLA-MLP), and SSA-MLP to approximate the concrete compressive strength (CSC) in civil engineering. The results show that SSA-MLP and GOA-MLP can be promising alternatives to laboratory and traditional CSC evaluative methods.

The main contributions can be summarized as follows:

- The integration of SSA with MLP assists it since it can professionally avoid local minima;
- The high speed of convergence of SSA helps reach the optimal MLP structure quickly;
- The proposed method achieves better results compared with other algorithms in terms of accuracy, error rate, and convergence curve;
- The proposed method can deal with different scenarios with varying complexity.

This paper has the following sections: Section 2 reviews some of the previous studies that investigate DF. Section 3 provides a description for MLP. Section 4 discusses the details of the new MLPSSA-DF method. Section 5 discusses the dataset, experiments, and results. Finally, Section 6 summarizes the findings of the whole paper.

2. Forensic's Background

The initiation of the DF investigations depends on the presence of some indications on the computer system. There are several indications that a computer system has experienced suspicious incidents and is a victim of cybercrime [25]. Table 1 shows these indications.

Table 1. DF indications.

Computer System Part	Indication
Booting	Slow
Turn-off	Slow
Response	Slow
Data	Deleted or destroyed
Processes	Unknown processes run
Disk space	Low
Pop-up windows	Many unexpected
Battery	Drains quickly
Wi-Fi	Unstable connection
Blue Screen	Displayed many times

The goal of conducting DF investigations is to answer a range of questions as quickly as possible. These include:

1. What are the motives for cybercrime?
2. Where is the cybercrime site?
3. When did the suspicious act happen?
4. What suspicious acts occurred on the computer that warrants the action Investigation?
5. What is the application program or tool that was used in the execution of the cybercrime?

DF investigations rely heavily on system files because they provide accurate and important information about the sequence of events involved in cybercrimes [26]. At the end of the DF investigations comes the question about how the cybercrime occurred, and this question depends on the answer to all the previous questions.

In the past, a general rule was devised regarding potential digital evidence based on documentary evidence considered admissible in court [27]. These steps include earning, identification, assessing, and admission. Years later, a specific methodology was developed for cybercrime research [28]. This methodology is an approach to all the proposed models that have been followed so far and includes the following steps:

1. Define;
2. Preserve;
3. Gather;
4. Test;
5. Analyze;
6. Present;
7. Check.

In [29], the authors expanded the model in [28] by including two additional steps, which are set up and approach strategies. The new model was called the abstract DF model. The additional two steps were carried out after the define step and before the preserving step. The setup step involves the tools that are used to deal with the suspicious acts. On the other hand, the approach step involves a formal strategy for investigating the effect on the technology and viewers. The problem with this model is that it is quite generic, and there is no clear method to test it.

Through the models that have been proposed in [27–29], we can say that the steps of the DF are compatible with the steps of machine learning. In general, a DF investigation relies heavily on reconstructing events and preparing accurate evidence.

In [30], the authors devised a new method that initially requires the collection of evidence to reconstruct the acts, followed by the development of initial hypotheses. These hypotheses are studied, analyzed, and examined. From here begins the actual process of DF investigations. It ends with finding the outcomes of the electronic case. A new technique has been developed in [31] to reconstruct the latest acts. It is based primarily on defining the condemnation of data objects and the relationships between them.

In [32], the authors used an ant system to take the place on Grid hosts. The Ant-based Replication and Mapping Protocol (ARMAP) is used to spread resource information with a decentralized technique, and its performance is assessed using an entropy index. The authors in [33] developed a new method to reconstruct acts. It uses a finite state machine. It shows the transition from one state to another based on some conditions put by the evidence. In [34], the authors proposed ontology for reconstructing acts by representing the acts as entities. The events are represented by developing a time model to show the instance change in the state instead of using a period. The shellbag-based technique was proposed in [35], which preserves information in the Windows Registry. This information is about the files, folders, and windows that appear, such as deleted, modified, and relocated files and folders.

Nowadays, many tools are used to collect acts and save them in a temporal repository. However, it can be difficult to analyze the raw data.

In [36], Hal investigated the potential of EXplainable artificial intelligence (XAI) to enhance the analysis of DF evidence using examples of the current state of the art as a starting point. Bhavsar [37] pointed out the challenges in this forensics standard to design the framework of the investigation process for criminal activities using the best digital forensic instruments. Dushyant [38] discussed the advantages and disadvantages of incorporating machine learning and deep learning into cybersecurity to protect systems from unwanted threats. In [39], Casino conducted a review of all the relevant reviews in the field of digital forensics. The main challenges are related to evidence acquisition and pre-processing, collecting data from devices, the cloud, etc.

Unlike previous studies in the literature, where the proposed algorithm was trained on general datasets or limited applications, the newly developed MLP-SSA is proposed for the first time in the context of DF. Different evaluations are performed to test the viability of the MLPSSA to be used as a robust classifier for DF.

3. DF Using MLP

The MLP neural network can be used for the classification of files that have been accessed, manipulated, changed, and deleted by application programs. Once act depends on using some features that are represented by footprints. Identifying the affected files by specious acts facilitates the process of event reconstruction in the file system.

3.1. Dataset

The dataset used in the experiments is collected from three resources: the audit log, file system, and registry. It ensures that if some features are missed in one resource, they may be found in another resource. The dataset represents a database for the collected features of the system file or the metadata. The record contains the values of the files' features that have been affected by a specific program. These are the footprints that are specific for each application program. It describes the system events or acts or metadata. As in supervised classification, the dataset has one column for class. In this dataset, the last column is the application program.

The collection of the features related to the system file using an application program is carried out using some programs such as the .Net program. This paper runs a program called VMware, which is used for collecting the features and building a training dataset. The main advantage of VMware is that it can get red from the useless programs and reduce their effects. The operating system used in this experiment is Windows 7 as it is a commonly used platform. The application programs used in the experiments are Internet Explorer, Acrobat Reader, MS-Word, MS-Excel (Microsoft Office version 2007), and VLC media player.

These applications are performed in different ways. First, the applications are loaded separately. Thus, if one application is completely loaded and then closed, the other program is loaded after it. The main issue in this execution of programs is that they do not do anything in the file system except loading and closing. Second, the applications also are loaded and executed and then closed one by one. In this case, the first three applications perform one act, which is saving a file. The last application visit (www.msn.com accessed on 20 April 2022) website. Third, as the first and second execution, the applications are loaded separately. However, different acts are executed by each application.

The acts of the first three applications include saving files, opening files, and creating new files. However, the last application performs a set of acts that include visiting secured/unsecured websites and sending/receiving emails with/without attachments. Fourth, in this execution, the applications are executed at the same time as opposed to previous executions. As in execution three, different acts take place. The number of examples of records in a database is 23,887. Table 2 shows the features on the dataset that is used to investigate the digital forensics.

Table 2. The features in a dataset.

Feature Number	Feature Name	Feature Value Example
1	Filename-length	255 characters
2	Filenamespace	_user/
3	Object Id	$ObjId$
4	Original-Volume-Id	94F8 − 9C08
5	Domain-Id	$dds :: domain :: DomainParticipant(DOMAIN\_ID, qos\_provider.participant_qos())$
6	Original-Object-Id	40dff02fc9b4d4118f120090273fa9fc

3.2. Preparing a Dataset

The performance of a machine learning model is greatly affected by the dataset. Therefore, preparing a dataset is an important stage for developing efficient and reliable models. It enhances the generalizing process. Different issues have been applied to a dataset to preprocess it before using it in the learning algorithm. First, restructuring of the dataset by distributing the values into sets in such a way the feature's states reduce.

Another important concern in this regard is that most machine learning algorithms deal with numerical values instead of text values. Therefore, in the used dataset, there is a need to assign numerical values to some feature values using the word2vect tool. Using this tool, an index is assigned to each word. Second, cleaning the dataset and git-rid of missing and outlier values. These have a positive impact on enhancing the generalization and generating less biased models. Third, normalizing the dataset by scaling the values of the features into a predefined range using the min-max method. This helps to deal with all features equally instead of making one feature overwhelm the others.

3.3. Salp Swarm Algorithm (SSA)

The inspiration for the SSA algorithm is from the salp aquatic animals. They have a specialized technique for obtaining food. The first salp in the swarm leads the other members in the sequence. This implies that other swarm members change their positions dynamically concerning the leader. Figure 1 shows a single salp on the right side and a swarm of salps on the left side.

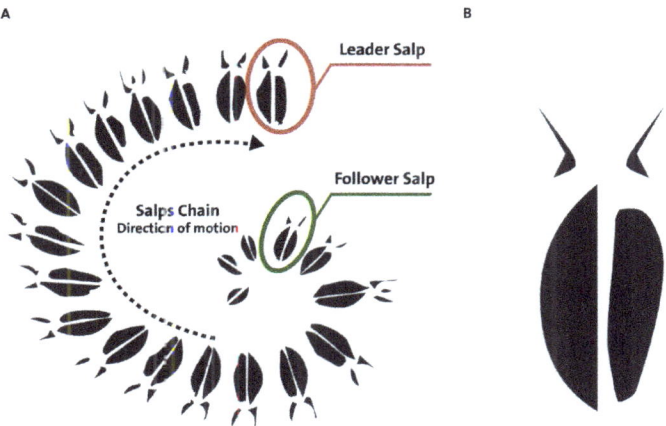

Figure 1. Single salp (**A**) and the salps chain (**B**).

Algorithm 1 (SSA) is an evolutionary algorithm that was developed by Mirjalili [20]. The swarm S of n salps can be represented in Equation (1), where Foo is the source of food. The population of salps is represented by a matrix. Each row is a salp or solution. The length of the salp is the number of features in a dataset (d). The number of salps (n) is the swarm's size. The first row in the matrix is the leader salp, and the other rows are for the follower salps.

$$S_i = \begin{bmatrix} s_1^1 & s_2^1 & \cdots & s_d^1 \\ s_1^2 & s_2^2 & \cdots & s_d^2 \\ \vdots & \vdots & \vdots & \vdots \\ s_1^n & s_2^n & \cdots & s_d^n \end{bmatrix} \quad (1)$$

Equation (2) illustrates the location of the first salp

$$s_j^1 = \begin{cases} Foo_j + cp_1((up\_bound_j - low\_bound_j)cp_2 \\ + low\_bound_j), \ cp_3 \geq 0.5 \\ Foo_j - cp_1((up\_bound_j - low\_bound_j)cp_2 \\ + low\_bound_j), \ cp_3 < 0.5 \end{cases} \quad (2)$$

where s_j^1 and Foo_j are the locations of leader salp and the source of food in the j_{th} dimension, respectively. In Equation (3), cp_1 gradually decreases and changes its value across cycles, and $curr$ and $last$ are the current and the last cycles, respectively. The other parameters cp_2 and c_3 in Equation (2) are randomly chosen from [0, 1]. cp_2 and cp_3 direct the next location in the j_{th} dimension to $+\infty$ or $-\infty$ and determine the step size. The up_bound_j and low_bound_j are the limits of the j_{th} dimension.

$$cp_1 = 2e^{-(\frac{4curr}{last})^2} \quad (3)$$

$$s_j^i = \frac{1}{2}(s_j^i + s_j^{i-1}) \quad (4)$$

In Equation (4), $i \geq 2$, and s_j^i is the location of the i_{th} salp at the j_{th} dimension.

Algorithm 1 SSA

In_variables: n is the size_swarm, d is the no_dimensions
Out_variables: The best salp (Foo)
Initial_variables s_i ($i = 1, 2, \ldots, n$), up_bound and low_bound
while (last cycle is not reached) **do**
    Compute each salp's fitness value
    Determine **Foo** as the optimal salp
    Update c_1 by Equation (3)
    for (each salp s_i) **do**
        if $l == 1$ **then**
            Change the location of the first salp by Equation (2)
        else
            Change the locations of the other salps by Equation (4)
        end if
        Change the locations of the salps using the lower and upper bound
    end for
end while
Return (**F**)

3.4. Multi-Layer Perceptron Neural Networks (MLP)

MLP is a type of feed-forward NN, which is used for training the data and discovering the hidden patterns in the training data. The pattern is then applied to the hidden instances of the dataset to obtain the results. Three layers are in the architecture of MLP: the first, the middle, and the last layers. Each layer consists of a set of computational nodes that simulate the human neurons. The MLP's complexity increases by adding more middle layers. The standard MLP contains a single hidden layer. Figure 2 shows a standard MLP, with a first layer that has n nodes, a single middle layer that has m nodes, and a last layer that has k nodes.

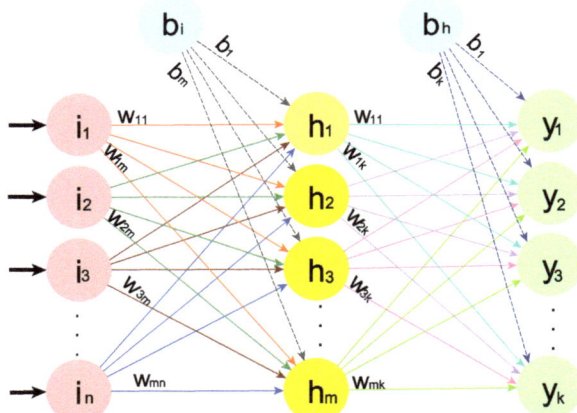

Figure 2. MLP components.

The MLP can be visualized as a directed graph that connects the first layer with the middle layer and the middle layer with the last layer. Each middle node is connected with the first layer with n weights and with r weights with the last layer. In addition, there are m biases. The main processes that take place in the hidden nodes are the summation as in Equation (5) and the activation as in Equation (6). The output of Equation (5) of node m is performed using Equation (5). After computing the sum, a transfer function is applied to the input as in Equation (6).

$$SumFun_m = \sum_{n=1}^{n} d_{nm} * P_n + c_m \tag{5}$$

where d_{nm} is the weight between the first node Pn and the middle node hm, and c_m is the bias m that enters the middle node m.

$$o_m = Sig(SumFun_m) \tag{6}$$

where o_m is the output node m; $m = 1, 2, \ldots, s$; Sig is the sigmoid function as in Equation (7)

$$Sig(SumFun_m) = \frac{1}{1 + e^{-SumFun_m}} \tag{7}$$

After collecting the results from all the m nodes, the final result O_m can be generated as shown in Equations (8) and (9).

$$SumFun_m = \sum_{n=1}^{s} d_{nm} * o_m + c_m \tag{8}$$

where d_{nm} is the weight between the node n in the middle layer and the node m in the last layer, and c_m is the bias m that enters to the output node m.

$$O_m = f(SumFun_m) \tag{9}$$

where O_m is the final result m; $m = 1, 2, \ldots, r$; Sig is the function applied in Equation (7).

4. SSAMLP-DF Model

MLP is a type of NN that uses the feed-forward NN architecture and backpropagation method to propagate data from input nodes to hidden nodes to output nodes. The following list shows the stages of the SSAMLP-DF model:

- Collecting data;
- Data pre-processing;
- MLP construction;
- MLP training phase;
- MLP testing phase.

There are many areas in which the MLP has been implemented. The main features of MLP that helped it to be commonly used include the nonlinear structure, adaptive update of its parameters, and the ability to generalize well compared with other algorithms. Initially, the dataset is divided into three parts: 70% to train the model, 15% to validate the model, and 15% to test the model and mitigate the overfitting problem. This is called the "Hold-Out" validation method. Overfitting is a popular machine learning problem in which the error rate in the training phase is too small, but it increases in the testing phase. Of course, this is not what is required.

The main target of MLP is to build an accurate model with a minimum error achieved in the testing phase. By applying the MLP to the training part of a dataset, the initial MLP structure is constructed. These include the MLP's layers and the number of nodes in each middle layer. The weights of the MLP are set to nonzero values. MLP trains a model until a specific condition is satisfied, such as reaching the maximum number of cycles or achieving a threshold error rate. If the trainer achieves the stopping condition, then the weight parameters of the generated model are kept. If the stopping condition is not met, then the MLP structure is updated by changing the number of nodes in the middle layer.

The MLP structure has a large number of nodes in the middle layer and is then progressively resourced across cycles until an acceptable performance is achieved. This method is called the punning method, and it is used to determine the number of nodes in the middle layer that is suitable to generate the desired performance. After the model is generated, the testing dataset is applied to this model, and the error rate is approximated. The basic measurement is the accuracy, which is calculated by dividing the correctly classified instances by the total number of instances in the testing part of the dataset. It is computed as follows:

$$ACC\text{-}DF = \frac{TP\text{-}DF + TN\text{-}DF}{TP\text{-}DF + TN\text{-}DF + FP\text{-}DF + FN\text{-}DF} \quad (10)$$

where $(TP\text{-}DF)$ is the number of files that are predicted to have tampered with a specific computer program, and it is tampered by a specific computer program. $(FN\text{-}DF)$ is the number of files that are tampered by a specific computer program and incorrectly predicted that they are not tampered by a specific computer program. $(FP\text{-}DF)$ is the number of files that are not tampered by a specific program and incorrectly predicted to be tampered by a specific program. $(TN\text{-}DF)$ is the number of files that are not tampered by a specific program and incorrectly predicted to be not tampered by a specific program. The generated DF-model is trained using MATLAB.

This section presents the proposed DF model by integrating the SSA and MLP algorithms. In this model, the SSA is used to train the MLP based on one hidden layer. Two important issues must be taken into account: the representation of the solution in the SSA and the fitness function. Each solution is represented as a one-dimensional array and its values represent a candidate MLP structure. The solution in the SSA-MLP model is partitioned into three parts: the weight parameters that connect the input nodes with the hidden nodes, the weight parameters that connect the hidden nodes with the output nodes, and the biases. The solution's length equals the number of connection weights in addition to the number of biases. Equation (11)

$$Solution's length = (P \times M) + (2 \times M) + 1 \quad (11)$$

where P is the number of input nodes, and M is the the number of the hidden nodes.

The fitness value in the SSA-MLP model is the mean square error (MSE). This is calculated by subtracting the predicted value from the actual value by the generated solutions (MLPs) for the training part instances of the dataset. MSE is shown in Equation (12), where R is the actual value, $\hat{R}$ is the value generated from prediction, and N is the number of examples in the training dataset.

$$MSE = \sum_{i=1}^{N}(R - \hat{R})^2 \quad (12)$$

The steps of the proposed SSA-MLP can be summarized as follows:

1. The candidate individuals of the SSA are initialized randomly. These represent the possible solutions of the MLP.
2. Evaluating the fitness of solutions. In this step, randomly generated values are assigned to the bits of solutions that represent the possible values of the connection weights and biases. These solutions are then assessed by the MLP. MSE is a popular function that is selected to evaluate the evolutionary-based MLP models. The main purpose of the training stage is to obtain the best network structure of the MLP that generates the minimum error rate or MSE value when implemented on a training part of the dataset
3. Update each solution by changing the position of search agents in the SSA algorithm.
4. Steps 2 and 3 are repeated until the stopping condition is satisfied. The last stage of the proposed model is generating the optimal MLP structure with the best weight parameters and biases. This MLP structure is tested on the testing and validation parts of the dataset. Therefore, the obtained error rate by MSE is the minimum.

Figure 3 shows the assignment of a salp to MLP. Figure 4 shows the general steps of the proposed DF-SSA-MLP model.

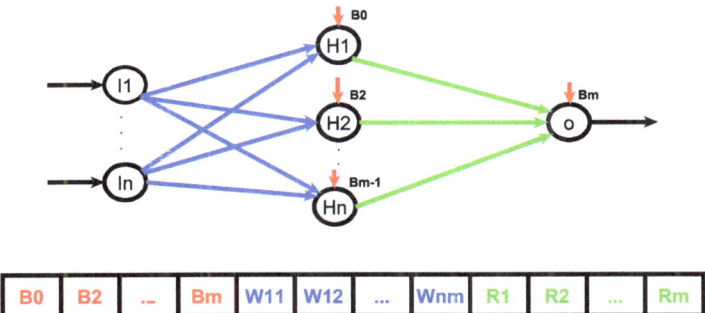

Figure 3. Assigning a salp to MLP.

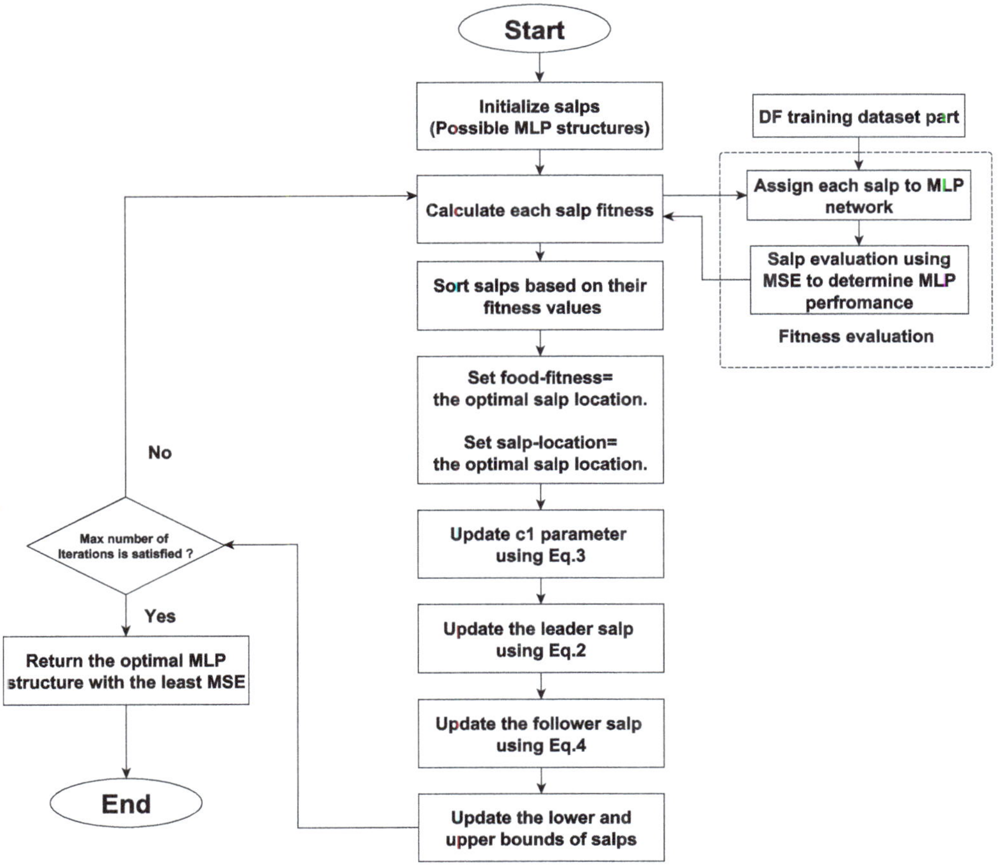

Figure 4. DF-SSA-MLP flowchart.

5. Experiments and Results

This section presents the classification results of the proposed SSAMLP versus other metaheuristic algorithms in terms of accuracy. Twenty-four experiments were established to evaluate the different algorithms in terms of the accuracy of results using different MLP structures. Furthermore, an analysis of their convergence curves is illustrated. The proposed approach is compared against seven metaheuristic algorithms integrated with MLP using the same experiment specifications. These algorithms are: Particle Swarm Optimization (PSO) [40], Ant Colony Optimization (ACO) [41], Genetics Algorithm (GA) [42], Differential Evolution algorithm (DE) [43], and BackPropagation [44].

These experiments performed the proposed methods on 14 benchmark mathematical functions used for minimization problems. Table 3 shows the values of the parameters for the applied algorithms that will be used to validate the proposed SSAMLP model.

Table 3. The parameters' values for the applied algorithms.

Algo	Parameter	Val
SSA	c1	[0–1]
	c2	[0–1]
GA	CrossProb	0.9
	Mut-prob	0.1
	SelecMeth	Roulette wheel
PSO	AccCons	[2.1, 2.1]
	InerWe	[0.9, 0.6]
DE	CrossProb	0.9
	DiffWe	0.5
ACO	IniPh (τ)	0.000001
	PhUpCons (Q)	20
	PhCons (q)	1
	GPhDec-rate (p_g)	0.9
	LPhDecRate (p_t)	0.5
	PhSen (α)	1
	VisSen (β)	5
ES	λ	10
	σ	1
PBIL	LeRate	0.05
	GPopMem	1
	BadPoMem	0
	EPara	1
	MutProb	0.1

Table 4 shows the specification of the performed experiments. These include the structure of the MLP related to the number of layers and the number of hidden nodes. The target is to study the effects of different MLP structures on the performance of the algorithms and determine the best MLP structure for all the studied algorithms.

Table 4. Experiments specifications.

Experiment Number	Number of Layers	Number of Hidden Nodes
1	1	8
2	1	7
3	1	6
4	1	5
5	1	4
6	1	3
7	1	2
8	1	1
10	2	7,1
11	2	6,2
12	2	5,3
13	2	4,4
14	2	6,1
15	2	5,2
16	2	4,3
17	2	5,1
18	2	4,2
19	2	3,3
20	2	3,2
21	2	4,1
22	2	3,1
23	2	2,2
24	2	2,1
25	2	1,1

Table 5 and Figure 5 show the accuracy results of the proposed SSAMLP model versus other algorithms that are integrated into the MLP algorithm.

Table 5. Accuracy results of the SSA-MLP against other algorithms based on different experiment specifications.

Exp #	SSAMLP	BPMLP	GAMLP	PSOMLP	ACOMLP	DEMLP	ESMLP	PRILMLP
1	91.34	87.33	90.56	90.21	84.28	83.20	80.58	82.79
2	92.88	87.98	91.34	91.66	85.06	83.78	80.99	84.39
3	93.41	88.05	91.97	86.00	83.90	81.45	81.77	84.67
4	94.22	88.47	93.11	92.71	85.13	83.10	81.93	84.80
5	95.84	89.50	94.61	93.98	86.09	84.32	82.77	85.39
6	95.76	89.48	92.78	92.67	85.00	84.22	82.55	85.23
7	95.00	88.09	92.71	92.36	84.53	84.01	82.07	85.14
8	94.78	84.67	90.89	91.45	82.77	83.99	81.20	84.33
9	94.38	82.66	88.60	91.55	82.30	83.90	80.99	85.80
10	93.88	81.01	82.45	90.66	81.86	81.77	80.43	82.26
11	93.76	81.00	82.19	90.61	81.23	80.98	78.51	80.87
12	92.17	80.90	80.77	88.50	80.90	79.76	77.56	80.28
13	91.89	80.74	80.36	87.94	80.89	79.32	76.11	79.90
14	90.04	80.03	80.12	84.44	80.13	77.65	75.89	78.94
15	90.01	80.00	79.45	84.21	80.10	76.30	76.26	77.85
16	89.99	79.99	79.39	83.97	80.01	76.10	75.97	77.48
17	89.87	79.43	79.34	83.56	79.56	75.98	75.36	77.31
18	89.60	79.20	78.22	82.30	79.36	74.34	75.12	77.22
19	88.45	79.10	77.83	82.28	79.20	74.29	74.84	74.09
20	88.34	78.33	77.51	82.14	78.20	73.29	72.91	73.67
21	88.23	78.11	75.23	81.90	77.28	72.11	72.37	71.88
22	87.88	77.20	74.33	81.36	77.02	72.00	72.21	71.51
23	87.21	76.37	74.07	81.23	76.98	71.98	71.00	71.47
24	87.11	75.30	73.44	81.11	76.09	71.86	70.66	71.22
25	87.02	75.19	73.19	81.02	76.00	70.96	70.34	70.18

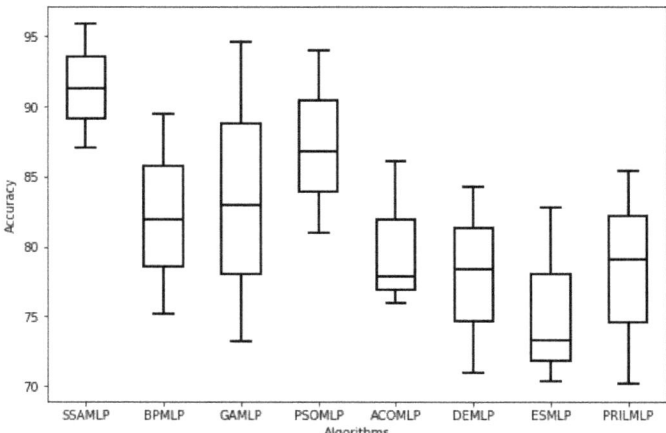

Figure 5. Boxplot representation for the proposed SSAMLP and other algorithms in terms of classification accuracy.

As can be seen in Table 5, the best MLP structure is achieved when the number of hidden nodes is four and the number of layers is one. The results show that in the first three experiments, the accuracy of all hybrid models increases dramatically by decreasing the number of hidden nodes in the MLP's structures. This can be explained because reducing the complexity of computations enhances the performance of DF prediction. The accuracy results increase until a specified limit, which is when the number of layers is one and the number of hidden nodes is four. This structure represents the best MLP structure for all algorithms in which the best DF prediction accuracy is achieved. After that, the accuracy started to decrease linearly by decreasing the number of nodes in the middle layer. The reason is that decreasing the number of nodes in the middle layer makes the model become simple, and the computations are not sufficient to produce an efficient model. Hence, it is not recommended to decrease the number of hidden nodes to less than four or increase it to more than four.

Choosing a medium number of computational nodes in the middle layer can produce the best classification model and achieve the best DF prediction performance. In the remaining experiments, the number of layers in the MLP structure does not benefit the prediction performance. Conversely, increasing the number of layers inversely affects the classification performance. This can increase the complexity of the classification model and cause a major machine learning problem, which is overfitting. This problem occurs because complex models that are unable to produce a general prediction pattern in the learning phase are generated. The produced model is complex and passes by a large number of learning instances. This reflects badly on the testing phase and causes degradation in the prediction performance. Figure 6 shows the convergence curves of the proposed SSAMLP versus other algorithms. The Y-axes is the error rate in terms of MSE. The X-axis is the number of iterations. It can be seen that SSAMLP achieved the best convergence curves, as it obtains the least error rates in the final iterations.

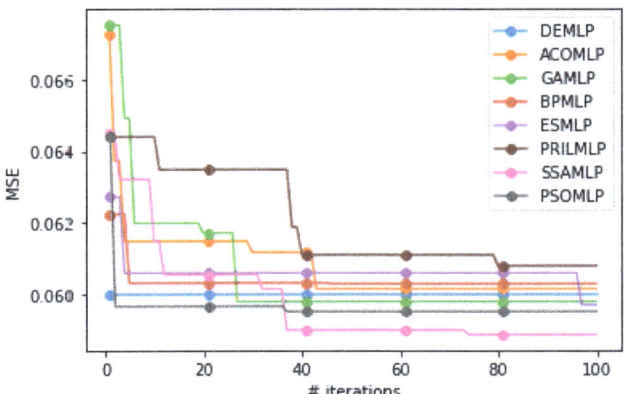

Figure 6. Convergence curves of the proposed SSAMLP and other algorithms.

Overall, the experiments confirm that MLP can be applied as a reliable prediction model for DF. Furthermore, it can be revealed that the number of features is sufficient to investigate the DF and identify the application programs that have affected the file system. The best accuracy achieved is 95.84%, which is somehow high. This indicates that there is about a 4.16 error rate. Although there is no recommended error rate threshold commonly reported for the DF models, it can be explained that this small error rate comes from the application programs that access the same files in the file system. This makes the extracted features of the application programs overlap for several files.

6. Conclusions and Future Works

Recently, cybercrime has increased significantly and tremendously. This made the need for DF urgent. The target of this paper is to propose the SSA in training MLPs. The few parameters, fast convergence, and strong ability to avoid local minima motivated our attempts to use SSA to train MLPs. This is a minimization problem in which the main objective is to select the optimal structure of MLP (best connection weight and bias parameters) to achieve the minimum MSE. The purpose is to apply the optimal MLP in determining the growing evidence by checking the historical acts on the file system to identify how the application programs affected these files.

The used dataset in the experiments has been collected based on applying for five different application programs and checking the footprint on the system files that are the results of system actions and log entries. There are four scenarios for applying the application programs to the system files. These represent simple, medium, and complex scenarios to access the file system. The dataset is used to train the hybrid MLP-SSA model and determine the best structure of MLP that produces the minimum MSE. The results of the experiments show that the proposed SSAMLP outperformed others compared with hybrid meta-heuristic algorithms in terms of accuracy, error rate, and convergence scale. Furthermore, SSAMLP proved its suitability to be used as a reliable model to investigate DF, with an accuracy of 95.84% when the number of middle layers is one and the number of hidden nodes is four.

To verify the proposed method, a set of meta-heuristic algorithms were applied to the same dataset, and their results were compared with the SSAMLP. The comparison results show the out-performance of the SSAMLP in the majority of cases.

For future works, it is worthy to train other types of MLPs using the SSA. The applications of the multiobjective SSA to train MLP in the context of DF are recommended as well.

Author Contributions: Conceptualization, R.A.K. and M.A.; methodology, R.A.K. and M.A.; software, A.A.; validation, M.A., A.A. and M.W.; formal analysis, R.A.K.; investigation, M.A.; data curation, M.W.; writing—original draft preparation, R.A.K.; writing—review and editing, R.A.K.; visualization, A.A.; supervision, M.A.; project administration, M.A.; funding acquisition, M.A. All authors have read and agreed to the published version of the manuscript.

Funding: The research reported in this publication was funded by the Deanship of Scientific Research and Innovation at Al-balqa Applied University, Al-Salt, Jordan. (Grant Number: DSR-2021#388).

Conflicts of Interest: The author declares that no conflict of interest.

References

1. Purnaye, P.; Kulkarni, V. A comprehensive study of cloud forensics. *Arch. Comput. Methods Eng.* **2022**, *29*, 33–46. [CrossRef]
2. Wu, H.; Zhou, J.; Tian, J.; Liu, J.; Qiao, Y. Robust Image Forgery Detection against Transmission over Online Social Networks. *IEEE Trans. Inf. Forensics Secur.* **2022**, *17*, 443–456. [CrossRef]
3. Li, B.; Ma, S.; Deng, R.; Choo, K.K.R.; Yang, J. Federated Anomaly Detection on System Logs for the Internet of Things: A Customizable and Communication-Efficient Approach. *IEEE Trans. Netw. Serv. Manag.* **2022**, *4*, 104–111. [CrossRef]
4. Alazab, M.; Alazab, M.; Shalaginov, A.; Mesleh, A.; Awajan, A. Intelligent mobile malware detection using permission requests and API calls. *Future Gener. Comput. Syst.* **2020**, *107*, 509–521. [CrossRef]
5. Alazab, M.; Venkatraman, S.; Watters, P.; Alazab, M.; Alazab, A. Cybercrime: The case of obfuscated malware. In *Global Security, Safety and Sustainability & e-DEMOCRACY*; Springer: Berlin/Heidelberg, Germany, 2011; pp. 204–211.
6. Alazab, M.; Alhyari, S.; Awajan, A.; Abdallah, A.B. Blockchain technology in supply chain management: An empirical study of the factors affecting user adoption/acceptance. *Clust. Comput.* **2021**, *24*, 83–101. [CrossRef]
7. Padma, K.; Don, K. Artificial Neural Network Applications in Analysis of Forensic Science. *Cyber Secur. Digit. Forensics* **2022**, *2*, 59–72.
8. Flatley, J. *Crime in England and Wales: Year Ending December 2015*; Office for National Statistics: London, UK, 2016.
9. Jin, Z.; Huang, J.; Wang, W.; Xiong, A.; Tan, X. Estimating Human Weight from A Single Image. *IEEE Trans. Multimed.* **2022**, *1*, 12–23. [CrossRef]
10. Mohammad, R.M. A neural network based digital forensics classification. In Proceedings of the 2018 IEEE/ACS 15th International Conference on Computer Systems and Applications (AICCSA), Aqaba, Jordan, 28 October–1 November 2018; IEEE: Piscataway, NJ, USA, 2018; pp. 1–7.
11. Jolfaei, A.; Usman, M.; Roveri, M.; Sheng, M.; Palaniswami, M.; Kant, K. Guest Editorial: Computational Intelligence for Human-in-the-Loop Cyber Physical Systems. *IEEE Trans. Emerg. Top. Comput. Intell.* **2022**, *6*, 2–5. [CrossRef]
12. Han, J.; Pei, J.; Kamber, M. *Data Mining: Concepts and Techniques*; Elsevier: Amsterdam, The Netherlands, 2011.
13. Dong, W.; Zeng, H.; Peng, Y.; Gao, X.; Peng, A. A deep learning approach with data augmentation for median filtering forensics. *Multimed. Tools Appl.* **2022**, *2*, 1–19. [CrossRef]
14. Carrier, B.; Spafford, E. An event-based digital forensic investigation framework. *Digit. Investig.* **2004**, *1*, 5–14.
15. Alhmoud, L.; Abu Khurma, R.; Al-Zoubi, A.; Aljarah, I. A Real-Time Electrical Load Forecasting in Jordan Using an Enhanced Evolutionary Feedforward Neural Network. *Sensors* **2021**, *21*, 6240. [CrossRef] [PubMed]
16. Pasti, R.; de Castro, L.N. Bio-inspired and gradient-based algorithms to train MLPs: The influence of diversity. *Inf. Sci.* **2009**, *179*, 1441–1453. [CrossRef]
17. Izci, D. An enhanced slime mould algorithm for function optimization. In Proceedings of the 2021 3rd International Congress on Human-Computer Interaction, Optimization and Robotic Applications (HORA), Online, 11–13 June 2021; IEEE: Piscataway, NJ, USA, 2021; pp. 1–5.
18. Askari, Q.; Saeed, M.; Younas, I. Heap-based optimizer inspired by corporate rank hierarchy for global optimization. *Expert Syst. Appl.* **2020**, *161*, 113702. [CrossRef]
19. Khurma, R.A.; Awadallah, M.A.; Aljarah, I. Binary Harris Hawks Optimisation Filter Based Approach for Feature Selection. In Proceedings of the 2021 Palestinian International Conference on Information and Communication Technology (PICICT), Gaza, Palestine, 28–29 September 2021; IEEE: Piscataway, NJ, USA, 2021; pp. 59–64.
20. Khurma, R.A.; Sabri, K.E.; Castillo, P.A.; Aljarah, I. Salp Swarm Optimization Search Based Feature Selection for Enhanced Phishing Websites Detection. In Proceedings of the EvoApplications, Virtual Event, 7–9 April 2021; pp. 146–161.
21. Yang, B.; Zhong, L.; Zhang, X.; Shu, H.; Yu, T.; Li, H.; Jiang, L.; Sun, L. Novel bio-inspired memetic salp swarm algorithm and application to MPPT for PV systems considering partial shading condition. *J. Clean. Prod.* **2019**, *215*, 1203–1222. [CrossRef]
22. Abu Khurma, R.; Almomani, I.; Aljarah, I. IoT Botnet Detection Using Salp Swarm and Ant Lion Hybrid Optimization Model. *Symmetry* **2021**, *13*, 1377. [CrossRef]
23. Al-Ahmad, B.; Al-Zoubi, A.; Abu Khurma, R.; Aljarah, I. An Evolutionary Fake News Detection Method for COVID-19 Pandemic Information. *Symmetry* **2021**, *13*, 1091. [CrossRef]
24. Ma, X.; Foong, L.K.; Morasaei, A.; Ghabussi, A.; Lyu, Z. Swarm-based hybridizations of neural network for predicting the concrete strength. *Smart Struct. Syst. Int. J.* **2020**, *26*, 241–251.
25. Czap, H. *Self-Organization and Autonomic Informatics (I)*; IOS Press: Amsterdam, The Netherlands, 2005; Volume 1.

26. Cho, G.S.; Rogers, M.K. Finding forensic information on creating a folder in logfile of ntfs. In Proceedings of the International Conference on Digital Forensics and Cyber Crime, Dublin, Ireland, 26–28 October 2011; Springer: Berlin/Heidelberg, Germany. 2011; pp. 211–225.
27. Pollitt, M. Computer forensics: An approach to evidence in cyberspace. In Proceedings of the National Information Systems Security Conference, Baltimore, MD, USA, 10–13 October 1995; Volume 2, pp. 487–491.
28. Palmer, G. A Road Map for Digital Forensic Research/DFRWS. 2001. Available online: https://dfrws.org/wp-content/uploads/2019/06/2001_USA_a_road_map_for_digital_forensic_research.pdf (accessed on 22 April 2022).
29. Reith, M.; Carr, C.; Gunsch, G. An examination of digital forensic models. *Int. J. Digit. Evid.* **2002**, *1*, 1–12.
30. Lee, H.C.; Palmbach, T.; Miller, M.T. *Henry Lee's Crime Scene Handbook*; Academic Press: Cambridge, MA, USA, 2001.
31. Rynearson, J.M. *Evidence and Crime Scene Reconstruction*; National Crime Investigation and Training: Glynco, Georgia, 2002.
32. Forestiero, A.; Mastroianni, C.; Spezzano, G. Building a peer-to-peer information system in grids via self-organizing agents *J. Grid Comput.* **2008**, *6*, 125–140. [CrossRef]
33. Chabot, Y.; Bertaux, A.; Nicolle, C.; Kechadi, M.T. A complete formalized knowledge representation model for advanced digital forensics timeline analysis. *Digit. Investig.* **2014**, *11*, S95–S105. [CrossRef]
34. Schatz, B.; Mohay, G.; Clark, A. Rich event representation for computer forensics. In *Proceedings of the Fifth Asia-Pacific Industrial Engineering and Management Systems Conference (APIEMS 2004)*; Queensland University of Technology Publications: Brisbane, Australia, 2004; Volume 2, pp. 1–16.
35. Zhu, Y.; Gladyshev, P.; James, J. Using shellbag information to reconstruct user activities. *Digit. Investig.* **2009**, *6*, S69–S77. [CrossRef]
36. Hall, S.W.; Sakzad, A.; Choo, K.K.R. Explainable artificial intelligence for digital forensics. *Wiley Interdiscip. Rev. Forensic Sci.* **2022**, *4*, e1434. [CrossRef]
37. Bhavsar, K.; Patel, A.; Parikh, S. Approaches to Digital Forensics in the Age of Big Data. In Proceedings of the 2022 9th International Conference on Computing for Sustainable Global Development (INDIACom), New Delhi, India, 23–25 March 2022; IEEE: Piscataway, NJ, USA, 2022; pp. 449–453.
38. Dushyant, K.; Muskan, G.; Gupta, A.; Pramanik, S. Utilizing Machine Learning and Deep Learning in Cybersecurity: An Innovative Approach. *Cyber Secur. Digit. Forensics* **2022**, *2*, 271–293.
39. Casino, F.; Dasaklis, T.K.; Spathoulas, G.; Anagnostopoulos, M.; Ghosal, A.; Borocz, I.; Solanas, A.; Conti, M.; Patsakis, C. Research trends, challenges, and emerging topics in digital forensics: A review of reviews. *IEEE Access* **2022**, *1*, 3–9. [CrossRef]
40. Marini, F.; Walczak, B. Particle swarm optimization (PSO). A tutorial. *Chemom. Intell. Lab. Syst.* **2015**, *149*, 153–165. [CrossRef]
41. Katiyar, S.; Ibraheem, N.; Ansari, A.Q. Ant colony optimization: A tutorial review. In Proceedings of the National Conference on Advances in Power and Control, Hong Kong, China, 8–12 November 2015; pp. 99–110.
42. Mirjalili, S. Genetic algorithm. In *Evolutionary Algorithms and Neural Networks*; Springer: Berlin/Heidelberg, Germany, 2019; pp. 43–55.
43. Deng, W.; Shang, S.; Cai, X.; Zhao, H.; Song, Y.; Xu, J. An improved differential evolution algorithm and its application in optimization problem. *Soft Comput.* **2021**, *25*, 5277–5298. [CrossRef]
44. Leung, H.; Haykin, S. The complex backpropagation algorithm. *IEEE Trans. Signal Process.* **1991**, *39*, 2101–2104. [CrossRef]

Review

Android Mobile Malware Detection Using Machine Learning: A Systematic Review

Janaka Senanayake [1,*], Harsha Kalutarage [1] and Mhd Omar Al-Kadri [2]

1 School of Computing, Robert Gordon University, Aberdeen AB10 7QB, UK; h.kalutarage@rgu.ac.uk
2 School of Computing and Digital Technology, Birmingham City University, Birmingham B4 7XG, UK; omar.alkadri@bcu.ac.uk
* Correspondence: j.senanayake@rgu.ac.uk

Abstract: With the increasing use of mobile devices, malware attacks are rising, especially on Android phones, which account for 72.2% of the total market share. Hackers try to attack smartphones with various methods such as credential theft, surveillance, and malicious advertising. Among numerous countermeasures, machine learning (ML)-based methods have proven to be an effective means of detecting these attacks, as they are able to derive a classifier from a set of training examples, thus eliminating the need for an explicit definition of the signatures when developing malware detectors. This paper provides a systematic review of ML-based Android malware detection techniques. It critically evaluates 106 carefully selected articles and highlights their strengths and weaknesses as well as potential improvements. Finally, the ML-based methods for detecting source code vulnerabilities are discussed, because it might be more difficult to add security after the app is deployed. Therefore, this paper aims to enable researchers to acquire in-depth knowledge in the field and to identify potential future research and development directions.

Keywords: Android security; malware detection; code vulnerability; machine learning

1. Introduction

In this technological era, smartphone usage and its associated applications are rapidly increasing [1] due to the convenience and efficiency in various applications and the growing improvement in the hardware and software on smart devices. It is predicted that there will be 4.3 billion smartphone users by 2023 [1]. Android is the most widely used mobile operating system (OS). As of May 2021, its market share was 72.2% [2]. The second highest market share of 26.99% is owned by Apple iOS, while the rest of the 0.81% is shared among Samsung, KaiOS, and other small vendors [2]. Google Play is the official app store for Android-based devices. The number of apps published on it was over 2.9 million as of May 2021. Of these, more than 2.5 million apps are classified as regular apps, while 0.4 million apps are classified as low-quality apps by AppBrain [3]. Android's worldwide popularity makes it a more attractive target for cybercriminals and is more at risk from malware and viruses. Studies have proposed various methods of detecting these attacks, and ML is one of the most prominent techniques among them [4]. This is because ML techniques are able to derive a classifier from a (limited) set of training examples. The use of examples thus avoids the need to explicitly define signatures in developing malware detectors. Defining signatures requires expertise and tedious human involvement and for some attack scenarios explicit rules (signatures) do not exist, but examples can be obtained easily. Numerous industrial and academic research has been carried out on ML-based malware detection on Android, which is the focus of this review paper.

The taxinomical classification of the review is presented in Figure 1. Android users and developers are known to make mistakes that expose them to unnecessary dangers and risks of infecting their devices with malware. Therefore, in addition to malware detection techniques, methods to identify these mistakes are important and covered in this paper (see Figure 1). Detecting malware with ML involves two main phases, which are analyzing Android Application Packages (APKs) to derive a suitable set of features and then training machine and deep learning (DL) methods on derived features to recognize malicious APKs. Hence, a review of the methods available for APK analysis is included, which consists of static, dynamic, and hybrid analysis. Similar to malware detection, vulnerability detection in software code involves two main phases, namely feature generation through code analysis and training ML on derived features to detect vulnerable code segments. Hence, these two aspects are included in the review's taxonomy.

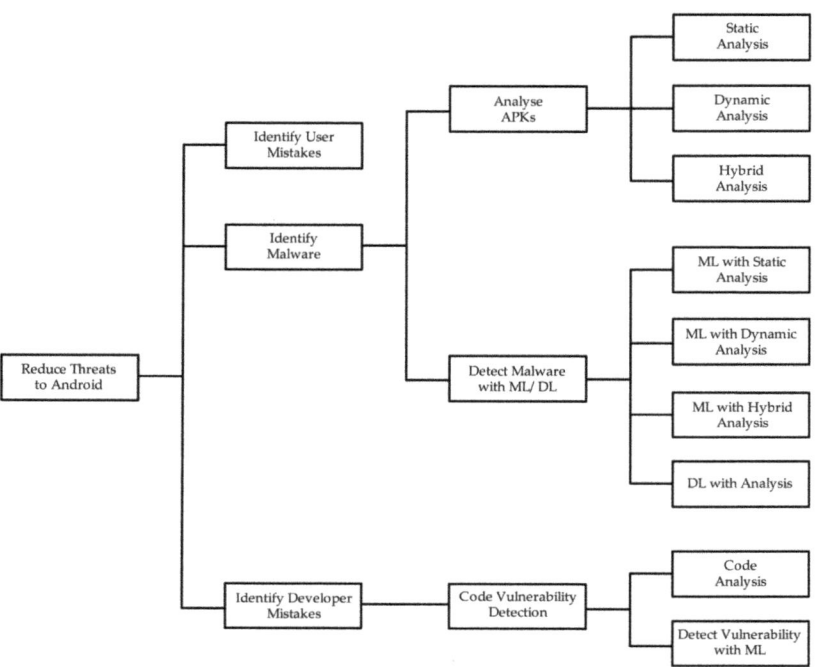

Figure 1. Taxonomy of the review.

The rest of this paper is organised as follows: Section 2 lays out the background to this study. Section 3 provides a detailed description of the review methodology, while Section 4 discusses related previous reviews on the topic. Section 5 discusses static, dynamic, and hybrid analysis techniques for Android malware detection and the application of ML and DL methods as well as a comparison of the methods used in the individual studies. Section 6 discusses ML methods to identify code vulnerabilities, with Section 7 exploring the results and discussions thereof. Finally, Section 8 concludes the paper.

2. Background

This section provides a high-level overview of the Android architecture and its built-in security as well as potential threat vectors for Android. It also provides an introduction to the ML process as it would be useful for non-ML background readers to understand the contents of this paper.

2.1. Android Architecture

Android is built on top of the Linux Kernel. Linux is chosen because it is open source, verifies the pathway evidence, provides drivers and mechanisms for networking, and manages virtual memory, device power, and security [5]. Android has a layered architecture [6]. The layers are arranged from bottom to top. On top of the Linux Kernal

Layer, the Hardware Abstraction Layer, Native C/C++ Libraries and Android Runtime. Java Application Programming Interface (API) Framework, and System Apps are stacked on top of each. Each layer is responsible for a particular task. For example, the Java API Framework provides Java libraries to perform a location awareness application-related activity such as identifying the latitude and the longitude.

Android-based applications and some system services use the Android Runtime (ART). Dalvik was the runtime environment used before the ART. Both ART and Dalvik were created for the Android applications-related projects. The ART executes the Dalvik Executable (DEX) format and the bytecode specification [7]. The other aspects are memory management and power management since the Android-based applications run on battery-powered devices with limited memory. Therefore, the Android operating system is designed in a way that any resource can be well managed [5]. For instance, the Android OS will automatically suspend the application in memory if an application is not in use at the moment. This state is known as the running state of the application life cycle. By doing this, it can preserve the power that can be utilised when the application reopens. Otherwise, the applications are kept idle until they are closed [8].

Built-In Security

Android comes with security already built in. It is a privileged separated operating system [9]. Sandboxing technique and the permission system in Android reduce some risks and bugs in the application. Sandboxing technique in Android isolates the running applications using unique identifiers which are based on the Linux environment [10]. Without having permissions granted from the user at the time of app installation or reconfiguration, apps cannot access system resources. If some of the permissions are not granted, then the application itself will not be usable. When a system update or upgrade happens, several improvements happen in terms of security and privacy. For example, Android 11, the latest stable Android version contains some changes related to security and privacy such as scoped storage enforcement, one-time permissions, permissions auto-reset, background location access, package visibility, and foreground services [11].

However, there are possibilities of malware attacks to exploit some vulnerabilities in the applications developed by various users, because the Google Play Store will not detect some vulnerabilities when publishing applications in the Play Store as in Apple App Store [12].

2.2. Threats to Android

While Android has good built-in security measures, there are several design weaknesses and security flaws that have become threats to its users. Awareness about those threats is also important to perform a proper malware detection and vulnerability analysis. Many research and technical reports have been published related to the Android threats [13] and classified Android threats based on the attack methodology. Social engineering attacks, physical access retrieving attacks, and network attacks are described under the ways of gaining access to the device. For the vulnerabilities and exploitation methods, man in the middle attacks, return to libc attacks, JIT-Spraying attacks, third-party library vulnerabilities, Dalvik vulnerabilities, network architecture vulnerabilities, virtualization vulnerabilities, and Android debug bridges and kernel vulnerabilities are considered.

The survey in [14] identified four types of attacks to Android; hardware-based attacks, kernel-based attacks, Hardware Abstraction Layer (HAL) based attacks, and application-based attacks. Hardware-based attacks such as Rowhammer, Glitch, and Drammer are related to sensors, touch screens, communication media, and DRAM. Kernel-based attacks such as Gooligan, DroidKungfu, Return-oriented Programming are related to Root Privilege, Memory, Boot Loader, and Device Driver. HAL-based attacks such as Return to User and TocTou are related to interfaces for cameras, Bluetooth, Wi-Fi, Global Positioning System (GPS), and Radio. Application-based attacks such as AdDetect, WuKong, and LibSift are related to third-party libraries, Intra-Library collusion, and privilege escalations.

Android applications are easily penetrable with proper knowledge of Android programming if suitable security mechanisms are not in place. In addition, Android marketplaces such as Google Play are not following extensive security protocols when new apps are published. For example, the Android game known as Angry Bird was hacked and the hacker managed to get into its APK file and embed a malicious code that sent text messages unknowingly by the user. The cost was 15 GPB to the user per message. More than a thousand users were affected [15].

2.2.1. Malware Attacks on Android

Malware attacks are the most common case that can be identified as a threat to Android. There are various definitions for malware given by many researchers depending on the harm they cause. The ultimate meaning of the malware is any of the malicious application with a piece of malicious code [16] which has an evil intent [17] to obtain unauthorised access and to perform neither legal nor ethical activities while violating the three main principles in security: confidentiality, integrity, and availability.

Malware related to smart devices can be classified into three perspectives as attack goals and behaviour, distribution and infection routes, and privilege acquisition modes [18]. Frauds, spam emails, data theft, and misuse of resources can be mentioned as the attack goals and behaviour perspective. Software markets, browsers, networks, and devices can be identified as the distribution and infection routes. Technical exploitation and user manipulation such as social engineering can be listed under the privilege and acquisition modes. Malware specifically related to the Android operating system is identified as Android malware [19] which harms or steals data from an Android-based mobile device. These are categorised as Trojans, Spyware, adware, ransomware, worms, botnet, and backdoors [20]. Google describes malware as potentially harmful applications. They classified malware as commercial and noncommercial spyware, backdoors, privilege escalation, phishing, types of frauds such as click fraud, toll fraud, Short Message Service (SMS) fraud, and Trojans [21].

App collusion also should be considered when studying malware. App collusion is two or more apps working together to achieve a malicious goal [22]. However, if those apps perform individually, there is no possibility of a malicious activity happening. It is a must to detect malicious inter-app communication and app permissions for app collusion detection [23,24].

2.2.2. Users and App Developers' Mistakes

The mistakes can happen knowingly or unknowingly from the developers as well as users. These mistakes may lead to threats arising to Android OS and its applications.

It has been identified that users are responsible for most security issues [25]. Some common mistakes done by the users will lead to serious threats in an Android application. At the time of installing Android applications, users will be asked to allow some permissions. However, all the users may not understand the purpose of each permission. They allow permission to run the application without considering the severity of it. Fraudulent applications might steal data and perform unintended tasks after getting the required permissions. It is possible to arise threats to the Android systems due to the mistakes performed by the app developers at the time of developing applications. In the publishing stage of the Android apps, Google Play will have only limited control over the code vulnerabilities in the applications. Sometimes developers are specifying unwanted permissions in the Android manifest file mistakenly, which encourages the user to grant the permissions if the permissions were categorised as not simple permissions [26]. Though the app development companies and some of the app stores are advising about following the security guidelines implemented at the time of development, many developers still fail to write secure codes to build secured mobile applications [27].

2.3. Machine Learning Process

ML is a branch of artificial intelligence that focuses on developing applications by learning from data without explicitly programming how the learned tasks are performed. The traditional ML methods make predictions based on past data. ML process lifecycle consists of multiple sequential steps. They are data extraction, data preprocessing, feature selection, model training, model evaluation, and model deployment [9]. Supervised learning, unsupervised learning, semisupervised learning, reinforcement learning and deep learning are the different subcategories of ML [28]. The supervised learning approach uses a labelled dataset to train the model to solve classification and regression problems depend on the output variable type (continuous or discreet). Unsupervised learning is used to identify the internal structures (clusters), the characteristics of a dataset, and a labelled dataset is not required to train the model. A mix of both supervised and unsupervised learning techniques are applied in semisupervised learning and used in a case of limited labelled data in the used dataset [29]. The learning model and the data used for training are inferred. The model parameters are updated with the received feedback from the environment in reinforcement learning where no training data is involved. This ML method proceeds as prediction and evaluation cycles [30]. DL is defined as learning and improving by analysing algorithms on their own. It works with models such as artificial neural networks (ANN) and consists of a higher or deeper number of processing layers [31].

3. Methodology

Android was first released in 2008. A few years later, the security concerns were discussed with the increasing popularity of Android applications [2]. More attention was received towards applying ML for software security in the last five years because many researchers continuously identify and propose novel ML-based methods [9]. This review was conducted according to the Preferred Reporting Items for Systematic reviews and Meta-Analysis (PRISMA) model [32]. Based on the objective of this study, first we formulated several research questions (see Section 3.1). Next, a search strategy was defined to identify the conducted studies which can be used to answer our research questions. The database usage and inclusion and exclusion criteria were also defined at this stage. The study selection criteria were defined to identify the studies aiming to answer the formulated research questions as the third stage. The fourth stage is defined as data extraction and synthesis, which describes the usage of the collected studies to analyse for providing answers to the research questions. We reviewed threats to the validity of the review and the mechanism to reduce the bias and other factors that could have influenced the outcomes of this study as the last step of the review process.

3.1. Research Questions

This systematic review aims to answer the following research questions.

RQ1: What are the existing reviews conducted in ML/DL based models to detect Android malware and source code vulnerabilities?
RQ2: What are code/APK analysing methods that can be used in malware analysis?
RQ3: What are the ML/DL based methods that can be used to detect malware in Android?
RQ4: What are the accuracy, strengths, and limitations of the proposed models related to Android malware detection?
RQ5: Which techniques can be used to analyse Android source code to detect vulnerabilities?

3.2. Search Strategy

The search strategy involves the outline of the most relevant bibliographic sources and search terms. In this review, we have used several top research repositories as main sources to identify studies. They were ACM Digital Libraries, IEEEXplore Digital Library, Science Direct, Web of Science and Springer Link. Google Scholar, and Research Gate were also used to identify research studies published in some quality venues. The search string that we used to browse through research repositories contained the following search terms: ("android malware") OR ("malware detection") OR ("machine learning") OR ("deep learning") OR ("static analysis") OR ("dynamic analysis") OR ("hybrid analysis") OR ("malware analysis") OR ("android vulnerability analysis") OR ("ML based malware detection") OR ("DL based malware detection").

3.3. Study Selection Criteria

Since mobile malware detection using ML techniques related trends increased from 2016, we limit our review to study related work from 2016 to May 2021. Initially through the research database search in the top research repositories, 109 research papers and from another sources 11 research papers were identified. From these 120 papers, 5 were excluded because of duplicate entries and another 5 were excluded because they were not available in public from those 110 articles. Due to data analysis issues and experiment issues in the given context, 4 articles were excluded though the full text is available. The remaining 106 articles were reviewed in this study. We performed the snowballing process [33], considering all the references presented in the retrieved papers and evaluating all the papers referencing the retrieved ones, which resulted in two additional relevant paper. We applied the same process as for the retrieved papers. The snowballing search was conducted in March 2021. Figure 2 shows a summary of the paper selection method for this systematic review.

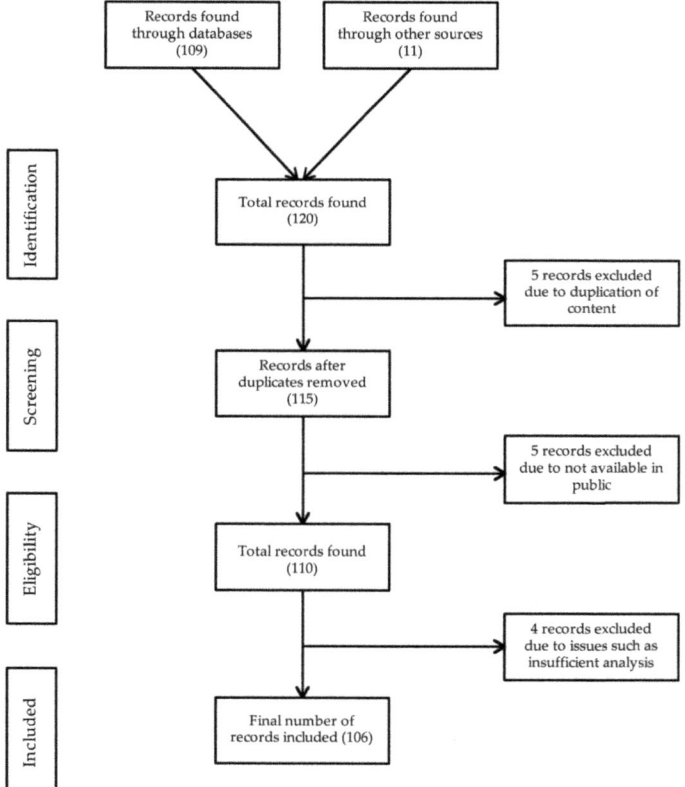

Figure 2. PRISMA method: collection of papers for the review.

3.4. Data Extraction and Synthesis

We extracted data from 9 studies to answer the RQ1, which is about the existing literature reviews related to Android malware detection using ML/DL models and Android vulnerability analysis. To map with RQ2, related studies were identified related to Android code/APK analysing techniques that can be used to analyse malware. The count for those studies was 22. To answer the RQ3 about ML/DL based techniques which can be used to detect malware, we extracted data from 13 different studies. Data from 36 research studies were extracted to find answers for the RQ4, which is about detection model accuracy, strengths, and weaknesses. The remaining 21 papers about Android source code vulnerability analysis and detection methods were used to answer the RQ5.

3.5. Threats to Validity of the Review

This review was conducted in a systematic approach explained above. We tried to minimise the bias and the other factors affecting the review study. Though we have conducted our review comprehensively, still there can be good papers which were not reviewed in this study since they are not available in the research repositories that we used. The period we were considering for the paper selection is from 2016 to May 2021, as the use of ML techniques for malware detection has increased significantly during this period due to recent advances in artificial intelligence. Therefore, if comprehensive studies were conducted before that, those studies were not captured in our work. When searching for the papers we considered the research papers written in the English language. Because of this limitation, our work may have overlooked some important works written in other languages such as Chinese, German, and Spanish.

4. Related Work

Previous reviews in [9,13,17,34–37] discussed various ML-based Android malware detection techniques and ways to improve Android security.

The review in [34] systematically reviewed the studies conducted in static analysis techniques used for Android applications from 2011 to 2015. The tools that can be used to perform Android code analysis using static analysis techniques were also summarised. Abstract representation, taint analysis, symbolic execution, program slicing, code instrumentation, and type/model checking were identified as fundamental analysis methods. Though this review correctly identified the most widely used approach to detect privacy and security related issues, the applicability of static analysis techniques for malware detection was not discussed. Apart from that, it did not take into account the recent research where novel analysis methods and malware detection methods were suggested. The study conducted in [35] provided a good systematic review mainly about static analysis techniques that can be used in Android malware detection. Four methods were identified as characteristic-based, opcode-based, program graph-based and symbolic execution-based. After that, it evaluated the capabilities of static analysis based Android malware detection methods on these four methods using the existing literature. The paper has identified ML and statistical models as possible methods by which Android malware can be identified. However, ML-based machine learning methods have not been thoroughly reviewed as the main focus is only on the static analysis techniques.

In [13], a survey was carried out using existing literature up to 2017 to identify malware detection techniques together with their advantages and disadvantages. Under static and dynamic analysis, they have grouped several approaches that can be used to identify Android malware. However, the analysis of this survey was not comprehensive as it focused on a limited number of studies. Based on the previous studies, a systematic review was conducted in [17]. According to it, there are five types of Android malware detection techniques. They are static detection, dynamic detection, hybrid detection, permission-based detection, and emulation based detection. They also summarised the reviewed work with the model accuracy of malware detection, but the approach of those studies was not discussed. The review conducted in [9] analysed several studies conducted until 2019 related to ML models which can be used to detect Android malware. The malware and APK analysis methods were not discussed in detail since the focus on identifying different ML models was the priority in this review. It is better to analyse the accuracies of the identified ML models. The novel ML/DL and other models which can be used to detect Android malware were also not in the focus of this review. The review in [36] provides a good analysis of static, dynamic, and hybrid detection techniques used in the existing research studies for Android malware detection. Along with that possibility of using machine learning models, several deep learning models are also discussed. However, this study did not comprehensively analyse the model accuracy of the machine learning methods for Android malware detection since this study focused more on discussing different malware detection approaches instead of considering the accuracy of those approaches. Hence, these works differ from our study.

In [37], a systematic review on DL-based methods for Android malware defence was discussed. Malware detection, malware family detection, repackaged/fake app detection, adversarial learning attacks and protections, and malicious behaviour analysis were identified as the malware defines objectives in this review together with the usage of DL models. Though they have identified the possible DL models, it is still better to analyse the accuracy and compare it with traditional ML methods and other hybrid approaches.

Apart from Android malware detection techniques, source code vulnerability analysis is also important to address security concerns in Android. The survey in [38] analysed several studies on ML-based and data mining approaches which can be used to identify software vulnerabilities until 2017. Though this survey provides a good analysis, they considered most of the research work in general software security. Therefore, the vulnerability analysis in Android code was not discussed. However, findings such as ML models' usage for vulnerability analysis are still beneficial for specific programming languages' related analysis.

However, several limitations have been identified in the above works, such as not covering recent proposals on ML methods to detect malware, narrow scopes, and lack of critical appraisals of suggested detection methods. The lack of a thorough analysis of ML/DL-based methods was also identified as a limitation of existing works. Android malware detection and Android code vulnerability analysis have a lot in common. ML methods used in one task can be customised for use in the other task. However, as per our understanding, there are no reviews that cover these two areas together. These shortcomings have been addressed in this work and therefore our work is unique.

5. Machine Learning to Detect Android Malware

Malware detection in Android can be performed in two ways; signature-based detection methods and behaviour-based detection methods [39]. The signature-based detection method is simple, efficient, and produces low false positives. The binary code of the application is compared with the signatures using a known malware database. However, there is no possibility to detect unknown malware using this method. Therefore, the behaviour-based/anomaly-based detection method is the most commonly used way. This method usually borrows techniques from machine learning and data science. Many research studies have been conducted to detect Android malware using traditional ML-based methods such as Decision Trees (DT) and Support Vector Machines (SVM) and novel DL-based models such as Deep Convolutional Neural Network (Deep-CNN) [40] and Generative adversarial networks [41]. These studies have shown that ML can be effectively utilised for malware detection in Android [9]. Most of these studies used datasets such as Drebin [42], Google Play [43], AndroZoo [44], AppChina [45], Tencent [46], YingYongBao [47], Contagio [48], Genome/MalGenome [49], VirusShare [50], IntelSecurity/MacAfee [51], MassVet [52], Android Malware Dataset (AMD) [53], APKPure [54], Anrdoid Permission Dataset [55], Andrototal [56], Wandoujia [57], Kaggle [58], CICMaldroid [59], AZ [60], and Github [61] to perform experiments and model training in their studies.

5.1. Static, Dynamic, and Hybrid Analysis

As mentioned earlier, analysing APKs to extract features is required to use some of the proposed ML techniques in the literature. To this end, three analysis techniques are identified as static, dynamic, and hybrid analysis method [62–64]. Static analysis can be performed by analysing the bytecode and source code (or re-engineered APK) instead of running it on a mobile device. Dynamic analysis detects malware by analysing the application while it is running in a simulated or real environment. However, there is a high chance of exposing the risks to a certain extent to the runtime environment in the dynamic analysis since malicious codes will be executed which can harm the environment. The hybrid analysis involves methods in both static and dynamic analysis.

Under the static analysis, four aspects were proposed [28] which are analysis techniques, sensitivity analysis, data structure, and code representation. Under the analysis techniques, Symbolic execution, taint analysis, program slicing, abstract interpretation, type checking, and code instrumentation were identified. For the sensitivity analysis, object, context, field, path, and flow were identified. For the data structure aspect, it is possible to list call graph (CG), Control Flow Graph (CFG), and Inter-Procedural Control Flow Graph (ICFG). Smali, Jimple, Wala-IR, Dex-Assembler, Java Byte code, or class were listed under the code representation aspect. Kernel, application, and emulator can be taken under inspection level aspect. Taint analysis and anomaly-based can be taken under the dynamic analysis approaches.

The feature extraction methods available in the static analysis consist of two types: Manifest Analysis and Code Analysis [65]. Features such as package name, permissions, intents, activities, services, and providers can be identified in Manifest Analysis. In the code analysis, features such as API calls, information flow, taint tracking, opcodes, native code, and cleartext analysis can be identified as possible features to extract. For the dynamic analysis, five feature extraction methods were identified. They were (1) Network traffic analysis for features like Uniform Resource Locators (URL), Internet Protocol (IP), Network protocols, certificates, and nonencrypted data, (2) Code instrumentation for features such as Java classes, intents, and network traffic, (3) System calls analysis, (4) System resources analysis for features such as processor, memory and battery usage, process reports, network usage, and (5) User interaction analysis for features such as buttons, icons, and actions/events. The study in [66] has explored the security of ML for Android malware detection techniques using a learning-based classifier with API calls extracted from converted smali files. Then a sophisticated secure learning method is proposed, which showed that it is possible to enhance the security of the system against a wide range of evasion attacks. This model is also applicable to anti-spam and fraud detection areas. This study can be further improved by exploring the possibilities of identifying attacks that can alter the training process.

5.2. Static Analysis with Machine Learning

Static analysis is the widely used mechanism for detecting Android malware. This is because malicious apps do not need to be installed on the device as this approach does not use the runtime environment [67].

5.2.1. Manifest Based Static Analysis with ML

Manifest based static analysis is a widely used static analysis technique. The model proposed in SigPID [68] discussed an Android permission-based malware detection mechanism. This model has identified only 22 permissions out of all the permissions listed in sample APKs that are significant by developing a three-level data purring method: permission ranking with negative rate, support based permission ranking, and permission mining with association rules. After that, the ML algorithms were employed to detect the malware. To this process, a binary format dataset of permissions, which was created using a database of malware and benign apps from Google Play was used. The support-vector machine (SVM) outperformed the other studied ML algorithms (Naïve Bayes (NB) and

(DT)) with over 90% accuracy. For the permission-based static analysis, this work was conducted comprehensively. However, it is better to check the other variables which are affecting the malware apart from permissions.

A malware detection method using Android manifest permission analysing was proposed in [69] with the use of static analyser and decompilation support of APKTool for the APK to code level extraction. AndroZoo repository was used as the dataset to train four different ML algorithms. Random Forest (RF), SVM, NB, and K-Means were used to perform the model validity process, and RF produced the highest accuracy for this model with 82.5% precision and 81.5% recall. However, the accuracy of this model is comparatively low with the other studies conducted in the same area. The close reason for that would be that this approach compares the permissions only.

The proposed work in [70] checked the possibility of using reduced dimension vector generating for malware detection. Based on that, malware detection using ML models with permission-based static analysis was performed. In the feature selection stage of this approach, the model removed the unnecessary features using a linear regression-based feature selection approach. Therefore, the classification model can run in real-time since the training time was decreased, with an accuracy of over 96%. The Multi-Layer Perception Model (MLP) algorithm outperformed NB, Linear Regression, k-nearest neighbors (KNN), C4.5, RF and Sequential Minimal Optimization (SMO). It is better to focus on hypermeter selections to also increase the performance of the classification. The model proposed in [71] performed a static analysis on Android apps. Android permissions and intents were used as the basic static features of malware classification while URLs, Emails, and IPs were used as the basic dynamic features. Initially, the APK files were decompiled using ApkTool. The extractor module of this extracted different types of information related to malware. After extracting the data through disassembling the dex files, the data were kept in a text files and they were used to create the feature vector. Then the ML algorithms RF, NB, Gradient Boosting (GB), and Ada Boosting (AB) were used to train and test the malware detection model with the usage of Drebin dataset and Google Play Store. After performing ML training and testing part for each of permission, intent, and network features individually it has identified that the above ML algorithms were performing with different accuracies. For permissions RF performed well with 0.98 precision and recall, for intents NB performed well with 0.92 precision and 0.93 recall, and for network both RF and AB performed similarly well with 0.97 precision and recall. Though this research concluded with such accuracies for malware detection it is still lacking the study of some other features like API calls, etc.

Android malware detection technique using feature weighting with join optimisation of weight mapping and classifier parameters model is proposed in JOWMDroid Framework in [72]. This model is a static analysis-based technique that selected a certain number of features out of the extracted features from the app which were related to malware detection. This process was done by decompiling the APK to manifest and class.dex files and prepared a binary feature matrix. Initial weight was calculated using Random Forest, SVM, Logistic regression (LR), and KNN ML models. Weight machine functions were designed to map the initial weight with final weights. As the last step, classifiers and weight mapping function parameters were jointly optimised by the Differential Evolutional algorithm. Drebin, AMD, Google Play, and APKPure datasets were used to train the model. Finally, it is identified that among weight unaware classifiers, RF performed better with 95.25% accuracy and for weight-aware classifiers, KNN and MLP performed better. However, with the integration of this JOWM-IO method, SVM and LR beat the RF with over 96% accuracy. If the correlation between features is also considered, the model accuracy for detecting malware will increase.

Table 1 comparatively summarises the above research studies related to manifest analysis based methods.

Table 1. Manifest based static Analysis with ML.

Year	Study	Detection Approach	Feature Extraction Method	Used Datasets	ML Algorithms/ Models	Selected ML Algorithms/ Models	Model Accuracy	Strengths	Limitations/Drawbacks
2018	[68]	Developing 3 level data purring method and applying ML models with SigPID	Manifest Analysis for Permissions	Google Play	NB, DT, SVM	SVM	90%	High effectiveness and accuracy	Considered only the permission analysis which may lead to omit other important analysis aspects
2021	[69]	Analysing permission and training the model with identified ML algorithm	Manifest Analysis for Permissions	Google Play, AndroZoo, AppChina	RF, SVM, Gaussian NB, K-Means,	RF	81.5%	The model was trained with comparatively different datasets	Did not consider other static analysis features such as OpCode, API calls, etc.
2021	[70]	Reducing dimension vector generation and based on that perform malware detection using ML models	Manifest Analysis for permissions	AMD, APKPure	MLP, NB, Linear Regression, KNN, C.4.5, RF, SMO	MLP	96%	Efficiency, applicability and understandability are ensured	Hyper-parameter selections are not made in the use
2021	[71]	Selecting feature using dimensionality reduction algorithms and using Info Gain method	Manifest Analysis for permissions and intents	Drebin, Google Play	RF, NB, GB, AB	RF, NB, AB	RF-98%, NB-92%, AB-97%	Analysed the features as individual components and not as a whole	Did not consider about other features such as API calls, Opcode etc
2021	[72]	Feature weighting with join optimisation of weight mapping with proposed JOWMDroid framework	Manifest Analysis for permission, Intents, Activities and Services	Drebin, AMD, Google Play APKPure	RF, SVM, LR, KNN	JOWM-IO method with SVM and LR	96%	Improved accuracy and efficiency	Correlation between features were not considered

5.2.2. Code Based Static Analysis with ML

Code based analysis is the other way of performing the static analysis to detect Android malware with ML. The model proposed in TinyDroid [39] analysed the latest malware listed in the Drebin dataset. Instruction simplification and ML are used in the model. Using the decompiled DEX files by converting APK to smali codes, the opcode sequence was abstracted. Then using that, features were extracted through N-gram and integrated with the exemplar selection method. In the exemplar selection method, for intrusion detection, a good representative of data was generated through a clustering algorithm, Affinity Propagation (AP). This is because in AP, the number of clusters determination or estimation is not required before running the application. Then the generated 2,3, and 4-gram sequences were fed into SVM, KNN, RF, and NB ML classifiers. RF algorithm was identified as the optimal algorithm for this scenario with 0.915 True Positive Rate, 0.106 False Positive Rate, 0.876 Precision, and 0.915 Recall for 2-gram sequence. High accuracy rates for the other 3 and 4-grams were also achieved compared to the studied ML algorithms. However, the proposed method still has issues such as using the malware samples taken only from few research studies and some organisation and lack of metamorphic malware samples. Therefore, some malware could remain undetected.

The approach proposed in [73] used the Drebin dataset with 5560 malware samples along with 361 malware from the Contagio dataset and 5900 benign apps from Google Play to propose another approach to detect malware by analysing API calls used in operand sequences. For the malware prediction model, the package level details were extracted from the API calls. The package n-grams were extracted from the package sequence, which represents application behaviour. Then they were combined with DT, RF, KNN, and NB ML algorithms to build a predictive model in this study and concluded that the RF algorithm performed with an accuracy of 86.89% after training the model on 2415 package n-grams. It is better to consider other information which contains in operands since it might affect the overall model. The relationship of system functions, sensitive permissions, and sensitive APIs were analysed initially in Anrdoidect [74]. A combination of system functions was used to describe the application behaviours and construct eigenvectors using the dynamic analysis technique. Based on the eigenvectors, effective methodologies of malware detection were compared along with the NB, J48 DT, and application functions decision algorithm and identified that the application functions' decision algorithm outperformed the others. There are still some improvements to be performed to this approach.

In MaMaDroid [75] model, API calls performed by apps were abstracted using static analysis techniques to classes, packages, or families. Then to determine the call graph of apps as Markov chain, the sequence of API calls was obtained. Then using ML algorithms, classification was performed using RF, KNN, and SVM and it was identified that RF had the highest accuracy among these three. However, in this method, dynamic analysis was not considered. The dynamic analysis is useful for an API calls analysis in a runtime environment to detect malicious applications.

Android malware detection approach using the method-level correlation relationship of application's abstracted API calls was discussed in [76]. Initially, the source codes of Android applications were split into methods, and abstracted API calls were kept. After that, the confidence of association rules between those calls was calculated. This approach provided behavioural semantic of the application. Then SVM, KNN, and RF algorithms were used to identify the behavioural patterns of the apps towards classifying as benign or malicious. Drebin and AMD datasets were used for this, and 96% accuracy was received with the RF algorithm. This method does not address the problems such as dynamic loading, native codes, encryptions, etc. though it has such high accuracy. If the dynamic analysis methods are also used, the accuracy of this model will increase to a further high level.

The model named SMART in [77] proposed a semantic model of Android malware based on Deterministic Symbolic Automation (DSA) to comprehend, detect, and classify malware. This approach identified 4583 malware that were not identified by leading anti-malware tools. Two main stages were included in this approach; malicious behaviour learning and malware detection and classification. In Stage 1, the model identified semantic clones among malware, and semantic models were constructed based on that. Then malicious features were extracted from DSA, and ML techniques were used to detect malware in Stage 2 after performing static analysing activities with bytecode analysis. Random Forest achieved the best classification results of 97% accuracy, and AB, C45, NB, and Linear SVM provided lower accuracy. Therefore, this work identified that DSA is possible to use for malware detection. DroidChain [78] proposed a static analysis model with behaviour chain model. The malware detection problem was transformed to a matrix model using the Wxshal algorithm to further analyse this approach. Privacy leakage, SMS financial charges, malware installation, and privilege escalation were proposed as malware models in this study using the behaviour chain model. In the static analysis part, using APKTool and DroidChain, Smali codes were extracted. Then the API call graph was generated using the Androguard [79] tool. After that, the incidence matrix was built, and the accessibility of the matrix to detect malware was calculated. The average accuracy of this model was 83%. This method can be improved to detect malware more accurately and efficiently by considering other static analysis features such as code analysis, permission analysis, etc.

The study conducted in [80] discussed testing malware detection techniques based on opcode sequence and API call sequence. The Hidden Markov Model (HMM) was trained in this and detection rates for models based on static, dynamic, and hybrid approaches were identified and it was concluded that the hybrid approaches are highly effective without performing static or dynamic analysis alone.

Tables 2 and 3 comparatively summarise the above research studies related to code analysis based methods, while Table 2 listed studies with model accuracy below 90% and Table 3 listed studies with model accuracy above 90%.

Table 2. Code based static Analysis with ML (Model Accuracy is below 90%).

Year	Study	Detection Approach	Feature Extraction Method	Used Datasets	ML Algorithms/ Models	Selected ML Algorithms/ Models	Model Accuracy	Strengths	Limitations/Drawbacks
2016	[78]	Transforming malware detection problem to matrix model using Wxshall algo and extracting Smali codes and generated the API call graph using Androguard	Code analysis for API Calls and code instrumentation for network traffic	MalGenome	Custom build ML based Wxshall algorithm, Wxshall extended algorithm	Wxshall extended algorithm	87.75%	Few false alarms	Required to expand the behaviour model and improve the efficiency
2017	[74]	Using the combination of system functions to describe the application behaviours and constructing eigenvectors and then using Androidetect	Code analysis for API calls and Opcodes	Google Play	NB, J48 DT, Application functions decision algorithm	Application functions decision algorithm	90%	Can identify the instantaneous attacks. Can judge the source of the detected abnormal behaviour. High performance in model execution	Did not consider some important static analysis features such as OpCode, API calls, etc.
2018	[39]	Using TinyDroid framework, n-Gram methods after getting the Opcode sequence from .smali after decompiling .dex	Code Analysis for Opcode	Drebin	NLP, SVM, KNN, NB, RF, AP	RF and AP with TinyDroid	87.6%	Lightweight static detection system. High performance in classification and detection	Malware samples were taken only from few research studies and some organisations which lack metamorphic malware samples
2018	[73]	Analysing Package level information extracted from API calls using decompiled Smali files	Code Analysis for API Calls and Information flow	Drebin, Contagio, Google Play	DT, RF, KNN, NB	RF	86.89%	Model performs well even when the length of the sequence is short	Other information contained in operands were not considered which affect to the overall model

Table 3. Code based static Analysis with ML (Model Accuracy is above 90%).

Year	Study	Detection Approach	Feature Extraction Method	Used Datasets	ML Algorithms/ Models	Selected ML Algorithms/ Models	Model Accuracy	Strengths	Limitations/Drawbacks
2016	[77]	Using Deterministic Symbolic Automaton and Semantic Modelling of Android Attack	Code Analysis for Opcode/Byte code	Drebin	AB, C4.5, NB, LinearSVM, RF	RF	97%	Use a combined approach of ML and DSA inclusion	Unable to detect new malware patterns since this will not perform complete static analysis
2017	[80]	Training Hidden Markov Models and comparing detection rates for models based on static data, dynamic data, and hybrid approaches	Code analysis for API calls and Opcode in static analysis and System call analysis	Harebot, Security Shield, Smart HDD, Winwebsec, Zbot, ZeroAccess	HMM	HMM	90.51%	Check the difference approaches available to detect ML	Did not consider other ML algorithms or other important features
2019	[75]	Determining the apps call graphs as Markov chain Then obtaining API call sequences and using ML models with MaMaDroid	Code Analysis for API calls	Drebin, oldbenign	RF, KNN, SVM	RF	94%	the system is trained on older samples and evaluated over newer ones	Requires a high memory to perform classification
2019	[76]	Calculating confidence of association rules between abstracted API calls which provides behavioural semantic of the app	Code Analysis for API calls	Drebin, AMD	SVM, KNN, RF	RF	96%	Efficient feature extraction process. Better stability of the system	Did not address the cases such as dynamic loading, native codes, encryption, etc.

5.2.3. Both Manifest and Code Based Static Analysis with ML

Some studies used both manifest and code based static analysis approaches to detect Android malware with ML. The implemented model in WaffleDetector [81], a static analysis approach to detect malware, was proposed by using a set of Android program features, sensitive permissions, and API calls with the utilization of Extreme Learning Machine (ELM). Tencent, YingYongBao, and Contagio datasets were used to train the algorithms. This method outperformed traditional binary classifiers (DT, Neural Network, SVM, and NB) with 97.06% accuracy. This approach still needs a few improvements, such as refining the combination of permissions and API calls.

The study conducted in [82] studied repackaged apps. The malware was identified from these repackaged apps with code-heterogeneity features. The codes of the apps were partitioned into subsets. Then the subsets were classified based on their behavioural features with Smalicode. Compared to the other nonpartitioning methods, this approach provides high accuracy with a False Negative Rate (FNR) of 0.35% and a False Positive Rate (FPR) of 2.97%. This method also used some Ensemble Learning mechanisms. It is better if the method improves the code heterogeneity mechanisms by using context and flow sensitivity.

Using the Drebin dataset, a method to detect Android malware using static analysis is discussed in [83]. Using this method with high accuracy of 98.7%, it was possible to detect malware using a sample of 10,865 applications. In this method, initially, the APK file was downloaded using the extracted download link from the APKPure website by using web mining techniques. Then the APK content was extracted using Apktool and generated the AndroidManifest.xml and classes.dex files. The application features were extracted from AndroidManifest.xml using the AAPT utility while decompiling classes.dex into a jar file using the dex2jar tool. Then the number of lines of code feature was extracted after extracting the java source files from the jar file using the jd-cmd tool. This static analysis approach was evaluated using ten different ML algorithms; KNN, SVM, Bayes Net, NB, LR, J48, RT, RF, AB, and BA. Out of them RF with 1000 decision trees outperformed the others with 0.987 precision, recall, and F-measure [83]. Though the model has high accuracy, it is better to study behavioural analysis of app behaviour by performing dynamic analysis.

In RanDroid [84] model, already classified malicious and benign apps were used to train the SVM, DT, RF, and NB ML algorithms. Initially, the APK files were decompiled using Androguard (a python-ased tool) [79]. Then the required features of permission, API calls, is_crypto_code, is_dynamic_code, is_native_code, is_reflection_code, is_database were extracted and transformed into binary vectors. Then it was trained using ML algorithm and identified that the DT was the most suitable algorithm for this static analysis approach with 97.7% accuracy. However, in this study, broadcast receivers, filtered intend, Control Flow Graph analysis, deep native code analysis, and dynamic analysis are not considered; they are identified as drawbacks.

In [85] a model named TFDroid has been proposed, which is a ML based malware detection by topics and sensitive data flow analysis using SVM with an accuracy of 93.7%. FlowDroid is a static analysis tool that was used in this approach to extract data flow in benign and malicious apps. The permission granularity was transformed using the data flow features. After that, a classifier was implemented for each category and performed the validation process. Google Play and Drebin datasets were used to train the model in this study. It is better to check the other possible ML algorithms' performance also. Since this study is related to data flow, it is better to perform dynamic analysis and introduce a hybrid model to increase the accuracy of detecting Android malware.

The DroidEnsemble [86] analyses the static behaviours of Android apps and builds a model to detect Android malware. In this approach, static features such as permissions, hardware features, filter intents, API calls, code patterns, and structural features of function call graphs of the application were extracted. Then after creating the binary vector, SVM, KNN, RF, and ML algorithms were performed to evaluate the performance of the features and their ensemble. The proposed methodology achieved detection accuracy of 95.8% and 90.68%, respectively, for static features and structural features. For ensemble of both types, the accuracy was increased to 98.4% with SVM. Sting features like API calls and structural features like function call graphs can be checked with dynamic analysis. Therefore, in this model, the malware detection accuracy would be increased when both static and dynamic analysis were integrated.

Table 4 comparatively summarised the above research studies related to both manifest and code based static analysis methods with ML.

Table 4. Both Manifest and Code based Static Analysis with ML.

Year	Study	Detection Approach	Feature Extraction Method	Used Datasets	ML Algorithms/ Models	Selected ML Algorithms /Models	Model Accuracy	Strengths	Limitations/Drawbacks
2017	[81]	Using customized method named Waffle Director	Manifest Analysis for Sensitive permissions and API calls	Tencent, YingYongBao, Contagio	DT, Neural Network, SVM, NB, ELM	ELM	97.06%	Fast Learning speed and Minimal human intervention	Combination of permissions and API calls are not refined
2017	[82]	Using a code-heterogeneity-analysis framework to classify Android repackaged malware by Smali code intermediate representation	Manifest Analysis for Intents. Permissions and API calls	Genome, VirusShare, Benign App	RF, KNN, DT, SVM	RF with custom model proposed	FNR- 0.35%, FPR- 2.96%	Provide in-depth and fine-grained behavioural analysis and classification on programs	Detection issues can happen when the malware use coding techniques like reflection and cannot handle if the encryption techniques used in DEX
2018	[84]	Extracting features and transforming into binary vectors and training using ML with RanDroid Framework	Manifest Analysis for Permissions, Code Analysis for API calls, op-code and native calls	Drebin	SVM, DT, RF NBs	DT	97.7%	Highly accurate to analyse permission, API calls, opcode an native calls toward malware detection	Broadcast receivers, filtered intend, Control Flow Graph analysis, deep native code analysis were not considered
2018	[86]	Creating the binary vector, apply ML models, evaluate performance of the features and their ensemble using DroidEnsemble	Manifest analysis for permissions, code analysis for API calls and system calls analysis	Google Play, AnZhi, LenovoMM, Wandoujia	SVM, KNN, RF	SVM	98.4%	Characterises the static behaviours of apps with ensemble of string and structural features.	Mechanism will fail if the malware contains encryption, anti-disassembly, or kernel-level features to evade the detector
2019	[83]	Extracting applications features from manifest while decompiling classes.dex into jar file and applying ML models	Manifest Analysis for permissions, activities and Code Analysis for Opcode	Drebin, playstore, Genome	KNN, SVM, BayesNet, NB, LR, J48, RT, RF, AB	RF with 1000 decision trees	93.7%	High efficiency, Lightweight analysis and fully automated approach	Did not consider about the API calls and other important features when analysing the DEX.
2019	[85]	Using FlowDroid for static analysis and proposing TFDroid framework to detect malware using sensitive data flow analysis	Manifest Analysis for permission and Code Analysis for information flow	Drebin, Google Play	SVM	SVM	95.7%	Analysed the functions of applications by their descriptions to check the data flow.	Did not consider the improving clustering techniques and applicability of other ML models

5.3. Dynamic Analysis with Machine Learning

The second analysis approach is dynamic analysis. Using this approach it is possible to detect malware with ML after running the application in a runtime environment. Android Malware detection using a network-based approach was introduced in [87]. In this approach, a detection application was developed. It contained three modules: network traces collection, network feature extraction, and detection. In the traces collection module, network activities of running applications were monitored and recorded the network traces periodically. The features extraction module extracted features of the network used by the applications. Those features were Domain Name System (DNS) based features, HyperText Transfer Protocol (HTTP) based features, Origin destination based features, and Transmission Control Protocol (TCP) based features. DT, LR, KNN, Bayes Network, and RF algorithm were used in the detection module. The RF algorithm provided the highest accuracy (98.7%) among them. However, this approach used network-based analysis. If the malware apps were using encrypted transfers, the malware detection accuracy would decrease. Therefore, the model also should consider such factors.

The proposed model in 6th Sense [88], using Markov Chain, NB, Logistic Model Tree (LMT) to detect malware using dynamic analysis is based on sensors available in a mobile device. A context-aware intrusion detection system is studied in this approach by collecting and observing changes in sensor data. This step happened when the applications were performing activities that enhanced security. This model distinguishes malware and benign applications. Three types of malware activities (triggering, leaking information, and stealing data) were identified using this approach via sensors available in the device. The collected data was divided as 75% for training and 25% for testing. For the Markov Chain-based detection technique, a training dataset was used to compute the state transitions and build a transition matrix. A training dataset was used with NB to determine the sensor condition changing frequency. For the other ML algorithms, all the data were defined as benign and malware. In this study, LMT outperformed others with 99.3% precision and 99.98% recall. Though this study is a comprehensive one, it is better if the tradeoffs such as frequency accuracy, battery frequency, etc. are considered.

The proposed method in [89] discussed dynamic analysis-based techniques which extract a set of dynamic permissions from APKs in different sources and run them in an emulator. Then it evaluates the model using NB, RF, Simple Logistic, DT, and K-Star ML models. After that, it is identified that Simple Logistic performs well with 0.997 precision and 0.996 recall. Some issues were in the dataset used in this model. For example, some benign and malicious apps were using the same permissions, and some apps crashed when running the application in an emulator. Therefore, if the dataset is fine-tuned more before use, this model provides even more accuracy.

In [90], a framework called Service Monitor was proposed, which is a lightweight host-based detection system that can detect malware on devices. This framework was built using dynamic analysis. Service Monitor monitored the way of requesting system services to create the Markov Chain Model. The Markov Chain is used as a feature vector to perform the classification tasks with ML algorithms: RF, KNN, and SVM. The RF method performed well with an accuracy of 96.7% after training the model with AndroZoo, Drebin, and Malware Genome datasets. Some benign apps also requested the system services in a similar way to malware. Therefore, this could lead to some misclassification of this model. To avoid that and enhance the classification accuracy, signature-based verification to the Service Monitor can be applied.

A mechanism named DATDroid was proposed in [91] which is a dynamic analysis based malware detection technique with an overall accuracy of 91.7% with 0.931 precision and 0.9 recall values with RF ML algorithm. As the initial stage, feature extraction was performed by collecting system calls, recording CPU and memory usage, and recording network packet transferring. Then in the feature selection stage, Gain Ratio Attribute Evaluator was applied. After that, the model training and validation were performed as the next stage to identify malicious and benign applications using APKPure and Genome Project datasets. In addition to the features studied in this, there can be an impact from features like HTTP, DNS, TCP/IP, and memory usage patterns towards identifying malware which should be discussed.

In [92], a framework which is named as MEGDroid, using the dynamic analysis was proposed to improve the event generation process in Android malware detection. In this method, it automatically extracted and represented information related to malware as a domain-specific model. Decompilation, model discovery, integration and transformation, analysis and transformation, and event production were the steps included in this model. The model was then used to analyse malware after training with the AMD dataset. This model extracted every possible event source from malware code and was developed as an Eclipse plugin. Based on the results, MEGDroid provides better coverage in malware detection through generating UI, whereas system events and monitoring the system calls are lacking in this approach.

Table 5 comparatively summarises the above research studies related to dynamic analysis based methods.

Table 5. Dynamic analysis based malware detection approaches.

Year	Study	Detection Approach	Feature Extraction Method	Used Datasets	ML Algorithms/Models	Selected ML Algorithms/Models	Model Accuracy	Strengths	Limitations/Drawbacks
2017	[87]	Extracting the DNS, HTTP, TCP, Origin based features of the network used by apps	Network traffic analysis for network protocols	Genome	DT, LR, KNN, Bayes Network, RF	RF	98.7%	Work with different OS versions, Detect unknown malware, and infected apps	If the malware apps using encrypted, not possible to detect malware properly
2017	[88]	Using Markov Chain-based detection technique, to compute the state transitions and to build transition matrix with 6thSense	System resources analysis for process reports and sensors	Google Play	Markov Chain, NB, LMT	LMT	95%	Highly effective and efficient at detecting sensor-based attacks while yielding minimal overhead	Tradeoffs such as frequency accuracy, battery frequency are not discussed which can affect the malware detection accuracy
2017	[89]	Using Dynamic based permission analysis using a run-time and detect malware using ML calculate the accuracy	Code instrumentation analysis Java classes and dynamic permissions	Pysingh, Android Botnet, DroidKin	NB, RF, Simple Logistic, DT, K-Star	Simple Logistic	99.7%	High Accuracy	Need to address the app crashing issue in the selected emulators in dynamic analysis
2019	[90]	Using dynamically tracks execution behaviours of applications and using ServiceMonitor framework	System call analysis	AndroZoo, Drebin and Malware Genome	RF, KNN, SVM	RF	96.7%	High accuracy and high efficiency	Not detecting difference in some system calls of malware and benign apps since signature based verification was not applied
2020	[91]	Extracting the features and permissions from Android app. Performing feature selection and proceed to classification with DATDroid	System call analysis, Code instrumentation for network traffic analysis and System resources analysis	APKPure, Genome	RF, SVM	RF	91.7%	High efficiency	Impact from features like HTTP, DNS, TCP, IP patterns are not considered
2021	[92]	Using decompilation, model discovery, integration and transformation, analysis and transformation, event production	Code instrumentation for java classes, intents	AMD	ML algorithms used in MEGDroid, Monkey, Droidbot	MEGDroid	51.6%	Considerably increases the number of triggered malicious payloads and execution code coverage	System calls are not monitored

5.4. Hybrid Analysis with Machine Learning

Hybrid analysis is the third approach which can be used in ML-based Android malware detection. The review in [93] identified three approaches of malware detection, which are the signature-based, anomaly-based, and topic modelling based approaches. ML algorithms such as DT, J48, RF, KNN, KMeans, and SVM can be applied to all these approaches. Signature-based malware was detected using ML algorithms after the feature extraction process. After the feature extraction, sensitive API calls were also analysed before applying ML algorithms. Documents were collected such as reviews, user documents, and app descriptions before following a similar approach as the signature-based method, initially in the topic modelling approach. It was identified that the behavioural based approach is better than the signature-based approach. If the topic modelling is combined with that approach, it was possible to achieve good results. The hybrid analysis method is created when the dynamic analysis method is integrated with the static analysis method. According to this study, the SVM classifier with the hybrid analysis method performed better than the other ML algorithms.

The model proposed in [94] discussed a methodology of using ML algorithms with static analysis and dynamic analysis. In the static analysis approach, malicious and benign applications' manifest data were taken as JSON files from MalGenome and Kaggale datasets to train the ML model. The trending apps were taken from well-known app stores. Androguard [79] was used to extract information from the APK files. After reverse engineering, decompiling, testing, and training with SVM, LR, KNN based ML models, a JSON file was prepared. According to this model, LR was identified as the most suitable ML algorithm, which has 81.03% accuracy. Many improvements are required to the proposed static analysis model since comparatively this has a low accuracy. However, the proposed dynamic analysis approach outperformed the static analysis approach with high accuracy of 93% of both precision and recall over the RF. In this approach, Droidbox was used to run APKs obtained from MalGenome and Android Wave Lock in a sandbox environment. Then a CSV file is obtained after converting the JSON file obtained by analysing the APK and after that the key features are extracted. As the last step, DT, RF, SVM, KNN, and LR

ML algorithms were used with extracted key features. Then accuracy and results were checked and the particular app was labelled as malware or benign. It would be better if this study explored the possibilities of using other ML algorithms also.

In [95], authors conducted an experiment using various ML technologies to analyse the relative effectiveness of the static and dynamic analysis method towards detecting malware. This study used the Drebin dataset and a custom dataset to train the ML algorithm to classify malware and benign apps. Altogether the whole dataset contains 103 malware and 97 benign apps. For the static analysis, the APK files were reverse-engineered by a tool available in Virustotal and extracted the permissions using a custom XML parser. Then binary feature vectors and permission vectors were created, and ML algorithms were applied. For dynamic analysis, applications were executed on separated Android Virtual Devices (AVDs). System calls and their frequencies were traced using the MonkeyRunner tool since the frequency representation of system calls contained behavioural information on apps. Usually, malware has higher frequencies compared to benign apps. After that, a feature vector of system calls was created, and ML algorithms were applied. The RF, J.48, Naïve Bayes, Simple Logistic, BayesNet Augmented Naïve Bayes (TAN), BayesNet K2, Instance Based Learner (IBk), SMO PolyKernel, and SMO NPolyKernel algorithms were used for both static and dynamic analysis. The best results of 0.96 for static analysis and 0.88 for dynamic analysis were achieved when RF with 100 trees was used. Permissions extracted from the AndroidManifest.xml file were considered for static analysis, and system calls extracted from the runtime were considered in the dynamic analysis.

The model proposed in [96] explained a hybrid analysis process to detect malware using ML algorithms with the accuracy of 80% when using the permissions analysis in static analysis approach and 60% accuracy when analysing by system calls. Malware samples were collected using a honeypot and search repositories such as Androditotal to train the model. However, this study lacks the consideration of other features' which affect malware detection that should also be considered to achieve a high accuracy model.

In [97], the model proposed a hybrid analysis-based efficient mechanism for Android malware detection, which used the malware genome dataset and the Drebin dataset to train the ML and DL models in the static analysis approach. CICMalDroid dataset for the dynamic analysis approach and 261 combined features were extracted for the hybrid analysis. To increase the performance, this model used dimension reduction using Principal Component Analysis (PCA). SVM, KNN, RF, DT, NB, MLP, and GB were used to train and test the model. Out of these ML/DL algorithms, GB outperformed the others in terms of accuracy (96.35%), but it took a comparatively long training time. Forty-six features from dynamic analysis results were also analysed. After performing combined hybrid analysis, GB again performed well with an accuracy of 99.36% and efficiency compared to the Random Forest and MLP. It is better to study the runtime environment and configuration more because this does not cover some areas.

The model described in [98] proposed a Tree TAN based hybrid malware detection mechanism by considering both static and dynamic features such as API calls, permissions, and system calls. LR algorithms were trained for these three features. Drebin, AMD, AZ, Github, and GP datasets were used in this and modelled the output relationships as a TAN to detect if the given app is malicious or benign with an accuracy of 0.97. There is a possibility of some malware remaining undetected from the model, which can be reduced using Reinforcement Learning techniques.

Tables 6 and 7 comparatively summarise the above research studies related to hybrid analysis based methods, where Table 6 listed studies with model accuracy below 90% and Table 7 listed studies with model accuracy above 90%.

Table 6. Hybrid analysis based malware detection approaches (model accuracy is below 90% or overall accuracy is not available).

Year	Study	Detection Approach	Feature Extraction Method	Used Datasets	ML algorithms/ Models	Selected ML algorithms/ Models	Model Accuracy	Strengths	Limitations/Drawbacks
2017	[96]	Using a set of Python and Bash scripts which automated the analysis of the Android data.	Manifest analysis for permissions and System call analysis for dynamic analysis	Andrototal	NB, DT	DT	80%	Model execution is efficient	Consider system call appearance rather than frequency and Lower number of samples used to train
2018	[95]	Using Binary feature vector and permission vector datasets were created using the analysis techniques and was used with the ML algorithms	Manifest analysis for permissions and system call analysis	Drebin	RF, J.48, NB, Simple Logistic, BayesNet TAN, BayesNet K2, SMO PolyKernel, IBK, SMO NPolyKernel	RF	Static-96%, Dynamic-88%	Compared with several ML algorithms	Accuracy depends on the 3rd party tool (Monkey runner) used to collect features.
2019	[94]	Preparing a JSON file after reverse engineering, decompiling, and analysing the APK by running in a sandbox environment and then extracting the key features and applied ML	Manifest analysis for permissions, code analysis for API calls and System call analysis	Ma Genome, Kaggle, Androguard [79]	SVM, LR, KNN, RF	LR for static analysis and RF for dynamic analysis	Static-81.03%, Dynamic-93%	Dynamic analysis performed was better than the static analysis approach in terms of detection accuracy	Did not perform a proper hybrid analysis approach to increase the overall accuracy

Table 7. Hybrid analysis based malware detection approaches (model accuracy is above 90%).

Year	Study	Detection Approach	Feature Extraction Method	Used Datasets	ML Algorithms/ Models	Selected ML Algorithms/ Models	Model Accuracy	Strengths	Limitations/Drawbacks
2017	[99]	Using import term extraction, clustering and applying genetic algorithm with MOCODroid	Code analysis for API calls and information flow and system call analysis	Virus-total, Google Play	Genetic algorithm, Multiobjective evolutionary algorithm	Multiobjective evolutionary classifier	95.15%	Possible to avoid the effects of the concealment strategies	Did not consider about other clustering methods.
2020	[97]	Extracted 261 combined features of the hybrid analysis with using the support of datasets and performed the ML/DL models	Manifest analysis for permissions and system call analysis	MalGenome, Drebin, CICMalDroid	SVM, KNN, RF, DT, NB, MLP, GB	GB	99.36%	Hybrid analysis is having higher accuracy comparing to static analysis and dynamic analysis individually	Runtime environment and configuration is not considered
2020	[98]	Using Conditional dependencies among relevant static and dynamic features. Then trained ridge regularised LR classifiers and modelled their output relationships as a TAN	Manifest analysis for permissions, code analysis for API calls and system call analysis	Drebin, AMD, AZ, Github, GP	TAN	TAN	97%	Highly accurate	Possibility of some malwares remain undetected
2021	[100]	Using exploit static, dynamic, and visual features of apps to predict the malicious apps using information fusion and applied Case Based Reasoning (CBR)	Manifest analysis for permissions and System call analysis	Drebin	CBR, SVM, DT	CBR	95%	Require limited memory and processing capabilities	Require to present the knowledge representation to address some limitations

5.5. Use of Deep Learning Based Methods

It is possible to use deep learning techniques also for detecting Android malware. In MLDroid, a web-based Android malware detection framework [101] was proposed by performing dynamic analysis. In this work, ML and DL methods were used with an overall 98.8% malware detection rate.

The model proposed in [102] disused a method to detect malware using a semantic-based DL approach and implemented a tool called DeepRefiner. This approach used the Long Short Term Memory (LSTM) on the semantic structure of Android bytecode with two layers of detection and validation. This method used the LSTM over Recurrent Neural Network (RNN) since RNN contains gradient vanish problem. Using this approach with an accuracy of 97.4% and a false positive rate of 2.54%, it was possible to detect malware. It was efficient and accurate compared with the traditional approaches. Since this approach uses the static analysis approach, some limitations can arise based on the runtime environment, which can be identified if this model uses the hybrid analysis approach.

MOCDroid [99] model discussed a multiobjective evolutionary classifier to detect malware in Android. It combined multiobjective optimisation with clustering to generate a classifier using third-party call group behaviours. This method produced an accuracy of 95.15%. Import term extraction, clustering, and applying a genetic algorithm were the three steps included in this process. Initially, the DEX files were uncompressed from the APK after using the decompression tool, and Java codes were obtained using the JADX tool [103]. Then the document term matrix was transformed. As the next step, K-Means clustering was applied since it was identified as the highest accuracy model for this, and the genetic algorithm was also applied. The results were compared with a random set of 10,000 benign and malicious apps with different antivirus engines. It is possible to consider other clustering methods to improve the accuracy of this method.

The work proposed in [104] discussed a method to detect Android malware using a deep convolutional neural network (CNN). Raw opcode sequence from disassembled Smali program was analysed using static analysers to classify the malware. The advantage of this method is automatically learning the feature indicative of malware. This work was inspired by n-gram based methods. To train the models Android Malware Genome project dataset [49] and Intel Security/MacAfee Lab dataset were used. The classification system of this provides 0.87 precision and recall accuracies. The accuracy of the malware detection can be increased when the dynamic analysis is also performed.

A deep learning-based static analysis approach was experimented with an accuracy of 99.9% and with an F1-score of 0.996 in [105]. This approach used a dataset of over 1.8 million Android apps. The attributes of malware were detected through vectorised opcode extracted from the bytecode of the APKs with one-hot encoding. After performing experiments on Recurrent Neural Networks, Long Short Term Memory Networks, Neural Networks, Deep Convents, and Diabolo Network models, it was identified that Bidirectional Long Short-Term Memory (BiLSTMs) is the best model for this approach. It is better to analyse the complete byte code using static analysis and check the app behaviour with dynamic analysis to build a more comprehensive malware detection tool based on deep learning techniques.

The DL-Droid framework based on deep learning techniques [106] proposed a new way of detecting Android malware with dynamic analysis techniques. This approach was having a detection rate of 97.8% by only including dynamic features. When the static features were also included in that, the detection rate would increase to 99.6%. The experiments were performed on real devices in which the application can run exactly the way the user experiences it. Further to this, some comparisons of detection performance and code coverage were also included in this work. Traditional ML classifier performances were also compared. This novel method outperformed the ML-based methods such as NB, SL, SVM, J48, Pruning Rule-Based Classification Tree (PART), RF, and DL. In addition to this work, seeking the possibilities to include intrusion detection mechanism in the DL-Droid would be a valuable addition.

The AdMat model proposed in [107] discussed a CNN on Matrix-based approach to detect Android malware. This model characterised apps and treated them as images. Then the adjacency matrix was constructed for apps, and it was simplified with the size of 219 × 219 to enhance the efficiency in data processing after transferring decompiled source code into call-graph of Graph Modelling Language (GML) format. Those matrices were the input images to the CNN, and the model was trained to identify and classify malware and benign apps. This model has an accuracy of 98.2%. Even though the model is highly accurate, there are limitations to this work, such as performing static analysis only, and the performance depends on the number of used features.

The model proposed in [108] discussed a DL-based method that uses CNN approach to analyse API sequence call, opcode, and permissions to detect Android malware in a zero-day scenario. The model achieved a weighted average detection rate of 91% and 81% on two datasets Drebin and AMD after the model was trained. The model can further improve if the dynamic analysis techniques are also considered.

With an accuracy of 95%, a multimodal analysis of malware apps using information fusion was presented in [100] which used hybrid analysis techniques. The study used CBR for training and validation purposes. SVM and DT were compared with the proposed model validation, but the classic ML algorithms were outperformed by the CBR-based method. If the work can represent the knowledge representation, some of the limitations can be addressed.

Tables 8 and 9 comparatively summarise the above research studies related to deep learning based malware detection methods, where Table 8 listed studies with model accuracy below 90% and Table 9 listed studies with model accuracy above 90%.

Table 8. Deep learning based Malware Detection Approaches (Model Accuracy is below 90% or overall accuracy is not available).

Year	Study	Detection Approach	Feature Extraction Method	Used Datasets	ML/DL Algorithms/Models	Selected DL Algorithms/Models	Model Accuracy	Strengths	Limitations/Drawbacks
2017	[104]	Using n-Gram methods after getting the Opcode sequence from .smali after dissembling .apk	Code Analysis for Opcodes	Genome, IntelSecurity, Macafee, Google Play	CNN, NLP	Deep CNN	87%	Automatically learn the feature indicative of malware without hand engineering	Assumption of all APKs are benign in Google Play dataset while all are malicious in malware dataset
2021	[108]	Using DL based method which uses Convolution Neural Network based approach to analyse features	Code Analysis for API calls, Opcode and Manifest Analysis for Permission	Drebin, AMD	CNN	CNN	91% and 81% on two datasets	Reduce over fitting and possible to train to detect new malware just by collecting more sample apps	Did not compared with other ML/DL methods

Table 9. Deep learning based malware detection approaches (model accuracy is above 90%).

Year	Study	Detection Approach	Feature Extraction Method	Used Datasets	ML/DL Algorithms/Models	Selected DL Algorithms/Models	Model Accuracy	Strengths	Limitations/Drawbacks
2018	[102]	Applying LSTM on semantic structure of bytecode with 2 layers of detection and validating with DeepRefiner	Code Analysis for Opcode/bytecode	Google Play, VirusShare, MassVet	RNN, LSTM	LSTM	97.4%	High efficiency with average of 0.22 s to the 1st layer and 2.42 s to the 2nd layer detection	Need to train the model regularly to update the training model on new malware
2020	[105]	Detecting Malware attributes by vectorised opcode extracted from the bytecode of the APKs with one-hot encoding before apply DL Techniques	Code Analysis for Opcode	Drebin, AMD, VirusShare	BiLSTM, RNN, LSTM, Neural Networks, Deep Convnets, Diabolo Network model	BiLSTMs	99.9%	Very high accuracy, Able to achieve zero day malware family without overhead of previous training	Did not analyse complete byte code
2020	[106]	Using DynaLog to select and extract features from Log files and using DL-Droid to perform feature ranking and apply DL	Code instrumentation analysis for java classes, intents, and systems calls	Intel Security	NB, SL, SVM, J48, PART, RF, DL	DL	99.6%	Experiments were performed on real devices High accuracy	Could have implemented the intrusion detection part also to make it more comprehensive malware detection tool
2021	[101]	Selecting features gained by feature selection approaches. Applying ML/DL models to detect malware	Code instrumentation for java classes, permissions and API calls at the runtime	Android Permissions Dataset, Computer and security dataset	farthest first clustering, Y-MLP, nonlinear ensemble decision tree forest, DL	DL with methods in MLDroid	95.8%	High accuracy and easy to retrain the model to identify new malware	Human interaction would be required in some cases. Can contain issues in the datasets
2021	[107]	Characterising apps and treating as images. Then constructing the adjacency matrix. Then applying CNN to identify malware with AdMat framework	Code Analysis for API calls, Information flow, and Opcode	Drebin, AMD	CNN	CNN	98.2%	High Accuracy and efficiency	Performance is depending on number of used features

6. Machine Learning Methods to Detect Code Vulnerabilities

Hackers do not just create malware. They also try to find loopholes in existing applications and perform malicious activities. Therefore, it is necessary to find vulnerabilities in Android source code. A code vulnerability of a program can happen due to a mistake at the designing, development, or configuration time which can be misused to infringe on the security [38]. Detection of code vulnerability can be performed in two ways. The first method is reverse-engineering the APK files using a similar approach discussed in Section 3. The second method is identifying the security flaws at the time of designing and developing the application [109]. The study conducted in [110] has identified five main categories of security approaches. They were secure requirements modelling, extended Unified Modeling Language (UML) based secure modelling profiles, non-UML-based secure modelling notations, vulnerability identification, adaption and mitigation, and software security-focused process. Under these categories, 52 security approaches were identified. All these approaches are used to identify software vulnerabilities at the time of designing and developing the applications. Based on the findings of the surveys and interviews conducted in [111] related to intervention for long-term software security, the importance of having an automated code analysis tool to identify vulnerabilities of the written codes has been identified. The empirical analysis conducted in [112] identified the static software metrics' correlation and the most informative metrics which can be used to find code vulnerability related to Android source codes.

6.1. Static, Dynamic, and Hybrid Source Code Analysis

Similar to analysing APKs for malware detection, there are three ways of analysing source codes. They are static analysis, dynamic analysis, and hybrid analysis. In static analysis, without executing the source code, a program is analysed to identify properties by converting the source to a generalised abstraction such as Abstract Syntax Tree (AST) [113]. The number of reported false vulnerabilities depends on the accuracy of the generalisation mechanism. The runtime behaviour of the application is monitored while using specific input parameters in dynamic analysis. The behaviour depends on the selection of input parameters. However, there are possibilities of undetected vulnerabilities [114].

In hybrid analysis, it provides the characteristics of both static analysis and dynamic analysis, which can analyse the source code and run the application to identify vulnerabilities while employing detection techniques [115].

The study conducted in [116] performed an online experiment where Android developers were the participants. Vulnerable code samples containing hard-coded credentials, encryptions, Structured Query Language (SQL) injections, and logging with sensitive data were given to the participants together with the guidance of static analysis tools and asked to indicate the appropriate fix. After analysing the experiment results, it has been identified that automated code vulnerability detection support is required for the developers to perform better when developing secure applications.

To analyse Android source code, Android Linters can be applied. Linters have been proposed to detect and fix these bad practices and they perform a static analysis based on AST or Universal AST (UAST) generation through written source codes [117]. The study in [118] discussed several Linters such as PMD, CheckStyle, Infer, and FindBugs, Detekt, Ktlint, and Android Lint discussed the usage of them. Android studio adopts the Android Lint, which identifies 339 issues related to correctness, security, performance, usability, accessibility, and internationalisation. In the proposed model in FixDroid [27], security-oriented suggestions along with their fixes were provided to the developer once the Android Lint identified security flaws. The FixDroid method can further be improved by employing ML techniques to produce highly accurate security suggestions.

However, just warning the developer about security issues in the code is not sufficient. There should be a mechanism to inform the developer about the severity level of the security issue also. By using app user reviews, OASSIS [119] proposed a method to prioritise static analysis warnings generated from Android Lint. Based on the review analysis using sentiment analysis, it was possible to identify the issues in Android apps. After receiving prioritised lint warnings, developers will able to take prompt actions. The study in [120] proposed a mechanism named as MagpieBridge to integrate static analysis into Integrated Development Environments (IDEs) and code editors such as Eclipse, IntelliJ,

Jupyter, Sublime Text, and PyCharm. However, the possibility of extending this to the Android platform should be discussed further.

In [121], using static and dynamic analysis, a vulnerability identification of Secure Sockets Layer (SSL)/Transport Layer Security (TLS) certificate verification in Android application was described. This experiment found that out of the analysed 2213 Android apps, 360 apps contain vulnerable codes using the proposed framework of DCDroid. Therefore, through SSL/TLS certificates, it is possible to identify some vulnerabilities.

6.2. Applying ML to Detect Source Code Vulnerabilities

It has been proven that ML methods can be applied on a generalised architecture such as AST to detect Android code vulnerabilities [38]. Most of the research was conducted using static analysis techniques to analyse the source code.

With the use of ML, vulnerability detection rules were extracted with static metrics as discussed in [122]. Thirty-two supervised ML algorithms were considered for most common vulnerabilities and identified that when the model used the J48 ML algorithm, 96% accuracy could be obtained in vulnerability detection. The model proposed in [123] discussed an automated mechanism to classify well-written and malicious code using a portable executable (PE) structure through static analysis and ML with an accuracy of 98.77%. The proposed methodology used RF, GB, DT, and CNN as ML models.

The study in [124] built a model to predict software vulnerabilities of codes using ML before releasing the code. After developing a source code representation using AST and intelligently analysing it, the ML models were applied. Popular datasets such as NIST SAMATE, Draper VDISC, and SATE IV Juliet Test Suite, which contain C, C++, Java, and Python source codes, were used to train the model. However, using this model, it was not possible to locate a specific place of vulnerability. It is identified as a drawback, and it has not proven that the same approach is possible to apply to other programming languages and frameworks. However, there is a possibility of using this approach for Android applications, which were developed using Java.

In [125], using C and C++ source codes, a vulnerability detection system was proposed using ML and deep feature representation learning. Apart from using the existing datasets, the Drapper dataset was compiled using Drebin and Github repositories with millions of open-source functions and labelled with carefully selected findings. The findings of the research were compared with Bag of Words (BOW), RF, RNN, and CNN models.

The study conducted in [126] developed a mechanism to classify subroutines as vulnerable or not vulnerable in C language using ML methods. The National Vulnerability Dataset (NVD) was used to collect C programming code blocks and their known vulnerabilities. After preparing the AST and preprocessing the data, feature extraction, feature selection, and classification tasks were performed and ML algorithms were applied.

The applicability of deep learning to detect code vulnerabilities was discussed in [127]. Comparison of using three DL algorithms CNN, LSTM, and CNN-LSTM were discussed in this study. The proposed model has an accuracy of 83.6% when applying the DL models. Using Deep Neural Networks, it was possible to predict vulnerable code components. The model in [128] evaluated it using some Java-based Android applications. In this mechanism, N-gram analysis and statistical feature selection for constructing features were performed. This model can classify vulnerable classes with high precision, accuracy, and recall.

In [129], a model was proposed to detect zero-day Android malware using a distinctive parallel classifier and a mechanism to identify oncoming highly elusive vulnerabilities in the source code with an accuracy of 98.27% with the use of Ml algorithms; PART, Ripple Down Rule Learner (RIDOR), SVM, and MLP.

ML-Based Vulnerability Detection Specifically for Android

There is less research conducted relating to Android vulnerability detection with ML. The methodology of the studies, which were conducted on general programming languages, could apply to the Android code vulnerability detection after training the model using specific code datasets and adjusting the generalisation mechanism.

The work conducted in [130] prepared a manually curated dataset that can be used to fix vulnerabilities of open-source software. The possibility of automatically identifying security-related commits in the relevant code repository has been proven since it has been successfully used to train classifiers.

In [131] repository of Android security vulnerabilities was created named AndroVul, which includes dangerous permissions, security code smells, and high-risk shell command vulnerabilities. In [132], a study was conducted to predicatively analyse the vulnerabilities in Internet of Things (IoT) related Android applications using statistical codes and applying ML. In this study, 1406 Android apps were taken with various risk levels, and six ML models (KNN, LR, RF, DT, SVM, and GB) were administered to examine security risk prediction. It is identified that RF performs well in the intermediate risk level. GB performs well at a very high-risk level compared to the other ML model-based approaches. The study conducted in [133] proposed an ML-based vulnerabilities detection mechanism to identify security flaws of Android Intents using hybrid analysis. Adaboost algorithm was used to perform the ML based analysis.

Tables 10 and 11 summarise selected studies from above which are related to Android vulnerability analysis. Table 10 lists the studies which have model accuracy below 90% and Table 11 lists the studies which have model accuracy above 90%.

Table 10. Android vulnerability detection mechanisms (Model accuracy is below 90%).

Year	Study	Code Analysis Method	Approach	Used ML/DL Methods/ Frameworks	Accuracy of the Model
2017	[127]	Dynamic Analysis	Collected 9872 sequences of function calls as features. Performed dynamic analysis with DL methods	CNN-LSTM	83.6%
2017	[133]	Hybrid Analysis	Decompiled the apk file. Performed static analysis of the manifest file to obtain the components/permissions. Dynamic analysis and fuzzy testing were conducted and obtained system status.	AB and DT	77%
2019	[115]	Hybrid Analysis	Reverse engineered the APK, Decoded the manifest files & codes and extracted meta data from it. Performed dynamic analysis to identify intent crashing and insecure network connections for API calls. Generated the report.	AndroShield	84%
2020	[124]	Hybrid Analysis	Performed intelligent analysis of generated AST. Checked ML can differentiate vulnerable and nonvulnerable.	MLP and a customised model	70.1%

Table 11. Android vulnerability detection mechanisms (model accuracy is above 90%).

Year	Study	Code Analysis Method	Approach	Used ML/DL Methods/ Frameworks	Accuracy of the Model
2017	[113]	Static Analysis	Generated the AST, navigated it, and computed detection rules. Identified smells when training with manually created dataset.	ADOCTOR framework	98%
2017	[128]	Static Analysis	Combined N-gram analysis and statistical feature selection for constructing features. Evaluated the performance of the proposed technique based on a number of Java Android programs.	Deep Neural Network	92.87%
2019	[129]	Hybrid Analysis	Decompiled the APK and selected the features and executed the APK and generated log files with system calls. Generated the vector space and trained with ML algorithms as parallel classifiers.	MLP, SVM, PART, RIDOR, MaxProb, ProdProb	98.37%
2020	[121]	Hybrid Analysis	In static analysis, vulnerabilities of SSL/TLS certification were identified. Results from static analysis about user interfaces were analysed to confirm SSL/TLS misuse in dynamic analysis.	DCDroid	99.39%
2021	[122]	Static Analysis	32 supervised ML algorithms were considered for 3 common vulnerabilities: Lawofdemeter, BeanMemberShouldSerialize, and LocalVariablecouldBeFinal	J48	96%
2021	[123]	Static Analysis	Classified malicious code using a PE structure and a method for classifying it using a PE structure	CNN	98.77%

7. Results and Discussion

Based on the reviewed studies in ML/DL based methods to detect malware, it is identified that 65% of studies related to malware detection techniques used static analysis, 15% used dynamic analysis, and the remaining 20% followed the hybrid analysis technique. This is illustrated in Figure 3. This high attractiveness of static analysis may be due to the various advantages associated with it over dynamic analysis, such as ability to detect more vulnerabilities, localising vulnerabilities, and offering cost benefits.

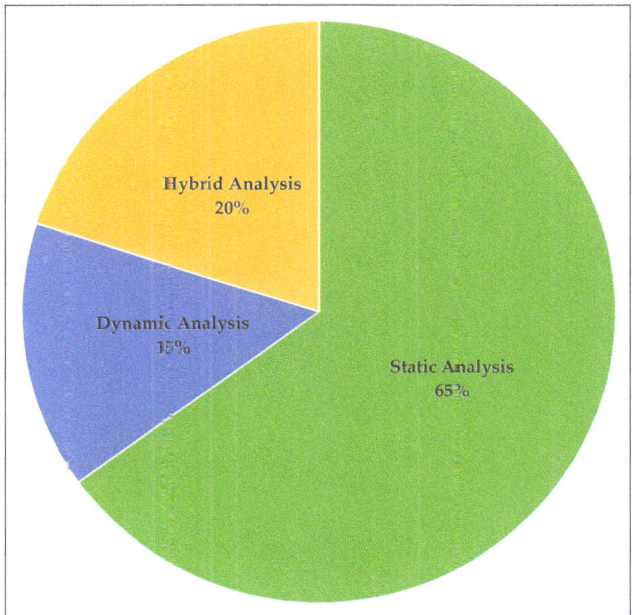

Figure 3. Malware analysis techniques used in the reviewed studies.

Many ML/DL based malware detection studies used the code analysis method as the feature extraction method. Apart from that, manifest analysis and system call analysis methods are the other widely used methods. Figure 4 illustrates those feature extraction methods used in the reviewed studies. It is possible to detect a substantial amount of malware after analysing decompiled source codes rather than analysing permissions or other features. That may be the reason for the high usage of code analysis in malware detection.

By using the feature extraction methods, permissions, API calls, system calls, and opcodes are the most widely extracted features. This is illustrated in Figure 5 along with the other extracted features in the reviewed studies. Many hybrid analysis methods extracted permissions as the feature to perform static analysis. It is easy to analyse permissions when comparing with the other features too. These could be reasons for the high usage of permissions as the extracted feature. Services and network protocols have low usage in feature extractions. The reason for this may be it is comparatively not easy to analyse those features.

The datasets used in ML/DL based Android malware detection studies to train the algorithms are illustrated in Figure 6. Drebin was the most widely used dataset in Android Malware Detection, and it was used in 18 reviewed studies. Google Play, MalGenome, and AMD datasets are the other widely used datasets. The reason for the highest usage of the Drebin dataset may be because it provides a comprehensive labelled dataset. Since Google Play is the official app store of Android, it may be a reason to have high usage for the dataset from Google.

It is identified that the RF, SVM, and NB are at the top of widely studied ML models to detect Android malware. The reason may be that the resource cost to run RF, SVM, or NB based models is low. Models like CNN, LSTM, and AB have less usage because to run such advanced models, good computing power is required, and the trend for DL-based models was also boosted in recent years. Table 12 summarises widely used ML/DL algorithms with their advantages and disadvantages. Figure 7 illustrates all of the studied ML/DL models with their usage in the reviewed studies.

The majority of the studies used hybrid analysis and static analysis as the source code analysis techniques in vulnerability detection in Android, as illustrated in Figure 8. To perform a highly accurate vulnerability analysis, the source code should be analysed and executed too. Therefore, this may be the reason to have hybrid analysis and static analysis as the widely used source code analysis methods to detect vulnerabilities.

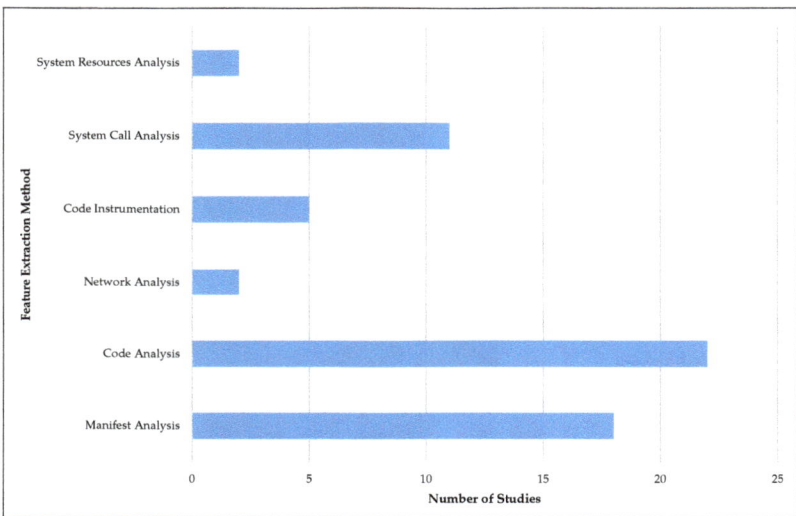

Figure 4. Feature extraction methods used in the reviewed studies.

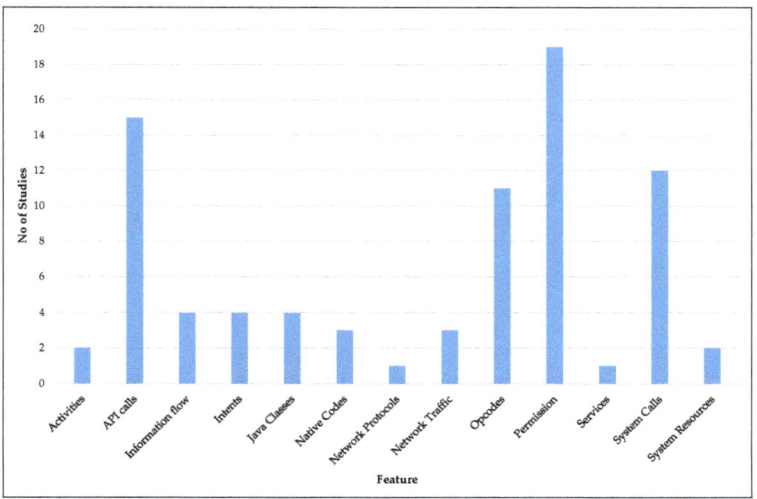

Figure 5. Extracted features in the reviewed studies.

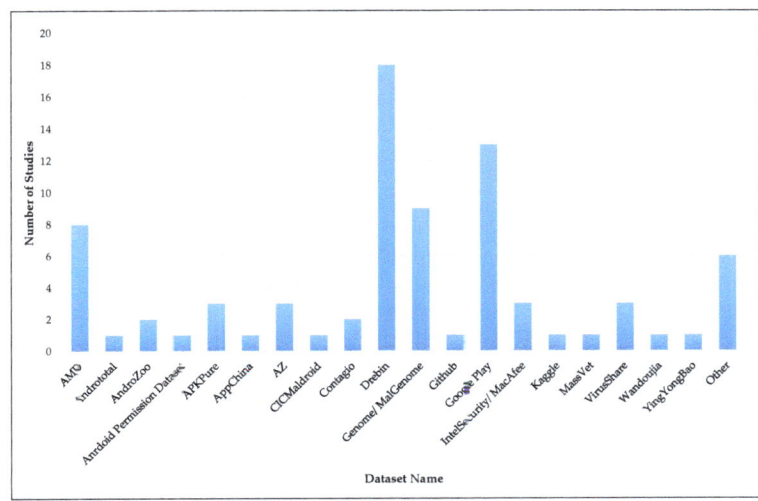

Figure 6. Usage of datasets.

Table 12. Commonly used ML/DL algorithms for Android malware detection.

Algorithm	Advantages	Disadvantages
DT	• Possible handle samples with missing values • Easy to understand	• Might cause the overfitting problem
NB	• Easily and quickly trainable	• Need to calculate prior probability • Not applicable if the feature variables are corelated
Regression Models	• Widely used in statistics based studies • Direct and Fast	• Not possible to deal well with high dimensional features
KNN	• Suitable to solve multiclassification problems	• Computation overhead is relatively high • Issues with the skewness of data
SVM	• Possible to solve high dimensional nonlinear small scale problems	• High overhead in data processing • Might face some issues when there are missing values in the sample
K-Means	• Easy to implement • Fast and simple	• Sensitive to outliers
RF	• Reduces overfitting • Normalising of data is not required	• Requires much time to train • Requires high computational power
Neural Networks	• Highly accurate • Strong fault tolerance	• Requires much time to train • Require a large number of data to train the model
LSTM	• Capable to remember facts for lengthy interval	• Requires high computational resources
CNN	• Reduce unimportant parameters by weight sharing and downsampling	• High computational cost
Ensemble Learning	• Accuracy is high	• Overhead on model training and maintenance

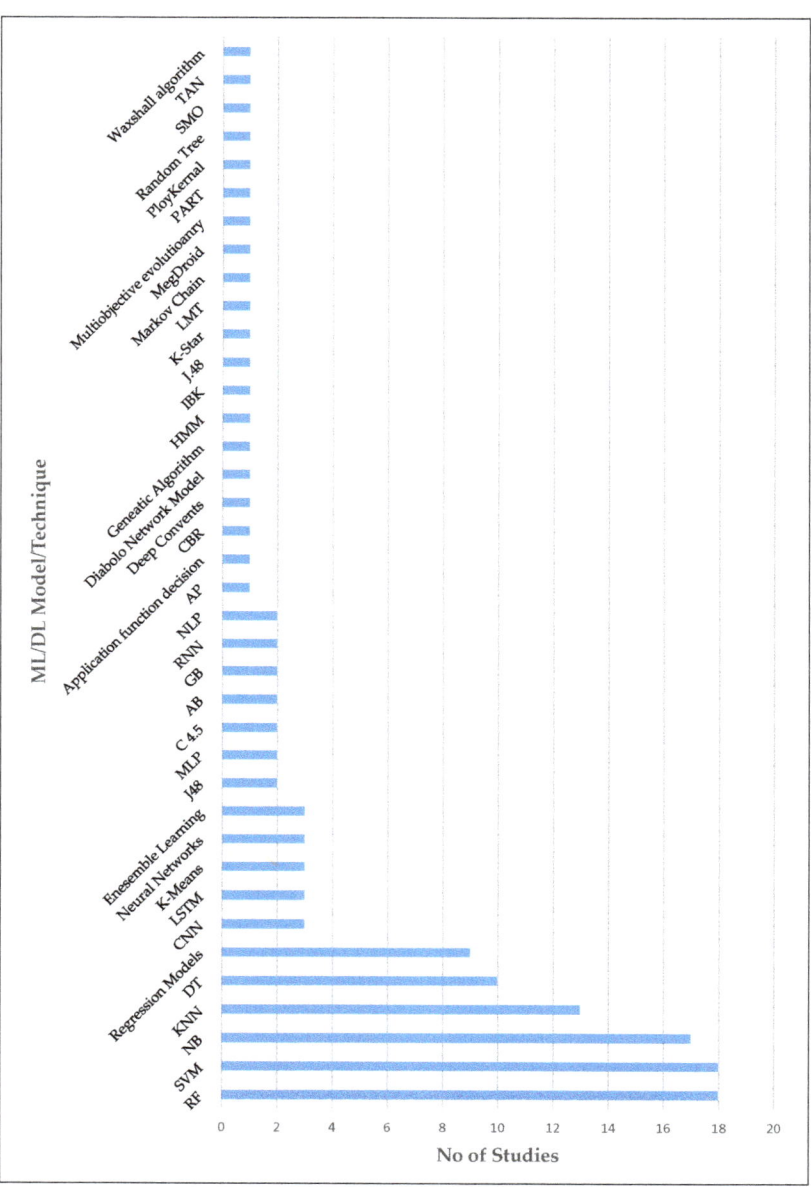

Figure 7. ML/DL models used in the reviewed studies.

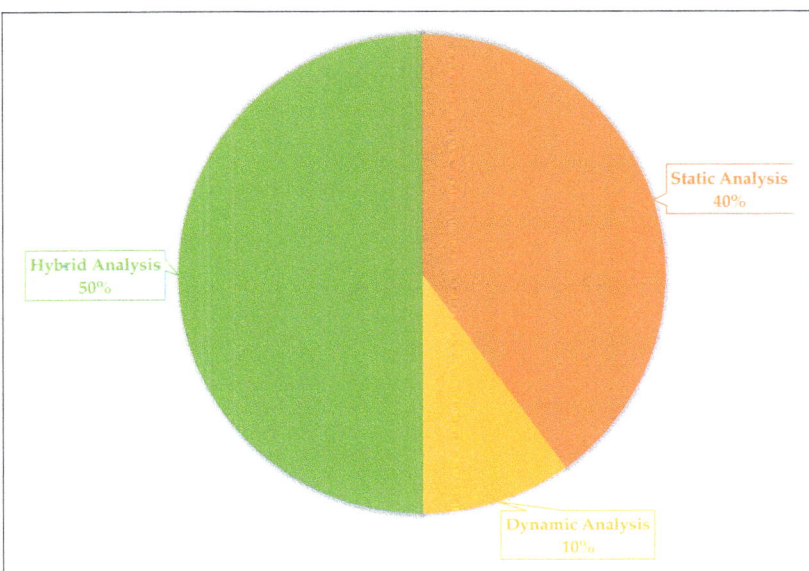

Figure 8. Android source code vulnerability analysis methods.

8. Conclusions and Future Work

Any smartphone is potentially vulnerable to security breaches, but Android devices are more lucrative for attackers. This is due to its open-source nature and the larger market share compared to other operating systems for mobile devices. This paper discussed the Android architecture and its security model, as well as potential threat vectors for the Android operating system. Based upon the available literature, a systematic review of the state-of-the-art ML-based Android malware detection techniques was carried out, covering the latest research from 2016 to 2021. It discussed the available ML and DL models and their performance in Android malware detection, code and APK analysis methods, feature analysis and extraction methods, and strengths and limitations of the proposed methods Malware aside, if a developer makes a mistake, it is easier for a hacker to find and exploit these vulnerabilities. Therefore, methods for the detection of source code vulnerabilities using ML were discussed. The work identified the potential gaps in previous research and possible future research directions to enhance the security of Android OS.

Both Android malware and its detection techniques are evolving. Therefore, we believe that similar future reviews are necessary to cover these emerging threats and their detection methods. As per our findings in this paper, since DL methods have proven to be more accurate than traditional ML models, it will be beneficial to the research community if more comprehensive systematic reviews can be performed by focusing only on DL-based malware detection on Android. The possibility of using reinforcement learning to identify source code vulnerabilities is another area of interest in which systematic reviews and studies can be carried out.

Author Contributions: Conceptualization, J.S., H.K. and M.O.A.-K.; methodology, J.S., H.K. and M.O.A.-K.; validation, J.S., H.K. and M.O.A.-K.; investigation, J.S.; Project administration, H.K.; writing—original draft preparation, J.S.; writing—review and editing, J.S., H.K. and M.O.A.-K.; visualization, J.S.; supervision, H.K. and M.O.A.-K.; All authors have read and agreed to the published version of the manuscript.

Funding: This research received no external funding.

Acknowledgments: We thank the Accelerating Higher Education Expansion and Development (AHEAD) grant of Sri Lanka, University of Kelaniya—Sri Lanka and Robert Gordon University—United Kingdom for their support.

Conflicts of Interest: The authors declare no conflict of interest.

References

1. Number of Mobile Phone Users Worldwide from 2016 to 2023 (In Billions). Available online: https://www.statista.com/statistics/330695/number-of-smartphone-users-worldwide/ (accessed on 19 May 2021).
2. Mobile Operating System Market Share Worldwide. Available online: https://gs.statcounter.com/os-market-share/mobile/worldwide/ (accessed on 19 May 2021).
3. Number of Android Applications on the Google Play Store. Available online: https://www.appbrain.com/stats/number-of-android-apps/ (accessed on 19 May 2021).
4. Gibert, D.; Mateu, C.; Planes, J. The rise of machine learning for detection and classification of malware: Research developments, trends and challenges. *J. Netw. Comput. Appl.* **2020**, *153*, 102526. [CrossRef]
5. Khan, J.; Shahzad, S. Android Architecture and Related Security Risks. *Asian J. Technol. Manag. Res. [ISSN: 2249–0892]* **2015**, *5*, 14–18. Available online: http://www.ajtmr.com/papers/Vol5Issue2/Vol5Iss2_P4.pdf (accessed on 19 May 2021).
6. Platform Architecture. Available online: https://developer.android.com/guide/platform (accessed on 19 May 2021).
7. Android Runtime (ART) and Dalvik. Available online: https://source.android.com/devices/tech/dalvik (accessed on 19 May 2021).
8. Cai, H.; Ryder, B.G. Understanding Android application programming and security: A dynamic study. In Proceedings of the 2017 IEEE International Conference on Software Maintenance and Evolution (ICSME), Shanghai, China, 17–22 September 2017; pp. 364–375. [CrossRef]
9. Liu, K.; Xu, S.; Xu, G.; Zhang, M.; Sun, D.; Liu, H. A Review of Android Malware Detection Approaches Based on Machine Learning. *IEEE Access* **2020**, *8*, 124579–124607. [CrossRef]
10. Gilski, P.; Stefanski, J. Android os: A review. *Tem J.* **2015**, *4*, 116. Available online: https://www.temjournal.com/content/41/14/temjournal4114.pdf (accessed on 19 May 2021).
11. Privacy in Android 11 | Android Developers. Available online: https://developer.android.com/about/versions/11/privacy (accessed on 19 May 2021).
12. Garg, S.; Baliyan, N. Comparative analysis of Android and iOS from security viewpoint. *Comput. Sci. Rev.* **2021**, *40*, 100372. [CrossRef]
13. Odusami, M.; Abayomi-Alli, O.; Misra, S.; Shobayo, O.; Damasevicius, R.; Maskeliunas, R. Android malware detection: A survey. In *International Conference on Applied Informatics*; Springer: Cham, Switzerland, 2018; pp. 255–266. [CrossRef]
14. Bhat, P.; Dutta, K. A survey on various threats and current state of security in android platform. *ACM Comput. Surv. (CSUR)* **2019**, *52*, 1–35. [CrossRef]
15. Tam, K.; Feizollah, A.; Anuar, N.B.; Salleh, R.; Cavallaro, L. The evolution of android malware and android analysis techniques. *ACM Comput. Surv. (CSUR)* **2017**, *49*, 1–41. [CrossRef]
16. Li, L.; Li, D.; Bissyandé, T.F.; Klein, J.; Le Traon, Y.; Lo, D.; Cavallaro, L. Understanding android app piggybacking: A systematic study of malicious code grafting. *IEEE Trans. Inf. Forensics Secur.* **2017**, *12*, 1269–1284. [CrossRef]
17. Ashawa, M.A.; Morris, S. Analysis of Android malware detection techniques: A systematic review. *Int. J. Cyber-Secur. Digit. Forensics* **2019**, *8*, 177–187. [CrossRef]
18. Suarez-Tangil, G.; Tapiador, J.E.; Peris-Lopez, P.; Ribagorda, A. Evolution, detection and analysis of malware for smart devices. *IEEE Commun. Surv. Tutor.* **2013**, *16*, 961–987. [CrossRef]
19. Mos, A.; Chowdhury, M.M. Mobile Security: A Look into Android. In Proceedings of the 2020 IEEE International Conference on Electro Information Technology (EIT), Chicago, IL, USA, 31 July–1 August 2020; pp. 638–642. [CrossRef]
20. Faruki, P.; Bharmal, A.; Laxmi, V.; Ganmoor, V.; Gaur, M.S.; Conti, M.; Rajarajan, M. Android security: A survey of issues, malware penetration, and defenses. *IEEE Commun. Surv. Tutor.* **2014**, *17*, 998–1022. [CrossRef]
21. Android Security & Privacy 2018 Year in Review. Available online: https://source.android.com/security/reports/Google_Android_Security_2018_Report_Final.pdf (accessed on 19 May 2021).
22. Kalutarage, H.K.; Nguyen, H.N.; Shaikh, S.A. Towards a threat assessment framework for apps collusion. *Telecommun. Syst.* **2017**, *66*, 417–430. [CrossRef]
23. Asavoae, I.M.; Blasco, J.; Chen, T.M.; Kalutarage, H.K.; Muttik, I.; Nguyen, H.N.; Roggenbach, M.; Shaikh, S.A. Towards automated android app collusion detection. *arXiv* **2016**, arXiv:1603.02308.
24. Asăvoae, I.M.; Blasco, J.; Chen, T.M.; Kalutarage, H.K.; Muttik, I.; Nguyen, H.N.; Roggenbach, M.; Shaikh, S.A. Detecting malicious collusion between mobile software applications: The Android case. In *Data Analytics and Decision Support for Cybersecurity*; Springer: Cham, Switzerland, 2017; pp. 55–97. [CrossRef]
25. Malik, J. Making sense of human threats and errors. *Comput. Fraud Secur.* **2020**, *2020*, 6–10. [CrossRef]
26. Calciati, P.; Kuznetsov, K.; Gorla, A.; Zeller, A. Automatically Granted Permissions in Android apps: An Empirical Study on their Prevalence and on the Potential Threats for Privacy. In Proceedings of the 17th International Conference on Mining Software Repositories, Seoul, Korea, 29–30 June 2020; pp. 114–124. [CrossRef]

27. Nguyen, D.C.; Wermke, D.; Acar, Y.; Backes, M.; Weir, C.; Fahl, S. A stitch in time: Supporting android developers in writing secure code. In Proceedings of the 2017 ACM SIGSAC Conference on Computer and Communications Security, Dallas, TX, USA, 30 October–3 November 2017; pp. 1065–1077. [CrossRef]
28. Garg, S.; Baliyan, N. Android Security Assessment: A Review, Taxonomy and Research Gap Study. *Comput. Secur.* **2020**, *100*, 102087. [CrossRef]
29. Van Engelen, J.E.; Hoos, H.H. A survey on semi-supervised learning. *Mach. Learn.* **2020**, *109*, 373–440. [CrossRef]
30. Alauthman, M.; Aslam, N.; Al-Kasassbeh, M.; Khan, S.; Al-Qerem, A.; Choo, K.K.R. An efficient reinforcement learning-based Botnet detection approach. *J. Netw. Comput. Appl.* **2020**, *150*, 102479. [CrossRef]
31. Shrestha, A.; Mahmood, A. Review of deep learning algorithms and architectures. *IEEE Access* **2019**, *7*, 53040–53065. [CrossRef]
32. Page, M.; McKenzie, J.; Bossuyt, P.; Boutron, I.; Hoffmann, T.; Mulrow, C.; Shamseer, L.; Tetzlaff, J.M.; Akl, E.A.; Brennan, S.E.; et al. The PRISMA 2020 statement: An updated guideline for reporting systematic reviews. *BMJ* **2020**, *372*. [CrossRef]
33. Wohlin, C. Guidelines for snowballing in systematic literature studies and a replication in software engineering. In Proceedings of the 18th International Conference on Evaluation and Assessment in Software Engineering, London, UK, 13–14 May 2014; pp. 1–10. [CrossRef]
34. Li, L.; Bissyandé, T.F.; Papadakis, M.; Rasthofer, S.; Bartel, A.; Octeau, D.; Klein, J.; Traon, L. Static analysis of android apps: A systematic literature review. *Inf. Softw. Technol.* **2017**, *88*, 67–95. [CrossRef]
35. Pan, Y.; Ge, X.; Fang, C.; Fan, Y. A Systematic Literature Review of Android Malware Detection Using Static Analysis. *IEEE Access* **2020**, *8*, 116363–116379. [CrossRef]
36. Sharma, T.; Rattan, D. Malicious application detection in android—A systematic literature review. *Comput. Sci. Rev.* **2021**, *40*, 100373. [CrossRef]
37. Liu, Y.; Tantithamthavorn, C.; Li, L.; Liu, Y. Deep Learning for Android Malware Defenses: A Systematic Literature Review. *arXiv* **2021**, arXiv:2103.05292.
38. Ghaffarian, S.M.; Shahriari, H.R. Software vulnerability analysis and discovery using machine-learning and data-mining techniques: A survey. *ACM Comput. Surv. (CSUR)* **2017**, *50*, 1–36. [CrossRef]
39. Chen, T.; Mao, Q.; Yang, Y.; Lv, M.; Zhu, J. TinyDroid: A lightweight and efficient model for Android malware detection and classification. *Mob. Inf. Syst.* **2018**, *2018*. [CrossRef]
40. Nisa, M.; Shah, J.H.; Kanwal, S.; Raza, M.; Khan, M.A.; Damaševičius, R.; Blažauskas, T. Hybrid malware classification method using segmentation-based fractal texture analysis and deep convolution neural network features. *Appl. Sci.* **2020**, *10*, 4966. [CrossRef]
41. Amin, M.; Shah, B.; Sharif, A.; Ali, T.; Kim, K.I.; Anwar, S. Android malware detection through generative adversarial networks. *Trans. Emerg. Telecommun. Technol.* **2019**, e3675. [CrossRef]
42. Arp, D.; Spreitzenbarth, M.; Hubner, M.; Gascon, H.; Rieck, K.; Siemens, C. Drebin: Effective and explainable detection of android malware in your pocket. In Proceedings of the 2014 Network and Distributed System Security Symposium, San Diego, CA, USA, 23–26 February 2014; doi:10.14722/ndss.2014.23247 [CrossRef]
43. Google Play. Available online: https://play.google.com/ (accessed on 19 May 2021).
44. AndroZoo. Available online: https://androzoo.uni.lu/ (accessed on 19 May 2021).
45. AppChina. Available online: https://tracxn.com/d/companies/appchina.com (accessed on 19 May 2021).
46. Tencent. Available online: https://www.pcmgr-global.com/ (accessed on 19 May 2021).
47. YingYongBao. Available online: https://android.myapp.com/ (accessed on 19 May 2021).
48. Contagio. Available online: https://www.impactcybertrust.org/dataset_view?idDataset=1273/ (accessed on 19 May 2021).
49. Zhou, Y.; Jiang, X. Dissecting android malware: Characterization and evolution. In Proceedings of the 2012 IEEE Symposium on Security and Privacy, San Francisco, CA, USA, 20–23 May 2012; pp. 95–109. [CrossRef]
50. VirusShare. Available online: https://virusshare.com/ (accessed on 19 May 2021).
51. Intel Security/MacAfee. Available online: https://steppa.ca/portfolio-view/malware-threat-intel-datasets/ (accessed on 19 May 2021).
52. Chen, K.; Wang, P.; Lee, Y.; Wang, X.; Zhang, N.; Huang, H.; Zou, W.; Liu, P. Finding unknown malice in 10 s: Mass vetting for new threats at the google-play scale. In Proceedings of the 24th USENIX Security Symposium (USENIX Security 15), Redmond, WA, USA, 7–8 May 2015; pp. 659–674
53. Android Malware Dataset. Available online: http://amd.arguslab.org/ (accessed on 19 May 2021).
54. APKPure. Available online: https://m.apkpure.com/ (accessed on 19 May 2021).
55. Anrcoid Permission Dataset. Available online: https://data.mendeley.com/datasets/b4mxg7ydb7/3 (accessed on 19 May 2021).
56. Maggi, F.; Valdi, A.; Zanero, S. Andrototal: A flexible, scalable toolbox and service for testing mobile malware detectors. In Proceedings of the Third ACM Workshop on Security and Privacy in Smartphones & Mobile Devices, Berlin, Germany, 8 November 2013; pp. 49–54. [CrossRef]
57. Wandoujia App Market. Available online: https://www.wandoujia.com/apps (accessed on 19 May 2021).
58. Google Playstore Appsin Kaggle. Available online: https://www.kaggle.com/gauthamp10/google-playstore-apps (accessed on 19 May 2021).
59. CICMaldroid Dataset. Available online: https://www.unb.ca/cic/datasets/maldroid-2020.html (accessed on 19 May 2021).
60. AZ Dataset. Available online: https://www.azsecure-data.org/other-data.html/ (accessed on 19 May 2021).

61. Github Malware Dataset. Available online: https://github.com/topics/malware-dataset (accessed on 19 May 2021).
62. Alqahtani, E.J.; Zagrouba, R.; Almuhaideb, A. A Survey on Android Malware Detection Techniques Using Machine Learning Algorithms. In Proceedings of the 2019 Sixth International Conference on Software Defined Systems (SDS), Rome, Italy, 10–13 June 2019; pp. 110–117. [CrossRef]
63. Lopes, J.; Serrão, C.; Nunes, L.; Almeida, A.; Oliveira, J. Overview of machine learning methods for Android malware identification. In Proceedings of the 2019 7th International Symposium on Digital Forensics and Security (ISDFS), Barcelos, Portugal, 10–12 June 2019; pp. 1–6. [CrossRef]
64. Choudhary, M.; Kishore, B. HAAMD: Hybrid analysis for Android malware detection. In Proceedings of the 2018 International Conference on Computer Communication and Informatics (ICCCI), Coimbatore, India, 4–6 January 2018; pp. 1–4. [CrossRef]
65. Kouliaridis, V.; Kambourakis, G. A Comprehensive Survey on Machine Learning Techniques for Android Malware Detection. *Information* **2021**, *12*, 185. [CrossRef]
66. Chen, L.; Hou, S.; Ye, Y.; Chen, L. An adversarial machine learning model against android malware evasion attacks. In *Asia-Pacific Web (APWeb) and Web-Age Information Management (WAIM) Joint Conference on Web and Big Data*; Springer: Cham, Switzerland, 2017; pp. 43–55. [CrossRef]
67. Lubuva, H.; Huang, Q.; Msonde, G.C. A review of static malware detection for Android apps permission based on deep learning. *Int. J. Comput. Netw. Appl.* **2019**, *6*, 80–91. [CrossRef]
68. Li, J.; Sun, L.; Yan, Q.; Li, Z.; Srisa-An, W.; Ye, H. Significant permission identification for machine-learning-based android malware detection. *IEEE Trans. Ind. Inform.* **2018**, *14*, 3216–3225. [CrossRef]
69. Mcdonald, J.; Herron, N.; Glisson, W.; Benton, R. Machine Learning-Based Android Malware Detection Using Manifest Permissions. In Proceedings of the 54th Hawaii International Conference on System Sciences, Maui, HI, USA, 5–8 January 2021; p. 6976. [CrossRef]
70. Şahin, D.Ö.; Kural, O.E.; Akleylek, S.; Kılıç, E. A novel permission-based Android malware detection system using feature selection based on linear regression. *Neural Comput. Appl.* **2021**, 1–16. [CrossRef]
71. Nawaz, A. Feature Engineering based on Hybrid Features for Malware Detection over Android Framework. *Turk. J. Comput. Math. Educ. (TURCOMAT)* **2021**, *12*, 2856–2864.
72. Cai, L.; Li, Y.; Xiong, Z. JOWMDroid: Android malware detection based on feature weighting with joint optimization of weight-mapping and classifier parameters. *Comput. Secur.* **2021**, *100*, 102086. [CrossRef]
73. Zhang, P.; Cheng, S.; Lou, S.; Jiang, F. A novel Android malware detection approach using operand sequences. In Proceedings of the 2018 Third International Conference on Security of Smart Cities, Industrial Control System and Communications (SSIC), Shanghai, China, 18–19 October 2018; pp. 1–5. [CrossRef]
74. Wei, L.; Luo, W.; Weng, J.; Zhong, Y.; Zhang, X.; Yan, Z. Machine learning-based malicious application detection of android. *IEEE Access* **2017**, *5*, 25591–25601. [CrossRef]
75. Onwuzurike, L.; Mariconti, E.; Andriotis, P.; Cristofaro, E.D.; Ross, G.; Stringhini, G. MaMaDroid: Detecting Android malware by building Markov chains of behavioral models (extended version). *ACM Trans. Priv. Secur. (TOPS)* **2019**, *22*, 1–34. [CrossRef]
76. Zhang, H.; Luo, S.; Zhang, Y.; Pan, L. An efficient Android malware detection system based on method-level behavioral semantic analysis. *IEEE Access* **2019**, *7*, 69246–69256. [CrossRef]
77. Meng, G.; Xue, Y.; Xu, Z.; Liu, Y.; Zhang, J.; Narayanan, A. Semantic modelling of android malware for effective malware comprehension, detection, and classification. In Proceedings of the 25th International Symposium on Software Testing and Analysis, Saarbrücken, Germany, 18–20 July 2016; pp. 306–317. [CrossRef]
78. Wang, Z.; Li, C.; Yuan, Z.; Guan, Y.; Xue, Y. DroidChain: A novel Android malware detection method based on behavior chains. *Pervasive Mob. Comput.* **2016**, *32*, 3–14. [CrossRef]
79. Androguard. Available online: https://pypi.org/project/androguard/ (accessed on 19 May 2021).
80. Damodaran, A.; Di Troia, F.; Visaggio, C.A.; Austin, T.H.; Stamp, M. A comparison of static, dynamic, and hybrid analysis for malware detection. *J. Comput. Virol. Hacking Tech.* **2017**, *13*, 1–12. [CrossRef]
81. Sun, Y.; Xie, Y.; Qiu, Z.; Pan, Y.; Weng, J.; Guo, S. Detecting Android malware based on extreme learning machine. In Proceedings of the 2017 IEEE 15th International Conference on Dependable, Autonomic and Secure Computing, 15th International Conference on Pervasive Intelligence and Computing, 3rd International Conference on Big Data Intelligence and Computing and Cyber Science and Technology Congress (DASC/PiCom/DataCom/CyberSciTech), Orlando, FL, USA, 6–10 November 2017; pp. 47–53. [CrossRef]
82. Tian, K.; Yao, D.; Ryder, B.G.; Tan, G.; Peng, G. Detection of repackaged android malware with code-heterogeneity features. *IEEE Trans. Dependable Secur. Comput.* **2017**, *17*, 64–77. [CrossRef]
83. Kabakus, A.T. What static analysis can utmost offer for Android malware detection. *Inf. Technol. Control* **2019**, *48*, 235–249. [CrossRef]
84. Koli, J. RanDroid: Android malware detection using random machine learning classifiers. In Proceedings of the 2018 Technologies for Smart-City Energy Security and Power (ICSESP), Bhubaneswar, India, 28–30 March 2018; pp. 1–6. [CrossRef]
85. Lou, S.; Cheng, S.; Huang, J.; Jiang, F. TFDroid: Android malware detection by topics and sensitive data flows using machine learning techniques. In Proceedings of the 2019 IEEE 2nd International Conference on Information and Computer Technologies (ICICT), Kahului, HI, USA, 14–17 March 2019; pp. 30–36. [CrossRef]

86. Wang, W.; Gao, Z.; Zhao, M.; Li, Y.; Liu, J.; Zhang, X. DroidEnsemble: Detecting Android malicious applications with ensemble of string and structural static features. *IEEE Access* **2018**, *6*, 31798–31807. [CrossRef]
87. Garg, S.; Peddoju, S.K.; Sarje, A.K. Network-based detection of Android malicious apps. *Int. J. Inf. Secur.* **2017**, *16*, 385–400. [CrossRef]
88. Sikder, A.K.; Aksu, H.; Uluagac, A.S. 6thsense: A context-aware sensor-based attack detector for smart devices. In Proceedings of the 26th USENIX Security Symposium (USENIX Security 17), Vancouver, BC, Canada, 16–18 August 2017; pp. 397–414. [CrossRef]
89. Mahindru, A.; Singh, P. Dynamic permissions based android malware detection using machine learning techniques. In Proceedings of the 10th Innovations in Software Engineering Conference, Jaipur, India, 5–7 February 2017; pp. 202–210. [CrossRef]
90. Salehi, M.; Amini, M.; Crispo, B. Detecting malicious applications using system services request behavior. In Proceedings of the 16th EAI International Conference on Mobile and Ubiquitous Systems: Computing, Networking and Services, Houston, TX, USA, 12–14 November 2019; pp. 200–209. [CrossRef]
91. Thangavelooa, R.; Jinga, W.W.; Lenga, C.K.; Abdullaha, J. DATDroid: Dynamic Analysis Technique in Android Malware Detection. *Int. J. Adv. Sci. Eng. Inf. Technol.* **2020**, *10*, 536–541. [CrossRef]
92. Hasan, H.; Ladani, B.T.; Zamani, B. MEGDroid: A model-driven event generation framework for dynamic android malware analysis. *Inf. Softw. Technol.* **2021**, *135*, 106569. [CrossRef]
93. Raphael, R.; Mathiyalagan, P. An Exploration of Changes Addressed in the Android Malware Detection Walkways. In Proceedings of the International Conference on Computational Intelligence, Cyber Security, and Computational Models, Coimbatore, India, 19–21 December 2019; Springer: Singapore, 2019; pp. 61–84. [CrossRef]
94. Jannat, U.S.; Hasnayeen, S.M.; Shuhan, M.K.B. Ferdous, M.S. Analysis and detection of malware in Android applications using machine learning. In Proceedings of the 2019 International Conference on Electrical, Computer and Communication Engineering (ECCE), Cox'sBazar, Bangladesh, 7–9 February 2019; pp. 1–7. [CrossRef]
95. Kapratwar, A.; Di Troia, F.; Stamp, M. *Static and Dynamic Analysis of Android Malware*; ICISSP: Porto, Portugal, 2017; pp. 653–662. [CrossRef]
96. Leeds, M.; Keffeler, M.; Atkison, T. A comparison of features for android malware detection. In Proceedings of the SouthEast Conference, Kennesaw, GA, USA, 13–15 April 2017; pp. 63–68. [CrossRef]
97. Hadiprakoso, R.B.; Kabetta, H.; Buana, I.K.S. Hybrid-Based Malware Analysis for Effective and Efficiency Android Malware Detection. In Proceedings of the 2020 International Conference on Informatics Multimedia, Cyber and Information System (ICIMCIS), Jakarta, Indonesia, 19–20 November 2020; pp. 8–12. [CrossRef]
98. Surendran, R.; Thomas, T.; Emmanuel, S. A TAN based hybrid model for android malware detection. *J. Inf. Secur. Appl.* **2020**, *54*, 102483. [CrossRef]
99. Martín, A.; Menéndez, H.D.; Camacho, D. MOCDroid: Multi-objective evolutionary classifier for Android malware detection. *Soft Comput.* **2017**, *21*, 7405–7415. 10.1007/s00500-016-2283-y. [CrossRef]
100. Qaisar, Z.H.; Li, R. Multimodal information fusion for android malware detection using lazy learning. *Multimed. Tools Appl.* **2021** 1–15. [CrossRef]
101. Mahindru, A.; Sangal, A. MLDroid—Framework for Android malware detection using machine learning techniques. *Neural Comput. Appl.* **2021**, *33*, 5183–5240. [CrossRef]
102. Xu, K.; Li, Y.; Deng, R.H.; Chen, K. Deeprefiner: Multi-layer android malware detection system applying deep neural networks. In Proceedings of the 2018 IEEE European Symposium on Security and Privacy (EuroS&P), London, UK, 24–26 April 2018; pp. 473–487. [CrossRef]
103. JADX. Available online: https://github.com/skylot/jadx/ (accessed on 19 May 2021).
104. McLaughlin, N.; Martinez del Rincon, J.; Kang, B.; Yerima, S.; Miller, P.; Sezer, S.; Safaei, Y.; Trickel, E.; Zhao, Z.; Doupé, A.; et al. Deep android malware detection. In Proceedings of the Seventh ACM on Conference on Data and Application Security and Privacy, Scottsdale, AZ, USA, 22–24 March 2017; pp. 301–308. [CrossRef]
105. Amin, M.; Tanveer, T.A.; Tehseen, M.; Khan, M.; Khan, F.A.; Anwar, S. Static malware detection and attribution in android byte-code through an end-to-end deep system. *Future Gener. Comput. Syst.* **2020**, *102*, 112–126. [CrossRef]
106. Alzaylaee, M.K.; Yerima, S.Y.; Sezer, S. DL-Droid: Deep learning based android malware detection using real devices. *Comput. Secur.* **2020**, *89*, 101663. [CrossRef]
107. Vu, L.N.; Jung, S. AdMat: A CNN-on-Matrix Approach to Android Malware Detection and Classification. *IEEE Access* **2021**, *9*, 39680–39694. [CrossRef]
108. Millar, S.; McLaughlin, N.; del Rincon, J.M.; Miller, P. Multi-view deep learning for zero-day Android malware detection. *J. Inf. Secur. Appl.* **2021**, *58*, 102718. [CrossRef]
109. Acar, Y.; Stransky, C.; Wermke, D.; Weir, C.; Mazurek, M.L.; Fahl, S. Developers need support, too: A survey of security advice for software developers. In Proceedings of the 2017 IEEE Cybersecurity Development (SecDev), Cambridge, MA, USA, 24–26 September 2017; pp. 22–26. [CrossRef]
110. Mohammed, N.M.; Niazi, M.; Alshayeb, M.; Mahmood, S. Exploring software security approaches in software development lifecycle: A systematic mapping study. *Comput. Stand. Interfaces* **2017**, *50*, 107–115. [CrossRef]
111. Weir, C.; Becker, I.; Noble, J.; Blair, L.; Sasse, M.A.; Rashid, A. Interventions for long-term software security: Creating a lightweight program of assurance techniques for developers. *Softw. Pract. Exp.* **2020**, *50*, 275–298. [CrossRef]

112. Alenezi, M.; Almomani, I. Empirical analysis of static code metrics for predicting risk scores in android applications. In Proceedings of the 5th International Symposium on Data Mining Applications, Cham, Switzerland, 29 March 2018; Springer: Cham, Switzerland, 2018; pp. 84–94. [CrossRef]
113. Palomba, F.; Di Nucci, D.; Panichella, A.; Zaidman, A.; De Lucia, A. Lightweight detection of android-specific code smells: The adoctor project. In Proceedings of the 2017 IEEE 24th International Conference on Software Analysis, Evolution and Reengineering (SANER), Klagenfurt, Austria, 20–24 February 2017; pp. 487–491. [CrossRef]
114. Pustogarov, I.; Wu, Q.; Lie, D. Ex-vivo dynamic analysis framework for Android device drivers. In Proceedings of the 2020 IEEE Symposium on Security and Privacy (SP), San Francisco, CA, USA, 18–21 May 2020; pp. 1088–1105. [CrossRef]
115. Amin, A.; Eldessouki, A.; Magdy, M.T.; Abdeen, N.; Hindy, H.; Hegazy, I. AndroShield: Automated android applications vulnerability detection, a hybrid static and dynamic analysis approach. *Information* **2019**, *10*, 326. [CrossRef]
116. Tahaei, M.; Vaniea, K.; Beznosov, K.; Wolters, M.K. Security Notifications in Static Analysis Tools: Developers' Attitudes, Comprehension, and Ability to Act on Them. In Proceedings of the 2021 CHI Conference on Human Factors in Computing Systems, Yokohama, Japan, 8–13 May 2021; pp. 1–17. [CrossRef]
117. Goaër, O.L. Enforcing green code with Android lint. In Proceedings of the 35th IEEE/ACM International Conference on Automated Software Engineering Workshops, Melbourne, VIC, Australia, 21–25 September 2020; pp. 85–90. [CrossRef]
118. Habchi, S.; Blanc, X.; Rouvoy, R. On adopting linters to deal with performance concerns in android apps. In Proceedings of the 2018 33rd IEEE/ACM International Conference on Automated Software Engineering (ASE), Montpellier, France, 3–7 September 2018; pp. 6–16. [CrossRef]
119. Wei, L.; Liu, Y.; Cheung, S.C. OASIS: Prioritizing static analysis warnings for Android apps based on app user reviews. In Proceedings of the 2017 11th Joint Meeting on Foundations of Software Engineering, Paderborn, Germany, 4–8 September 2017; pp. 672–682. [CrossRef]
120. Luo, L.; Dolby, J.; Bodden, E. MagpieBridge: A General Approach to Integrating Static Analyses into IDEs and Editors (Tool Insights Paper). In Proceedings of the 33rd European Conference on Object-Oriented Programming (ECOOP 2019). Schloss Dagstuhl-Leibniz-Zentrum fuer Informatik, Dagstuhl, Germany, 15–19 July 2019. [CrossRef]
121. Wang, Y.; Xu, G.; Liu, X.; Mao, W.; Si, C.; Pedrycz, W.; Wang, W. Identifying vulnerabilities of SSL/TLS certificate verification in Android apps with static and dynamic analysis. *J. Syst. Softw.* **2020**, *167*, 110609. [CrossRef]
122. Gupta, A.; Suri, B.; Kumar, V.; Jain, P. Extracting rules for vulnerabilities detection with static metrics using machine learning. *Int. J. Syst. Assur. Eng. Manag.* **2021**, *12*, 65–76. [CrossRef]
123. Kim, S.; Yeom, S.; Oh, H.; Shin, D.; Shin, D. Automatic Malicious Code Classification System through Static Analysis Using Machine Learning. *Symmetry* **2021**, *13*, 35. [CrossRef]
124. Bilgin, Z.; Ersoy, M.A.; Soykan, E.U.; Tomur, E.; Çomak, P.; Karaçay, L. Vulnerability Prediction From Source Code Using Machine Learning. *IEEE Access* **2020**, *8*, 150672–150684. [CrossRef]
125. Russell, R.; Kim, L.; Hamilton, L.; Lazovich, T.; Harer, J.; Ozdemir, O.; Ellingwood, P.; McConley, M. Automated vulnerability detection in source code using deep representation learning. In Proceedings of the 2018 17th IEEE International Conference on Machine Learning and Applications (ICMLA), Orlando, FL, USA, 17–20 December 2018; pp. 757–762. [CrossRef]
126. Chernis, B.; Verma, R. Machine learning methods for software vulnerability detection. In Proceedings of the Fourth ACM International Workshop on Security and Privacy Analytics, Tempe, AZ, USA, 21 March 2018; pp. 31–39. [CrossRef]
127. Wu, F.; Wang, J.; Liu, J.; Wang, W. Vulnerability detection with deep learning. In Proceedings of the 2017 3rd IEEE International Conference on Computer and Communications (ICCC), Chengdu, China, 13–16 December 2017; pp. 1298–1302. [CrossRef]
128. Pang, Y.; Xue, X.; Wang, H. Predicting vulnerable software components through deep neural network. In Proceedings of the 2017 International Conference on Deep Learning Technologies, Chengdu, China, 2–4 June 2017; pp. 6–10. [CrossRef]
129. Garg, S.; Baliyan, N. A novel parallel classifier scheme for vulnerability detection in android. *Comput. Electr. Eng.* **2019**, *77*, 12–26. [CrossRef]
130. Ponta, S.E.; Plate, H.; Sabetta, A.; Bezzi, M.; Dangremont, C. A manually-curated dataset of fixes to vulnerabilities of open-source software. In Proceedings of the 2019 IEEE/ACM 16th International Conference on Mining Software Repositories (MSR), Montreal, QC, Canada, 26–27 May 2019; pp. 383–387. [CrossRef]
131. Namrud, Z.; Kpodjedo, S.; Talhi, C. AndroVul: A repository for Android security vulnerabilities. In Proceedings of the 29th Annual International Conference on Computer Science and Software Engineering, Toronto, ON, Canada, 4–6 November 2019; pp. 64–71.
132. Cui, J.; Wang, L.; Zhao, X.; Zhang, H. Towards predictive analysis of android vulnerability using statistical codes and machine learning for IoT applications. *Comput. Commun.* **2020**, *155*, 125–131. [CrossRef]
133. Zhuo, L.; Zhimin, G.; Cen, C. Research on Android intent security detection based on machine learning. In Proceedings of the 2017 4th International Conference on Information Science and Control Engineering (ICISCE), Changsha, China, 21–23 July 2017; pp. 569–574. [CrossRef]

Review

Business Email Compromise Phishing Detection Based on Machine Learning: A Systematic Literature Review

Hany F. Atlam [1,2,*] and Olayonu Oluwatimilehin [1]

1. School of Computing and Engineering, University of Derby, Derby DE22 3AW, UK
2. Computer Science and Engineering Department, Faculty of Electronic Engineering, Menoufia University, Menouf 32952, Egypt
* Correspondence: h.atlam@derby.ac.uk

Abstract: The risk of cyberattacks against businesses has risen considerably, with Business Email Compromise (BEC) schemes taking the lead as one of the most common phishing attack methods. The daily evolution of this assault mechanism's attack methods has shown a very high level of proficiency against organisations. Since the majority of BEC emails lack a payloader, they have become challenging for organisations to identify or detect using typical spam filtering and static feature extraction techniques. Hence, an efficient and effective BEC phishing detection approach is required to provide an effective solution to various organisations to protect against such attacks. This paper provides a systematic review and examination of the state of the art of BEC phishing detection techniques to provide a detailed understanding of the topic to allow researchers to identify the main principles of BEC phishing detection, the common Machine Learning (ML) algorithms used, the features used to detect BEC phishing, and the common datasets used. Based on the selected search strategy, 38 articles (of 950 articles) were chosen for closer examination. Out of these articles, the contributions of the selected articles were discussed and summarised to highlight their contributions as well as their limitations. In addition, the features of BEC phishing used for detection were provided, as well as the ML algorithms and datasets that were used in BEC phishing detection models were discussed. In the end, open issues and future research directions of BEC phishing detection based on ML were discussed.

Keywords: business email compromise (BEC); email phishing; phishing detection; machine learning (ML); systematic literature review

Citation: Atlam, H.F.;
Oluwatimilehin, O. Business Email
Compromise Phishing Detection
Based on Machine Learning: A
Systematic Literature Review.
Electronics 2023, *12*, 42.
https://doi.org/10.3390/
electronics12010042

Academic Editor: Suleiman Yerima

Received: 20 November 2022
Revised: 15 December 2022
Accepted: 19 December 2022
Published: 22 December 2022

Copyright: © 2022 by the authors.
Licensee MDPI, Basel, Switzerland.
This article is an open access article
distributed under the terms and
conditions of the Creative Commons
Attribution (CC BY) license (https://
creativecommons.org/licenses/by/
4.0/).

1. Introduction

The popularity of Internet-based public resources, such as cloud computing, social networks and online money processing, has significantly raised the danger of cyberattacks against enterprises. Since email has become one of the effective worldwide standards for commercial communication, cybercriminals attack email networks to undertake cyberattacks against companies for financial gain [1]. A Business Email Compromise (BEC) attack, often known as a CEO attack, is one of the most significant spear phishing attacks. BEC attacks are defined as sophisticated email phishing schemes that target businesses doing mundane tasks, such as money transfers [1]. Social engineering has shown to be a highly effective component of BEC attacks, which are designed to deceive corporations and their employees throughout the world. According to the Federal Bureau of Investigation (FBI) [2], victims worldwide lost more than USD 26 billion to BEC attacks between June 2016 and July 2019. In 2018, almost AUD 60 million was reported lost in Australia using this strategy. In addition, the United States (39%), the United Kingdom (26%), Australia (11%), Belgium and Germany (3%), Canada, the Netherlands, Hong Kong, Singapore, and Japan (2%), were the top 10 victim nations for BEC attacks in 2018–2019.

The reason why cyberattacks are becoming increasingly prevalent is that launching a cyberattack is simpler, cheaper, and less dangerous than launching a physical attack. The only requirements for committing a cybercrime are an Internet connection and a computer. In addition, the anonymity given by the Internet makes it difficult to trace and find attackers and bring them to justice [3].

BEC attacks are prevalent and have not been detected by conventional defence strategies, such as spam filters. Without a harmful payload, BEC attacks are difficult to detect with conventional screening equipment. BEC attacks are gaining popularity due to their effectiveness and difficulties in monitoring or detecting them [1]. Unlike other attacks using banking trojans or other forms of criminal ransomware, which may require a higher level of technical skill to execute, BEC attacks do not require an exceptionally high level of technical skill to execute; other than having the first name, last name, and email address of whoever they wish to address the email to, they do not need much analysis [2]. Hence, more investigation on BEC attacks is required to identify possible solutions for them.

BEC is a relatively new and fast-evolving attack in the phishing domain with less than ten years since its first identification in 2013 by the FBI. The novelty of this type of attack has led to several challenges regarding how much of its attack pattern and structure has been fully understood by experts to build an effective phishing detection model. In addition, how to ensure the resources needed to identify a BEC attack and the measure used to detect it do not become outdated due to the fast-changing pattern of this type of attack is important. These challenges make detecting BEC attacks using conventional defence strategies a very difficult task to achieve. To overcome these challenges, Machine Learning (ML) has been proposed by various researchers as an effective way to detect BEC attacks in a timely manner. Instead of using conventional phishing detection techniques that detect and block emails based on their origin, as well as applying common block listed locations which require significant time and effort to maintain, ML-based phishing detection techniques can identify and even predict advanced attacks by analysing large datasets to spot similarities, correlations, and trends. For instance, ML can be used to build a phishing detection model based on profiles where ML can be used to build a profile by analysing emails using features such as date, time, geo-location from where a person is accessing emails, relation graph which captures with whom the person interacts, etc. Then, the ML-based model will scan every incoming email against the profile and raise an alert for BEC in case of any deviation. ML-based techniques leveraged by modern email security platforms have become more effective, in which most techniques can detect around 98% of advanced phishing attacks [4].

This paper aims to provide a comprehensive systematic literature review that investigates and evaluates the state of the art of BEC phishing attacks, one of the primary attack domains that has a significant impact on organisations and has resulted in the loss of billions every year. Based on the selected search strategy, 38 articles out of 950 were chosen for further analysis. Out of the collected and analysed articles, articles were selected based on the manner of detection using ML algorithms, and additional assessment was obtained from the articles to comprehend what feature criteria were used for detection. In addition, a summary of the selected papers' contributions was provided. Compared to other surveys, to the best of the authors' knowledge, this is the first work to provide a systematic literature review of BEC phishing attacks. Most existing surveys focus on providing a general investigation and discussion of phishing attacks without focusing on BEC attacks and how creating effective BEC phishing detection models is now a necessity for various organisations around the world. This paper also provides a detailed discussion of BEC phishing attacks to allow researchers to have a complete overview of this type of attack, its detection methods, features, and challenges, which can allow them to develop optimised and sustainable techniques for detecting it effectively.

The contribution of this paper can be summarised as follows:

- Investigating and reviewing recent research on BEC detection by highlighting the merits of each study.
- Identifying common ML algorithms for the development of BEC phishing detection models.

- Determining common features used in BEC phishing detection models.
- Identifying common datasets used in BEC phishing detection models.
- Presenting challenges and future research directions of BEC phishing attacks.

The rest of the paper is organised as follows. Section 2 presents an overview of BEC attacks; Section 3 describes the research methodology used to produce the systematic literature review; Section 4 describes the analysis of data; Section 5 describes how this systematic review answers suggested research questions; Section 6 presents challenges and future research directions; and Section 7 is the conclusion.

2. An Overview of BEC Attack

Phishing is a type of email-based fraud and attack. Phishing happens when an attacker sends a bogus email that seems to originate from a reputable and approved source. The objective of the message is to deceive users into downloading malware on their devices and divulging sensitive information. Spear phishing is a targeted kind of phishing. Phishing and spear phishing both utilise email to target victims, but spear phishing delivers a personalised message to a particular individual. Before sending the email, the criminal searches the interests of the intended victim. It is important to understand that phishing emails nowadays are mostly used to acquire credentials [3].

BEC is one of the most significant spear-phishing attacks. This section provides an overview of this type of BEC attack to highlight the BEC lifecycle, types, and techniques that are used for detecting it.

2.1. BEC Attack

BEC is a form of attack that has evolved over the years from a simply compromised vendor email to requests for sensitive information, such as by targeting the real estate sector, and fraudulent requests for large amounts of gift cards. For a BEC attack to be successful, hackers first need to gain access to legitimate vendor email accounts. The most common method for accomplishing this is via phishing emails sent to the company's staff. The credentials of a worker who unknowingly lets themselves be compromised are a springboard for an attack [3].

The FBI created the term "business email compromise", or in short BEC, in 2013 when it first began tracking this issue. However, the strategy might be regarded as the natural progression of huge spamming campaigns that came before it. These promotions originated with what is now commonly referred to as Nigerian prince or lottery schemes. These email frauds were noticeable for their lack of professionalism—misspellings, grammatical errors, and implausible tales—and were easy to recognise and disregard. However, the offenders swiftly acquired technological expertise and today deploy some extremely sophisticated approaches [4].

2.2. BEC Lifecycle

BEC attacks are usually harder to spot than other phishing attacks, as they can play out in various ways. Figure 1 shows common steps for performing a successful BEC attack [4].

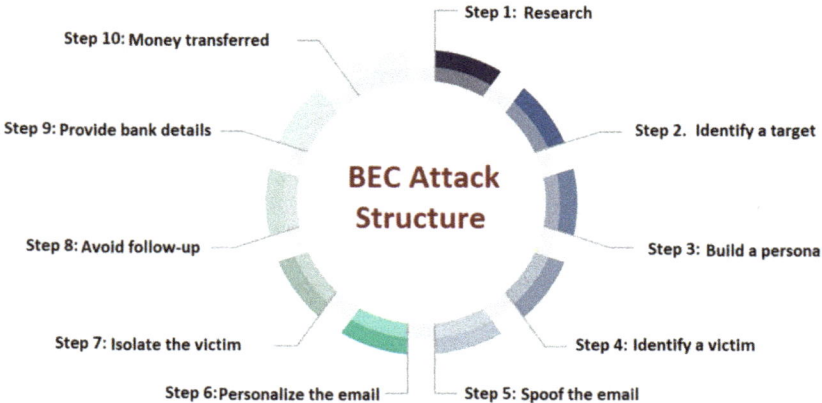

Figure 1. Common steps of performing a BEC attack.

The description of each step is as follows:

- Step 1—Research: Potential victims' vulnerabilities and openings are scouted by attackers.
- Step 2—Identify a target: Based on the research, the criminals decide what angle they will try to exploit and which organisation they will target.
- Step 3—Build a Persona: Through a web search, the criminals can identify board members in the target organisation.
- Step 4—Identify a victim: Next, they look for an individual at the target organisation whom they want to trick.
- Step 5—Spoof the email: The attack starts with an email that appears to come from the senior leader. The crooks first spearfish the executive to get their credential, then log in as them to send their email.
- Step 6—Personalise the email: The attacker puts all their research and persona-building work to good use, crafting an email that appears to come from the senior leader.
- Step 7—Isolate the Victim: Isolation is a popular technique to pressure the victim and stop them from checking with others.
- Step 8—Avoid Follow-up: The attacker does not want the victim checking in with the senior leader, so they discourage the victim by making the senior leader seem unavailable, such as by saying they are out of the office.
- Step 9—Provide bank details: The bank account detail is one of the attacker's biggest expresses, so they will only share after they have hooked the victim with their spoofed email.
- Step 10—Money transferred: The game is over; the money has been sent to the attacker and will never be seen again. Soon, someone will notice a big hole in the bank account, and that is when the alarm will go on.

2.3. Types of BEC Attacks

According to the FBI [4], there are five types of BEC attacks which include the following:

1. Email Account Compromise: This attack is targeted at small firms that use email to organise their financial transactions. The specifics of a recent transaction can be gleaned by breaking into an employee's email account and stealing the invoice. Attackers call a vendor and explain the situation, persuading the vendor that the final payment could not be processed. A new account, which the scammers would have set up to steal the money, is gently requested by them [4].

2. Lawyer Impersonation: This type of attack is fraud committed mostly against major corporations and the law firms that serve them. Attackers pose as a lawyer for a client of the company's law practice and ask for a quick transfer for the payment of an outstanding debt. To protect the transaction from being leaked, they convince the employee that the subject is private by making clear that they are demanding that the employee should not discuss it with anyone. Oftentimes, attackers plan it towards the conclusion of the work week to put the employee under more pressure to respond swiftly [4].
3. Data Theft: As part of a BEC attack, a top executive's email account is compromised and a request for critical company information is made. This is an example of a BEC attack that does not include any money laundering. It is often a prelude to a far more serious cyberattack, as this type of attack focuses on finance and human resources [5].
4. Vendor Email Compromise: It is common in businesses with overseas vendors. Attackers assume the role of vendors, demand payment for a fictitious invoice, and then transfer the funds to a fake account [4].
5. CEO Fraud: Attackers pretend to be a company's CEO or executive. As the CEO, they ask a worker in the accounting or finance division to transfer money to an account under the control of the attackers.

2.4. Phishing and BEC Techniques

This section highlights techniques of phishing attacks generally and BEC attack specifically.

2.4.1. Phishing-Related Techniques

Typically, a phishing attack includes sending an email that contains a spoof URL link that leads to a web page. The following are common phishing techniques:

- Direct Link: Links are usually accessible in the body of the email. In addition, they may contain hidden links or image links that lead to phishing or another dangerous website [6].
- PDF Files: Email attachments that include a PDF file are a typical phishing method because they make the recipients believe that the attached document is essential, such as an eye-catching business proposal or an urgent invoice. Even if the PDF file does not include any malicious code, the material inside the PDF file might be crafted to direct the recipients to phishing sites [7].
- HTML: The phishing email might also include a malicious HTML file. Even though HTML files are rarely utilised in commercial transactions, these attachments can nonetheless deceive unwary users. The target will be sent to a malicious URL if he or she clicks and downloads the attachment [8].
- File-hosting Services: One of the most prevalent methods used by attackers to trick users into visiting phishing sites is the misuse of file-hosting services [9].
- Malware-related Techniques: Keyloggers and Remote Access Tools (RATs) are the most utilised malware for BEC. Malware, unlike phishing attempts, may grab all computers' saved login credentials before delivering them to the attackers. Hacking forums are flooded with new keyloggers and RATs, offering cybercriminals easy access to sophisticated yet undetectable malware. BEC attacks have been plagued by a wide range of software, the most frequent of which being adware including AgentTesla, CyborgLogger, DarkComet, DiamondFox, Dracula Logger, iSpy Keylogger, Knight Logger, and LuminosityLink [10].

2.4.2. BEC Techniques

- **Spoofed BEC Messages:** The email domain may be manipulated to make the email appear to be legitimate in this method. Email header spoofing is used by attackers to produce fraudulent emails that appear to originate from a legitimate source. In the "From" address area, they use the true domain of the target company [4].

- **BEC Basic Header Trickery:** Another tactic adopted by attackers is to use a faked organisation's true domain in the "From" address box. Adding another domain to the "Reply-To" address box tricks the users into thinking they have received a reply. A reply to a field controlled by the attacker results in a reply being sent back to the attacker [1].
- **BEC Business Domain Similarity Attacks:** The "From" domain used in this attack is the same as the target domain. This gives the appearance that emails are coming from a higher authority and are being sent to employees asking for a quick response. To a foreign client or supplier, it may be necessary to make an immediate transfer of funds [2].
- **Executive Name Forgery:** In an attack known as ID Spoofing, the hacker inserts a fake executive name into the "From" box of an email. This title-spoofing technique takes advantage of the display label to spoof the names of corporate executives (CxOs) [2].
- **BEC Encoded Message Attack:** To avoid being detected by email gateways, BEC attackers sometimes utilise encoding ploys that alter the characters in the message. When a BEC message is opened in the email account of a client, it seems to be a regular email. A hex editor scan, on the other hand, reveals the real colours of the image [4].
- **BEC Attacks using Long Scenario to Lure Victims:** Using long, individualised emails to request sworn confidentiality from recipients because of legal repercussions of a vital business necessity is another typical practice. Law firms are frequently mentioned in these emails to remind their receivers that they need to follow legal and commercial requirements carefully in order not to divulge confidential information. The attacker will next seek access to the company's bank accounts and financial records, which they will study to make further demands for money transfers [5].
- **BEC Emails Demanding Gift Cards:** Criminals request iTunes, Amazon, and Walmart gift cards from their victims. Instead of requesting a wire transfer, this attack asks for the credentials of a gift certificate that the victim has received in person. It uses a standard message structure to demand first priority [4].
- **BEC Attackers Targeting Schools and Academic Institutes:** Emails from the school's principal or top management asking for wire transfers or gift vouchers from school workers are another common BEC technique [4].

2.4.3. Feature Selection Techniques

When developing an ML-based phishing detection model in the real world, it is usually never the case that all variables in the dataset are significant. Adding duplicate variables diminishes the model's capacity for generalisation and affects the overall accuracy of the detection model. Additionally, when more variables are added to the model, its total complexity grows. According to the Law of Parsimony of 'Occam's Razor,' the optimal answer for a problem is the one that requires the fewest assumptions. Thus, feature selection becomes an essential component in ML-based model development [9].

Feature selection is the method of reducing the input variables of a ML-based model by using only relevant data and getting rid of noise in the data. Its main goal is to clean up a model by getting rid of irrelevant or unnecessary data. Due to the complexity of some predictive modelling issues, considerable memory is often needed during model creation and training. In addition, certain models' functionality can deteriorate if the input variables are not pertinent to the target variable. In ML, the strategies for feature selection are categorised into two main categories: supervised and unsupervised. The supervised feature selection methods are applied to labelled data to discover the most important variables for improving the performance of supervised models. In other words, they use the target variables to identify the variables which can increase the efficiency of the model. Unsupervised feature selection methods are applied to unlabelled data in which the outcome is not considered while making the feature selection [10]. Figure 2 shows the categories of feature selection methods.

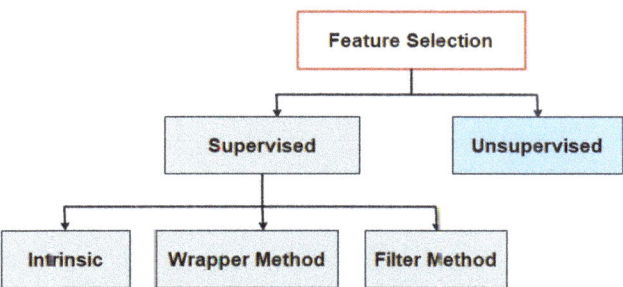

Figure 2. Feature selection methods.

The supervised methods are further divided into three methods, including filter, wrapper, and intrinsic methods [10,11].

- **Filter Method**: In this method, features are eliminated according to how they correlate with the output. Correlation is used to determine if the features are positively or negatively correlated to the output labels and then the features are dropped accordingly. Examples include Fisher's Score, Variance Threshold, Correlation Coefficient, Chi-Square Test, etc.
- **Wrapper Method**: Wrappers need a way to explore the space of all possible subsets of features, evaluating their quality by learning and evaluating a classifier with that subset of features. The feature selection procedure is determined by a particular ML algorithm that employs a greedy search strategy by comparing all potential feature combinations to the evaluation criterion. Wrapper approaches often produce more accurate predictions than filter methods. Wrapper methods include Forward Feature Selection, Backward Feature Elimination, Exhaustive Feature Selection, etc. [12].
- **Intrinsic Method**: This method, also called the embedded method, combines the advantages of wrapper and filter methods by including feature interactions while retaining an acceptable computing cost. This approach is iterative in the sense that it takes care of each iteration of the model training process and meticulously extracts the features that contribute the most to training for each iteration. Examples include Random Forest Importance and LASSO Regularisation (L1).

2.4.4. Evaluation Metrics for BEC Detection

Determining the effectiveness of BEC phishing detection models is significant to compare different models and identify the most effective model for each context. Based on numerous studies reviewed in the literature [5,9,10], the effectiveness of BEC phishing detection models is computed based on four main evaluation metrics, including accuracy, precision, recall, and F-measure. A description of how these evaluation metrics is computed is discussed below:

- **True Positive (TP)**: This represents the percentage of phishing emails in the training dataset that are correctly classified by a phishing detection model. Formally, if the number of phishing emails in the dataset is denoted by P and the number of correctly classified phishing emails by the phishing detection model is denoted by NP, the formula of TP is as follows:

$$TP = \frac{NP}{P} \qquad (1)$$

- **True Negative (TN)**: This represents the percentage of legitimate emails that are correctly classified as legitimate by a phishing detection model. If we denote the number of legitimate emails that are correctly classified as legitimate as NL and the total number of legitimate emails as L, the formula of TN is as follows:

$$TN = \frac{NL}{L} \qquad (2)$$

- **False Positive (FP):** This is the percentage of legitimate emails that are incorrectly classified by a phishing detection model as phishing emails. If we denote the number of legitimate emails that are incorrectly classified as phishing as Nf and the total number of legitimate emails as L, the formula of FP is as follows:

$$FP = \frac{Nf}{L} \qquad (3)$$

- **False Negative (FN):** This represents the percentage of the number of phishing emails that are incorrectly classified as legitimate by a phishing detection model. If we denote the number of phishing emails that are classified as legitimate by the algorithm as Npl and the total number of phishing emails in the dataset is denoted as P, the formula of FN is as follows:

$$FN = \frac{Npl}{P} \qquad (4)$$

Using TP, TN, FP, and FN, the four evaluation metrics, including accuracy, precision, recall, and F-measure, can be computed as follows:

- **Accuracy**: It represents the average number of successfully categorised emails throughout the entire dataset using the following formula:

$$Accuracy = \frac{TP + TN}{TP + FP + FN + TN} \qquad (5)$$

- **Precision**: It measures the exactness of a classifier, i.e., what percentage of emails that the classifier has labelled as BEC phishing are actually BEC phishing emails, and it is represented by this formula:

$$Precision = \frac{TP}{TP + FP} \qquad (6)$$

- **Recall:** It measures the completeness of a classifier's results, i.e., what percentage of phishing emails the classifier has labelled as phishing, and it is represented by this formula:

$$Recall = \frac{TP}{TP + FN} \qquad (7)$$

- **F-measure:** This is also known as $F1$ score and is defined as the harmonic mean of Precision and Recall, and it is calculated based on this formula:

$$F1 - Score = \frac{2TP}{2TP + FP + FN} \qquad (8)$$

3. Research Methodology

The purpose of a systematic literature review is to define, analyse, and interpret all available research relevant to a research topic, a specific subject, or a set of interesting occurrences. While several experts have offered solutions to detecting BEC attacks, the threat environment is expanding and becoming more dangerous despite their efforts. This systematic literature review investigates existing BEC attack techniques and detection methods, as well as various studies presented by researchers using different ML algorithms employed in the detection process and these studies' conclusions.

Conducting a systematic literature review consists of five stages, as shown in Figure 3. The objective of the first stage is to formulate the research questions that the current review will answer. This is followed by determining the inclusion and exclusion criteria to ensure that the selected articles are the best and most pertinent concerning the research objectives. The third stage is to specify which research databases will be searched to find relevant articles. In the fourth stage, the findings are analysed, and, in the fifth stage, the outcomes of each study topic are discussed.

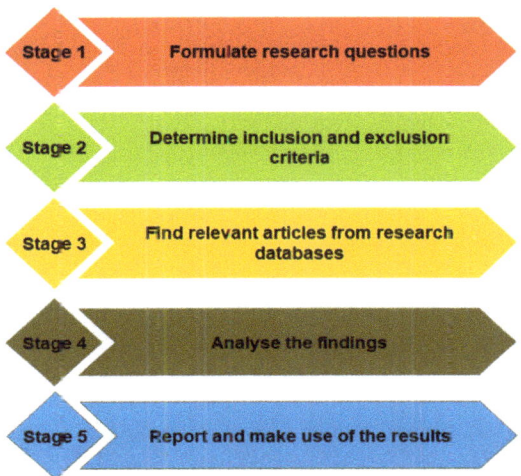

Figure 3. Stages of conducting a systematic literature review.

This methodology was utilised to allow readers to understand the stages used to complete this literature review systematically. Before beginning to evaluate many sources, we defined our research questions so that the review would be more focused. Next, the selection criteria were used to narrow down the retrieved publications to those relevant to the study's objectives. The digital libraries that were used to compile these articles are also offered as data sources. Article selection based on relevance was also covered. The presented methodology offers various benefits to show the steps taken by researchers to reach their study's intended results.

Although this methodology has been used in several systematic literature studies, there are some limitations, including the fact that it narrows the focus of the review/study and, hence, may not provide readers with all the facts they need to fully understand the subject matter at hand. In addition, data collection was limited to only six sources for collecting relevant publications in our study, which could limit the number of publications reviewed. Although these sources are the most reliable sources identified in various systematic literature studies, this could be considered a limitation as not all sources were investigated to identify relevant articles related to the study objectives. In addition, this study reviewed only articles published between 2012 and 2022. Although this study provides readers a review of state-of-the-art articles published in the last ten years, the search methodology limits the number of publications that can be reviewed in the study.

3.1. Research Questions

This paper seeks to address the following research questions:
- **RQ1**: What is the most recent and peer-reviewed literature regarding BEC phishing attacks?
- **RQ2**: What are the common ML algorithms used for developing ML-based BEC detection models?
- **RQ3**: What are the common datasets used in creating BEC detection models?

- **RQ4**: What are the conventional features used in developing an effective BEC detection model?

3.2. Inclusion and Exclusion Criteria

Inclusion and exclusion criteria were used to choose the applicable research. The primary purpose of these criteria was to answer the research questions and assure the creation of an effective literature review. The inclusion criteria were as follows:

- Peer-reviewed and scientific papers.
- Relevant to the specific research questions.
- Topic mainly on BEC phishing attack.
- English-language articles.
- Published between 2012 and 2022.

The exclusion criteria were as follows:

- Article concerning all other phishing attacks, including clone attacks, whaling, vishing, etc.
- Unpublished articles, non-peer-reviewed articles, and editorial articles.
- Articles that are not fully available.
- Non-English articles.
- Duplicates of already included articles.

3.3. Data Sources

Digital libraries were used to conduct the searches. The electronic databases used in this systematic literature review included the following:

- IEEE Xplore.
- PubMed.
- Elsevier ScienceDirect.
- Google Scholar.
- ACM Digital Library.
- SpringerLink.

The papers pertinent to the subject and study questions were gathered using keyword searches. The search terms used included the following:

- BEC phishing attack.
- Spoofed BEC messages.
- BEC basic header trickery
- BEC business domain similarity attacks.
- Executive name forgery.
- BEC encoded message attack.
- BEC emails demanding gift cards.

3.4. Selection of Relevant Articles

This step involved choosing relevant and recent studies on BEC phishing attacks among the 950 articles gathered from various online digital libraries. The process of selecting relevant publications was divided into three phases:

- **Phase 1**: Publications found during the search and those already in the collection were sorted using the inclusion and exclusion criteria. The scope of the search was narrowed to include only articles published between 2012 and 2022 that dealt with the topic of BEC phishing attacks.
- **Phase 2**: The titles and abstracts of the articles collected from several digital libraries were reviewed to determine how well they addressed the topic and the questions posed in this research work.
- **Phase 3**: During this stage, we focused on eliminating duplicates among the six digital libraries used for our publication collection.

4. Analysis of Results

The inclusion and exclusion criteria were applied to the collected publications in three phases, as indicated earlier. A total of 887 articles were removed based on the evaluation by simply reading the titles and abstracts and their relevance to the research questions. Furthermore, duplication across various online digital databases (25 publications) was removed, as shown in Figure 4.

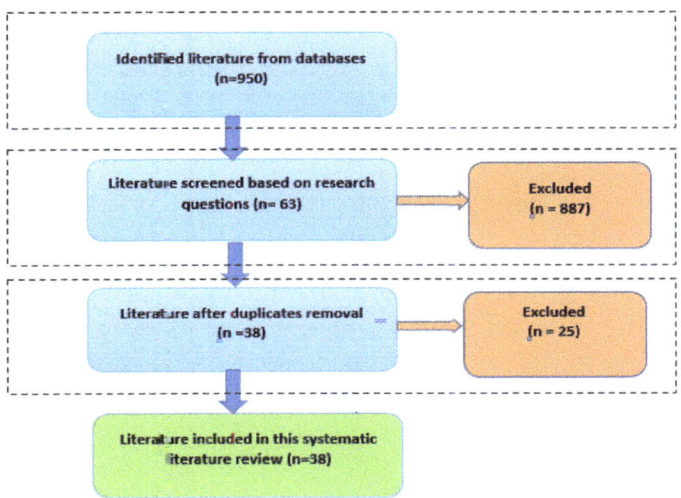

Figure 4. Flow diagram of the search.

The search that was executed in six different well-known online databases enabled us to collect most of the publications that are relevant to BEC phishing attacks. The results of the collected publications from each online database and the resultant number of publications after applying the three selection phases are shown in Table 1. The results show that Google Scholar and IEEE are the richest data sources of publications related to BEC phishing attacks.

Table 1. The number of search results per database after applying the three selection phases.

Data Source	Phase 1	Phase 2	Phase 3
Google Scholar	734	26	17
IEEE Xplore	90	15	10
PubMed	4	2	1
Elsevier ScienceDirect	42	4	2
ACM Digital Library	40	2	1
SpringerLink	40	14	7
Total	**950**	**63**	**38**

Additionally, the number of publications related to BEC phishing attacks per year is shown in Figure 5. The evidence suggests an upsurge in the study of BEC phishing attacks since 2017. However, many scientists still consider this to be a frontier. Research on BEC attacks has received consistent attention since 2017, as shown by the number of publications in 2017, 2018, 2019, 2020, 2021, and 2022.

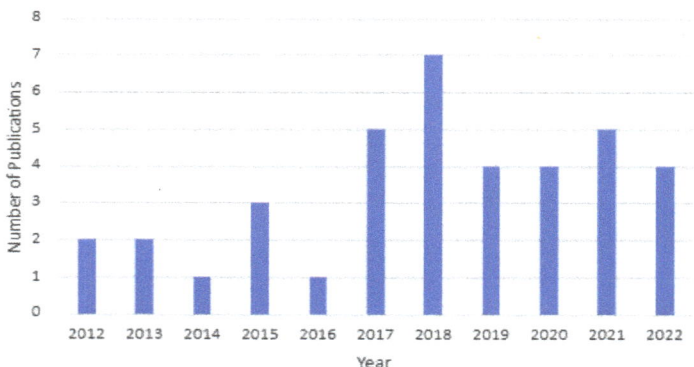

Figure 5. Number of selected articles published per year from 2012 to 2022.

Furthermore, the papers on BEC phishing attacks that were retrieved are separated by year and type (either as a journal or conference publication), as shown in Figure 6. Conference and journal articles both yield similar numbers of outcomes that meet our study objectives. In addition, Table 2 lists the ID, citation, publication category, and publication year for each of the examined articles. All of the papers that were read and retrieved were originally presented at academic conferences or published in scholarly publications.

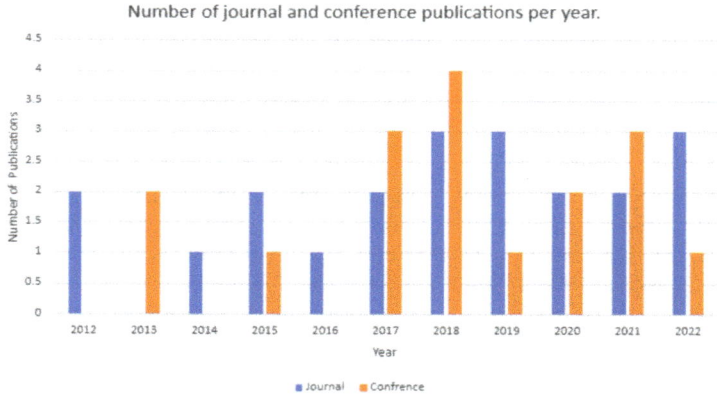

Figure 6. Number of journal and conference publications per year from 2012 to 2022.

Table 2. Retrieved publications that are related to the research questions.

Publication ID	Citation	Publication Type	Publication Year
1	Chakraborty and Mondal [11]	Journal	2012
2	Qasem, Shamsuddin, and Zain [12]	Journal	2012
3	Dhanaraj and Karthikeyani [13]	Conference	2013
4	Shams and Mercer [14]	Conference	2013
5	Laorden et al. [15]	Journal	2014
6	Rathod and Pattewar [16]	Conference	2015
7	Zhu, Dong, and Liu [17]	Journal	2015
8	Daeef et al. [18]	Journal	2015
9	Yasin and Abuhasan [19]	Journal	2016
10	Zweighaft [20]	Journal	2017

Table 2. Cont.

Publication ID	Citation	Publication Type	Publication Year
11	Rawal et al. [21]	Journal	2017
12	Zeng [22]	Conference	2017
13	Moradpoor, Clavie, and Buchanan [23]	Conference	2017
14	Niu et al. [24]	Conference	2017
15	Peng, Harris, and Sawa [25]	Conference	2018
16	Baykara and Gurel [26]	Conference	2018
17	Sahoo [27]	Conference	2018
18	Hiransha, Unnithan, and Kp [28]	Journal	2018
19	Singh, Pamula, and Shekhar [29]	Conference	2018
20	El Aassal et al. [30]	Journal	2018
21	Nidhin et al. [31]	Journal	2018
22	George Fomunyam [32]	Journal	2019
23	Oña et al. [33]	Conference	2019
24	Maleki and Ghorbani [34]	Journal	2019
25	Yang et al. [35]	Journal	2019
26	Garces, Cazares, and Andrade [36]	Conference	2020
27	Rendall, Nisioti, and Mylonas [37]	Journal	2020
28	Alam et al. [38]	Conference	2020
29	Alotaibi, Al-Turaiki, and Alakeel [39]	Journal	2020
30	Salahdine, and Kaabouch [40]	Conference	2021
31	Ripa, Islam, and Arifuzzaman [41]	Conference	2021
32	Dutta [42]	Journal	2021
33	Mughaid et al. [43]	Journal	2021
34	Mridha et al. [44]	Conference	2021
35	Li, Zhang, and Wu [45]	Conference	2022
36	Butt et al. [8]	Journal	2022
37	Magdy and Mikhail [46]	Journal	2022
38	Dewis and Viana [10]	Journal	2022

5. Discussion

Many researchers are still investigating BEC phishing attacks to identify better and more effective ways to counteract this growing threat. This paper serves as an excellent place for such researchers to begin understanding this paradigm by reviewing prior research that may be relevant to their study questions. To demonstrate how the reviewed papers have addressed our research questions, a discussion of the retrieved/analysed publications is provided in this section.

RQ1: What is the most recent and peer-reviewed literature regarding BEC phishing attacks?

To answer this research question, the retrieved/analysed publications that are related to BEC phishing attacks will be discussed. Table 3 summarises the contributions of each publication.

Table 3. Summary of recent studies in the literature regarding BEC phishing attacks.

Citation	Summary of Contribution	Limitations
Chakraborty and Mondal [11]	This paper analyses three DT classification algorithms for BEC phishing mail filtration. Logistic Model Tree classifier (LMT) produces the highest accuracy of 90%.	The accuracy achieved by the three DT classification algorithms is still low, and the model needs to be evaluated against a real-life dataset.

Table 3. *Cont.*

Citation	Summary of Contribution	Limitations
Qasem, Shamsuddin, and Zain [12]	This paper proposes a new hybrid multi-objective learning algorithm including MPPSON, MEPGAN, and MEPDEN to achieve a compact RBFN model with good prediction accuracy and prominent structure simultaneously for the process of detecting BEC attacks.	The proposed algorithms still suffer from slow convergence and long training times. There is a need to develop a sophisticated solution to overcome it.
Dhanaraj and Karthikeyani [13]	This paper uses Bayesian filtering to verify the sender's ID. The paper employs basic methods to differentiate humans from robot senders. The paper also creates new filters to detect emails requesting payment. The paper also uses N-gram language models, which assume that a word's location in a sequence depends on the previous N-1 words.	The accuracy achieved is still low. More investigation into filtering methods is required to improve the effectiveness and processing time of the proposed models.
Shams and Mercer [14]	This paper presents a unique BEC and spam categorisation algorithm based on email content language and readability. Although it only works for English emails, it may be useful for other languages.	There is a need for additional testing to evaluate the effectiveness of classifiers generated by stacking multiple algorithms.
Laorden et al. [15]	This paper presents a study on the effectiveness of anomaly detection applied to BEC and phishing filtering techniques. The study identifies that more than 85% of received emails are BEC/spam/phishing.	The threshold selection needs to be automated to improve filtering results.
Rathod and Pattewar [16]	This paper presents a Bayesian machine learning algorithm classification for accurately detecting BEC and phishing emails in HTML format that have been pre-processed to remove HTML tags, and stop words are applied. The Bayesian classifier could categorise real-world Gmail data with an accuracy of 96.46%.	Malicious URL detection needs to be incorporated to improve the effectiveness of the proposed content-based detection model.
Zhu, Dong, and Liu [17]	This study find that the number of neighbours, the distance measure, and the decision rule are the primary factors influencing categorisation performance. Several distance functions and other KNN parameters that were examined in this research are integrated into a support vector machine (SVM) and neural network, along with a dimension reduction and closest-neighbour index construction.	Feature selection, dimension, and class numbers need to be investigated further to improve classification effectiveness.
Daeef et al. [18]	In this paper, BEC and spare phishing are discussed as major threats to stealing user data. The paper analyses phishing email classifiers based on the email header and body.	A dynamic phishing dataset is needed to test the effectiveness of the proposed BEC phishing detection model.
Yasin and Abuhasan [19]	This paper presents an intelligent classification model for detecting phishing emails using knowledge discovery, data mining, and text processing techniques. The model was built using an intelligent pre-processing phase that extracts a set of features from the email's header, body, and term frequency.	The accuracy achieved by the proposed classification algorithm is still low, and the model needs to be evaluated against real-life datasets by considering the email's body and term frequency.
Zweighaft [20]	This paper presents the concept of BEC as a type of spear phishing in which fake or fraudulent emails are sent to employees of a specific business. The paper also presents an approach to securing access to content with multi-factor authentication and staff training.	The study provides some suggestions to detect BEC phishing attacks without evaluating the effectiveness of the suggested techniques.
Rawal et al. [21]	This paper classifies BEC and phishing emails using a SVM and a Random Forest (RF) classifier and achieves an accuracy of 99.87% with these algorithms.	The dataset used does not reflect real-life scenarios. A new dataset of real-life and dynamic emails is required.

Table 3. Cont.

Citation	Summary of Contribution	Limitations
Zeng [22]	This paper provides a predictive analytical technique that learns legitimate from harmful emails to overcome BEC attacks using static analysis.	Additional discriminating features need to be added to the predictive model to improve the accuracy of the detection model.
Moradpoor, Clavie, and Buchanan [23]	This paper presents an Artificial Neural Network (ANN) model algorithm classification for accurately detecting BEC and phishing emails. The ANN model contains ten hidden layers, five input features, one output layer, and one output feature. Confusion matrix, receiver operating characteristic (ROC), network performance, and error histogram are used to record, portray, and analyse the data. The ANN model achieves an accuracy of 96.46%.	Email word vectors are not enough to produce an effective detection model. Word embedding methods are required to vectorise full documents, rather than email word vectors, to improve the effectiveness of the detection model.
Niu et al. [24]	This paper proposes a novel SVM with Cuckoo Search (CS) detection model. The proposed approach detects more phishing emails than SVM with default parameters.	The proposed model utilizes only a single ML algorithm, so more algorithms are needed to evaluate the effectiveness of the model.
Peng, Harris and Sawa [25]	This paper presents attackers who pose as social networks, banks, IT administrators, or e-commerce sites. These emails may entice consumers to download malware or input personal information on a dangerous website. This study uses semantic analysis to check each phrase in an attacker's text.	The dataset used does not reflect the need for text emails, not graphics. A new dataset of text emails is required
Baykara and Gurel [26]	This paper builds an "Anti-Phishing Simulator" to identify BEC attacks using various ML techniques.	This method needs a real-life scenario to evaluate the effectiveness of the model.
Sahoo [27]	This paper uses data mining approaches to examine emails and avoid BEC phishing attacks.	The study provides some suggestions without evaluating their effectiveness.
Hiransha et al. [28]	This paper explains how to spot phishing emails including BEC attacks and avoid falling into their traps in a comprehensive manner.	The dataset used does not reflect real-life scenarios. A new real-life dataset is required.
Singh, Pamula, and Shekhar [29]	This paper evaluates the performance of Non-Linear SVM-based classifiers with two different kernel functions (Linear Kernel and Gaussian Kernel) over the SpamAssasin Public Corpus dataset.	The proposed model needs to be evaluated against more ML algorithms and different datasets.
El Aassal et al. [30]	This paper proposes a Multinomial Naive Bayes (MNB)-based ML algorithmic classification model for detecting BEC and phishing content in emails. The proposed MNB models can classify BEC and phishing emails with 96.8% accuracy.	The proposed model detects phishing only from the email content without considering the email header.
Nidhin et al. [31]	This paper proposes a supervised classifier for BEC phishing and legitimate emails. The paper classifies authentic and fraudulent emails using Naive Bayse (NB), Logistic Regression (LR), Decision Tree (DT), RF, Adaboost, and SVM.	A feature selection technique is not discussed as part of the proposed classifiers. In addition, the dataset used does not reflect real-life scenarios
George Fomunyam [32]	This paper identifies that ML algorithms are more effective than traditional mechanisms in detecting and categorising spam letters from internet fraudsters. This study presents new ways to combat cybercriminals' fraudulent BEC phishing attacks.	This study provides some suggestions to detect BEC phishing without evaluating its effectiveness.

Table 3. Cont.

Citation	Summary of Contribution	Limitations
Oña et al. [33]	This paper develops a new method for identifying BEC attacks and countering them. The paper provides a comprehensive discussion about how to integrate autonomous learning, feature selection, and ANNs using Scrum. This technique can identify and counteract an email server-based phishing attack.	The proposed methods have been not evaluated against either benchmark datasets or real-life data.
Maleki and Ghorbani [34]	This paper proposes a K-means-based ML algorithmic classification model for detecting BEC content in emails. The proposed model can classify BEC attacks with 92% accuracy.	Dynamic feature selection is not provided. In addition, the dataset used does not reflect real-life scenarios.
Yang et al. [35]	This paper provides a BEC detection model by analysing email headers, URLs, and scripts. A total of 500 authentic and 500 phishing emails were used in the experiments. The proposed method achieves 99% true positive, 9% false positive, and 91.7% precision.	The dataset used does not reflect real-life scenarios. A new dataset of real-life and dynamic emails is required.
Garces et al. [36]	The paper investigates phishing web attack anomalies and how integrating ML techniques with data analytics can be a very effective solution to detect BEC attacks faster.	A real-life dataset is needed to check the effectiveness of the model for taking proactive decisions to minimise the impact of an attack.
Rendall, Nisioti, and Mylonas [37]	In this paper, the researchers investigate the use of a multi-layered detection framework in which a potential phishing domain is classified multiple times by models using different feature sets.	The proposed framework lacks the use of a sophisticated data fusion process as part of JDL level 2 (situation refinement).
Alam et al. [38]	This paper provides a BEC phishing detection model using the DT and RF techniques.	More ML-based algorithms are needed to evaluate the effectiveness of their model.
Alotaibi et al. [39]	This paper presents a convolutional ANN for BEC email phishing detection. The approach can help enterprises protect against phishing email attacks.	A feature selection technique is not discussed as part of the proposed BEC phishing detection model.
Salahdine, El Mrabet, and Kaabouch [40]	This paper proposes utilising SVM, LR, and ANN algorithms for detecting BEC and phishing content in emails. The proposed models achieve an accuracy of 94.5%, 77.3%, and 92.9% in ANN, SVM and LR, respectively.	The proposed models detect phishing only from the email content without considering the email header.
Ripa, Islam, and Arifuzzaman [41]	This paper proposes utilising RF, SVM, KR, KNN, and DT algorithms for detecting BEC and phishing content in emails. The proposed models achieve an accuracy of 96.8%, 96.6%, 92.28%, 94.09%, and 96.47% in RF, SVM, LR, KNN and DT, respectively.	A dynamic and real-life phishing dataset is needed to test the effectiveness of the proposed BEC phishing detection models.
Dutta [42]	This paper proposes a recurrent ANN and short-term memory algorithm for detecting BEC and phishing content in emails. The proposed model achieves an accuracy of 94.8%.	The accuracy achieved by the proposed algorithm is still low, and a better accuracy is needed for the detection model.
Mughaid et al. [43]	This paper discusses ML methods and new technological solutions for mitigating BEC attacks. This study utilizes ML-based algorithms to classify emails as BEC/phishing or non-BEC/non-phishing.	The feature selection algorithm needs to be refined to keep up with attackers' evolving toolkits.
Mridha et al. [44]	This paper proposes RF and ANN-based algorithmic classification models for detecting BEC and phishing URLs accurately. The proposed RF and ANN models can classify BEC and phishing URL legitimacy labels with 99% accuracy.	The proposed model needs a GUI-based web browser extension framework to provide better precision for the detection model.

Table 3. Cont.

Citation	Summary of Contribution	Limitations
Li, Zhang, and Wu [45]	This paper proposes a BEC/phishing email detection method based on the persuasion principle.	A feature selection technique is not discussed as part of the proposed model.
Butt et al. [8]	This paper proposes utilising SVM, LSTM, RF, LR, ANN, and NB algorithms for detecting BEC and phishing content in emails. The proposed models achieve an accuracy of 99.6%, 98%, 94.5%, 93.9%, 95%, and 97% in SVM, LSTM, RF, LR, ANN and NB, respectively.	The proposed models detect phishing only from the email content without considering the email header.
Magdy, Abouelseoud and Mikhail [46]	This paper introduces a deep learning model to detect BEC attacks. The proposed classifier is designed with an eye on validation accuracy, achieving fast and competitive performance, and promoting its use in practical applications.	A mechanism for improving linguistic processing with content-based features is needed to improve the effectiveness of the detection model.
Dewis and Viana [10]	This paper proposes a BEC/phishing responder as a solution that uses a hybrid ML approach combining natural language processing and deep learning to detect phishing and BEC emails. It has achieved an average accuracy of 99% with the LSTM model for text-based datasets.	The dataset used does not reflect real-life scenarios. A new dataset of real-life and dynamic emails is required.

Table 3 provides a summary of the contributions of the retrieved publications that are related to BEC phishing attacks. Looking at the various reviewed studies illustrates that there are some similarities and differences among various researchers. Some researchers presented a comparative study of various supervised and unsupervised ML techniques to provide an effective BEC phishing detection model that provides the highest accuracy, precision, recall, and F-measure to detect phishing emails. For example, Butt et al. [8], Ripa, Islam, and Arifuzzaman [41], and Chakraborty and Mondal [11] created a comparative study using various ML algorithms, including DT, SVM, LSTM, RF, LR, ANN, NB, KR, and DT, to identify a ML algorithm that provides the highest accuracy on a specific dataset. Other researchers provided hybrid ML-based techniques that combine two or more algorithms with changes in variables to provide better accuracy for the BEC phishing detection model. For instance, Dewis and Viana [10] proposed a hybrid ML-based approach combining NLP and deep learning to detect BEC phishing emails. Their LSTM model has achieved an average accuracy of 99% for text-based datasets. In addition, Qasem, Shamsuddin, and Zain [12] proposed a new hybrid multi-objective learning algorithm combining MPPSON, MEP-GAN, and MEPDEN to achieve a compact RBFN model with good prediction accuracy and prominent structure while detecting BEC attacks.

Furthermore, some researchers focused more on the detection algorithm by investigating the best ML algorithm to implement their BEC phishing detection model, while other researchers focused more on the feature selection techniques to identify the best features that ensure the creation of a high-accuracy BEC phishing detection model. For example, Rendall, Nisioti, and Mylonas [37] used a multi-layered detection system where a potential phishing domain is classified multiple times by models using different feature sets, while the studies by Salahdine, El Mrabet, and Kaabouch [40] and Ripa, Islam, and Arifuzzaman [41] focused more on identifying the best ML algorithm for the detection model by comparing their effectiveness against various datasets.

Evaluating the proposed BEC phishing detection models by various researchers also revealed another difference among the retrieved publications: some researchers utilised publicly available datasets, while other researchers utilised real-world and dynamic datasets that they created in specific circumstances to evaluate the effectiveness of their detection models. For example, Garces and Cazres [36], Ripa et al. [41], Alam et al. [38], Dewis and Viana [10], Mridha et al. [44], and Nidhin et al. [31] evaluated the effectiveness of their phishing detection model using Kaggle dataset, one of the most common datasets in phishing detection domain, while other researchers created their own datasets, such as Baykara and Gurel [26], Rawal et al. [21], etc.

RQ2: What are the common ML algorithms for developing ML-based BEC detection models?

The technique used to identify a BEC attack is important to the process of identifying such an attack. It is possible to utilise a wide variety of algorithms to guarantee accuracy, although their detection effectiveness differs. This section lists various techniques that have been used by various researchers to build a BEC phishing detection model, as shown in Table 4. The most common techniques include the following:

- **Naive Bayes (NB):** This classifier uses the Bayes theorem to classify data samples. The Bayes theorem asserts that, given a hypothesis H and some evidence E, the probability of each category is computed, and then the highest probability category is the output [47,48].
- **Support Vector Machines (SVM):** This classifier is a speedy and efficient supervised approach for text classification algorithms. The input training set generates a hyperplane, a two-dimensional line that best separates categories. This hyperplane is the decision boundary. In BEC detection, the input is a set of criteria, such as the presence or absence of specified words; the output, 1 or −1, indicates whether the email is a BEC attack. For instance, certain phrases can be used to detect whether an email is a BEC attack [48,49].
- **Logistic Regression (LR):** A binary logistic model uses one or more predictor variables to estimate the probability of a binary response (features). It allows stating that a risk factor boosts the possibility of a specific consequence by a certain percentage [47].
- **Decision Tree (DT):** This technique utilises a tree-like model of actions and their effects, including chance event outcomes, resource costs, and utility. It is one approach to present a conditional-only algorithm. DT is used in operation research, especially decision analysis, to discover the most probable method to attain a goal. It is also a popular machine learning technique [50].
- **Random Forest (RF):** This technique blends many DT outputs to obtain a single outcome. Its simplicity of use and versatility have spurred its popularity as a classification and regression tool [51].
- **Artificial Neural Network (ANN):** This technique is a collection of algorithms that attempt to replicate the way the human brain works to discover hidden patterns and connections within a dataset. Neuronal systems, whether biological or synthetic, are what we mean when we talk about neural networks [8].
- **Natural Language Processing (NLP):** This is a subfield of computer science and, more specifically, a subfield of artificial intelligence (AI) that focuses on providing computers with the capacity to comprehend written and spoken language [52].

From Table 5, we can see that DT, SVM, ANN, NB, and Logistic algorithms have all been utilised in at least 10 of the 38 studies, indicating that researchers have found their results to be consistent enough to justify reusing these algorithms. In addition, it is important to highlight that certain algorithms, such as DT, SVM, NB, ANN, and Logistic algorithms, have a broad user base and are widely utilised by researchers and data scientists. As a result, they have well-updated libraries, and further enhancements are available to make them more compatible with several datasets due to their continuous use [53,54].

Table 4. List of algorithms used in the literature and their abbreviations.

Algorithm	Abbreviation
Linear Regression	Linear
Logistic Regression	Logistic
Decision Tree	DT
Support Vector Machine	SVM
Naive Bayes	NB
K-Nearest Neighbours algorithm	KNN
Random Forest	RF
Dimensionality Reduction Algorithm	DRA
Artificial Neural Network	ANN
Voted Perceptron	VP
Natural Language Processing	NLP
XGBoost Classifier	XGB
Group Method of Data Handling	GMDH
Probabilistic Neural Network	PNN
Genetic Programming	GP
Multilayer Perceptron	MP
Principal Component Analysis	PCA
Gaussian Kernel	GK

These algorithms have been used extensively by various researchers due to their effectiveness and accuracy in detecting BEC phishing attacks with various networks and datasets. Identifying the effective algorithm within each communication network should be the right course of action for an effective and successful BEC detection model. In addition, creating a hybrid technique that integrates two or more of these algorithms can yield an effective technique that can provide a better BEC phishing detection model.

In addition, if we look at supervised and unsupervised ML algorithms that have been utilised by researchers to build BEC phishing detection models, we find that most researchers prefer using supervised ML algorithms. For example, supervised ML algorithms, including DT, SVM, NB, and Logistic algorithms, were used in at least 10 of the 38 reviewed articles, while unsupervised algorithms, such as PCA, were utilised in only two of the 38 reviewed articles. Researchers prefer utilising supervised ML algorithms in BEC phishing detection models since unsupervised learning typically uses clustering algorithms to group email categories, such as BEC emails. However, a clustering algorithm would typically categorise many common categories (e.g., social emails and marketing emails), but since BEC emails are so rare, it results in low precision and many false positives. Therefore, supervised learning algorithms are more suitable for detecting BEC attacks at high precision.

It is also important to note that certain algorithms, such as GMDH, PNN, GP, MP, PCA, and GK, were used less than other algorithms; these algorithms were used in only one or two articles out of the 38 selected articles, making them less well known than the first group of algorithms. These algorithms may still not be known for various researchers to try and identify their effectiveness in various communication networks, but these algorithms can be the basis for creating an effective hybrid BEC phishing detection model in the near future.

Table 5. ML algorithms utilised by various researchers to build BEC phishing detection models.

Citation	SVM	RF	NLP	NB	ANN	DT	Linear	Logistic	VP	XGB	DRA	PNN	GP	MP	PCA	GK	KNN	GMDH
Fomunyam [32]	-	-	-	✓	-	-	-	-	-	-	-	-	-	-	-	-	-	-
Peng and Sawa [25]	-	-	✓	✓	-	-	-	-	-	-	-	-	-	✓	-	-	-	-
Baykara and Gurel [26]	-	-	-	✓	-	-	-	-	-	-	-	-	-	-	-	-	-	✓
Sahoo [27]	-	-	-	-	-	✓	-	-	-	-	-	-	-	-	-	-	-	-
Garces and Cazres [36]	-	-	-	-	✓	-	-	-	-	-	✓	-	-	-	-	-	-	-
Oña et al. [33]	-	✓	-	✓	✓	-	-	✓	-	-	-	-	-	-	-	-	-	-
Li and Wu [45]	✓	-	-	-	-	-	-	-	-	-	-	-	-	-	-	-	-	-
Salahdine et al. [40]	✓	✓	-	✓	✓	✓	-	✓	-	-	-	-	-	-	-	-	-	-
Rawal et al. [21]	✓	✓	-	✓	-	-	-	-	-	-	-	-	-	-	-	-	-	-
Zeng [22]	✓	✓	-	✓	-	-	-	✓	✓	-	-	-	-	-	-	-	-	-
Ripa et al. [41]	✓	-	✓	✓	-	-	-	✓	-	-	-	✓	-	✓	-	-	-	-
Hiransha et al. [28]	-	-	-	✓	-	-	-	-	-	-	-	-	-	-	-	-	-	-
Butt et al. [8]	-	✓	-	✓	-	✓	✓	-	-	-	-	-	-	-	-	-	-	-
Rendall et al. [37]	✓	-	-	-	-	✓	-	-	-	-	-	-	-	-	-	-	-	-
Alam et al. [38]	-	-	-	✓	✓	-	-	-	-	-	-	-	-	-	-	✓	✓	-
Singh et al. [29]	-	-	✓	-	✓	-	-	-	-	✓	-	-	-	-	-	-	-	-
El Aassal et al. [30]	✓	✓	✓	-	✓	-	-	-	✓	-	-	-	-	-	-	-	-	-
Moradpoor et al. [23]	-	-	-	-	-	-	-	-	-	-	-	-	-	-	-	-	-	-
Dutta [42]	✓	-	-	✓	-	-	-	✓	-	-	-	-	-	-	-	-	-	-
Yasin et al. [19]	-	-	✓	-	✓	-	-	✓	-	-	-	-	-	-	-	-	-	-
Mughaid et al. [43]	-	-	-	-	-	-	-	-	-	-	-	-	-	-	-	-	-	-
Magdy et al. [46]	-	-	-	✓	✓	✓	-	-	-	-	-	-	-	-	-	-	-	-
Dhanaraj et al. [13]	-	-	-	-	-	✓	✓	✓	-	-	-	-	-	-	-	-	-	-
Chakraborty et al. [11]	-	-	-	-	✓	-	-	-	-	-	-	-	-	-	-	-	-	-
Zhu and Liu [17]	-	-	-	-	-	✓	-	-	-	✓	-	-	-	-	-	-	-	-
Qasem et al. [12]	✓	-	-	-	-	-	-	-	-	-	-	-	-	-	-	-	-	-
Shams et al. [14]	-	-	-	✓	-	-	-	-	-	-	✓	-	-	-	-	-	-	-
Maleki et al. [34]	-	-	-	-	-	-	✓	-	✓	-	-	-	-	-	-	✓	✓	✓
Dewis and Viana [10]	-	-	-	-	-	-	-	-	-	✓	-	-	-	-	-	-	-	-
Mridha et al. [44]	-	-	✓	-	✓	-	-	-	-	-	-	-	-	-	-	-	-	-
Daeef et al. [18]	✓	✓	-	✓	-	-	-	-	-	-	-	-	-	-	✓	-	-	-
Nidhin et al. [31]	-	-	-	-	-	-	-	-	-	-	-	-	-	-	-	-	-	-
Yang et al. [35]	✓	✓	-	✓	-	✓	-	✓	✓	-	-	-	-	-	-	-	-	-
Alotaibi et al. [39]	-	-	-	-	✓	-	✓	-	-	-	-	-	-	-	-	-	-	-
Niu et al. [24]	✓	-	-	-	✓	-	-	-	-	-	-	-	-	-	-	-	-	-

RQ3: What are the common datasets used in creating BEC detection models?

Building a ML-based technique requires having a dataset for training and testing the suggested model to identify its effectiveness and accuracy. There are many common datasets that have been used by various researchers to build BEC phishing detection models. Table 6 summarises the datasets used by various researchers to build BEC phishing detection models. This paper highlights the fact that many datasets were created previously and get regular updates from their creators. For example, the Nazario dataset gets regular updates with fresh sample data, with the latest being in 2021. Other examples are the Spam email dataset, which was compiled in 2010, the Phishing corpus in 2005, the Enron spam in 2006, and the Spamassassin dataset in 2002, all of which get regular updates. The titles of the datasets used by the 38 publications are further categorised in Table 6.

In addition, out of the 38 studies, more than half used customised datasets in their study. In light of the ever-shifting nature of BEC attacks, most of the researchers acquired email samples from actively running servers and organisations' email systems. To keep up with the latest trends and conduct an in-depth study of emerging BEC attack routes and methodologies, a dynamic and continuously updated dataset that captures a wide range of emails is required. This further supports the argument that new customised datasets from working contexts are more widely utilised in research than the standard datasets provided.

RQ4: What are the conventional features used in developing an effective BEC detection model?

There are three primary locations from which features used to detect BEC phishing attacks are often extracted: header, body, and URLs. The URL is a subset of both the header and the body; thus, it is not surprising that it is a frequently utilised detection feature.

- **Header Features:** The header of an email is the section of the message that includes the sender's and the recipient's email addresses, as well as the message's topic. The technical details necessary for prompt email delivery are included in each message's unique header. Internet header refers to the part of an email that contains the sender's and the recipient's email addresses, the topics, and the dates. An email's header also contains useful technical information including the sender's email address, the receiver's email address, and a unique Message ID.
- **Body Features:** The body of an email is the key section of the message. Most features for detection are crafted based on the text or are conceptually comparable to a defined dictionary word search or a set bag of words.
- **URL Features:** These are features derived from the links in the body of an email and the authorised sender domain, which are extracted to be analysed based on criteria and defined structure and would serve as a good indicator for detecting BEC attacks. Table 7 shows the features used by various researchers to build effective BEC detection models.

Table 6. Datasets utilised by various researchers to build BEC phishing detection models.

Citation	Spam Archive Corpus [39]	Nazario [45]	Enron Corpus. [34]	Kaggle [41]	Spam Assasine [45]	Spam Enron [46]	Avocado Corpus [28]	Phishing Corpus [46]	Custom
Fomunyam [32]	✓								
Peng and Sawa [25]									✓
Baykara and Gurel [26]									✓
Sahoo [27]			✓						
Garces and Cazres [36]		✓							✓
Oha et al. [33]				✓					
Li and Wu [45]		✓							✓
Salahdine et al. [40]							✓		✓
Rathod et al. [16]				✓			✓		
Rawal et al. [21]									✓
Zeng [22]									✓
Ripa et al. [41]				✓	✓		✓		
Hiransha et al. [28]									✓
Butt et al. [8]									✓
Rendall et al. [37]		✓		✓	✓				
Alam et al. [38]									
Singh et al. [29]						✓			
El Aassal et al. [30]									
Yasin et al. [19]									✓
Mughaid et al. [43]				✓	✓			✓	✓
Magdy et al. [46]									✓
Dhanaraj et al. [13]									✓
Chakraborty et al. [11]									✓
Zhu and Liu [17]									
Qasem et al. [12]									
Shams et al. [14]									
Laorden et al. [15]									
Maleki et al. [34]			✓						
Dewis and Viana [10]									✓
Mridha et al. [44]									
Daeef et al. [18]									
Nidhin et al. [31]									✓
Yang et al. [35]	✓							✓	
Alotaibi et al. [39]									✓
Niu et al. [24]									

Table 7. Common features used by various researchers to build BEC detection models.

Citation	Header	Body	URL
George Famunyam [32]	✓	✓	-
Peng, Harris and Sawa [25]	✓	✓	-
Baykara and Gurel [26]	✓	✓	✓
Sahoo [27]	✓	✓	✓
Garces, Cazares, and Andrade [36]	-	✓	✓
Oña et al. [33]	✓	✓	-
Li, Zhang, and Wu [45]	✓	-	✓
Salahdine, El Mrabet, and Kaabouch [40]	✓	-	✓
Rathod and Pattewar [16]	✓	✓	-
Rawal et al. [21]	-	✓	✓
Zeng [22]	✓	✓	-
Ripa, Islam, and Arifuzzaman [41]	-	-	✓
Hiransha, Unnithan, and Kp [28]	✓	-	✓
Butt et al. [8]	✓	✓	-
Rendall, Nisioti, and Mylonas [37]	✓	-	✓
Alam et al. [38]	✓	-	✓
Singh, Pamula, and Shekhar [29]	-	✓	-
El Aassal et al. [30]	-	✓	-
Moradpoor, Clavie, and Buchanan [23]	-	✓	✓
Dutta [42]	-	✓	✓
Yasin and Abuhasan [19]	-	✓	✓
Mughaid et al. [43]	-	✓	-
Magdy, Abouelseoud, and Mikhail [46]	-	✓	-
Dhanaraj and Karthikeyani [13]	✓	-	-
Chakraborty and Mondal [11]	✓	✓	-
Zhu, Dong, and Liu [17]	-	✓	-
Qasem, Shamsuddin, and Zain [12]	-	✓	-
Shams and Mercer [14]	-	✓	-
Laorden et al. [15]	-	✓	-
Maleki and Ghorbani [34]	✓	✓	✓
Dewis and Viana [10]	-	✓	✓
Mridha et al. [44]	✓	✓	-
Daeef et al. [18]	✓	✓	-
Nidhin et al. [31]	✓	✓	-
Yang et al. [35]	✓	-	-
Alotaib, Al-Turaiki, and Alakeel [39]	✓	✓	-
Niu et al. [24]	✓	✓	✓

From Table 7, there is a large number of researchers utilising the body and header features, with a total of 28 researchers utilising the body and a total of 23 researchers utilising the header feature for BEC phishing detection. Furthermore, a total of 14 researchers used a combination of header and body features.

Moreover, the body feature is utilised by various researchers, as BEC is mainly focused on crafting a good email body to deceive corporations and their employees in which the content used in the BEC attacks includes a good and official mode or tone of writing to achieve the required level of deception. In addition, the header provides a good source for determining the authenticity of emails as most of the information, such as the sender's email address, SCL, and other vital components, which can serve as a good indicator for a malicious BEC email, will also be easily spotted from the header.

6. Challenges and Future Directions

BEC phishing attacks are particularly dangerous because they do not contain malicious links or dangerous email attachments. They are used to impersonate or compromise corporate or publicly accessible email accounts of executives or high-level employees, who are involved in finance or who wire transfer payments, to conduct fraudulent transfers, costing billions of dollars in damages. Detecting BEC phishing attacks is getting harder since hackers change their tactics regularly to deceive email recipients and BEC detection tools. There are several open issues and future research directions that still need to be investigated to provide an effective BEC phishing detection model, including the following:

- **Dynamic Feature Selection:** Feature selection is a practical way for data visualisation and a technique to increase the classification accuracy of classifiers. Feature selection aims to find the smallest subset of features with the highest amount of information. Applying dynamic feature selection for BEC phishing detection is significant to enable the detection model to determine the appropriate set of features from the list of features extracted for a specific situation in an automatic manner in order to build an effective BEC phishing detection system. This creates an adaptive feature selection method that dynamically selects features for prediction at any given time [55,56]. In some cases, some essential features that have high weights when computing similarity distances may not be beneficial to the detection outcome due to changes in users' behaviour and attack scenarios. Hence, allowing the phishing detection system to select appropriate features dynamically will provide the missing piece to allow the creation of adaptive and effective BEC phishing detection models [57]. Adopting dynamic feature selection can solve many issues of current BEC detection models and provide higher accuracy in BEC phishing detection.
- **Dataset Availability:** Datasets are designed to be used as a benchmark for ML-based phishing detection systems. The availability and dependability of datasets is another obstacle in utilising ML algorithms to build BEC phishing detection models. The availability of datasets is crucial to the design and effectiveness of any ML detector/classifier. Before developing a model, one must guarantee that appropriate amounts of data are available [58]. In addition, ML algorithms are data-hungry in which the more data are available, the better efficiency and performance it produces. However, there are no available datasets that imitate real-life scenarios in the BEC phishing detection domain. Although there are some datasets, such as Kaggle, Nazario, and Phishing Corpus, these datasets are becoming nearly obsolete as they contain static features that are no longer used in advanced BEC phishing attacks. There is a need for creating large datasets that capture real-life scenarios of different systems, corporations, and networks to enable researchers to evaluate their novel ML-based BEC phishing models as well as to provide optimisation to existing techniques.
- **NLP and Deep Learning**: Deep learning is a subset of representation learning where the model can automatically find the representations and features required for the classification task from the raw data. Deep-learning algorithms can provide better accuracy in BEC phishing detection by training on larger datasets, while traditional ML algorithms tend to reach a performance plateau quite quickly. Deep-learning algorithms are more effective in data classification processes because they have several hidden layers. Since, in BEC phishing detection, email/URL pairs are intrinsically made up of text elements, it is natural to use NLP techniques. The successful integra-

tion of NLP with deep learning can develop better accuracy in BEC phishing detection models [59]. More studies are required to investigate the integration of NLP with deep learning that allows us to combine the best of both approaches to create a better BEC phishing detection system.

- **Explainable AI for BEC Phishing Detection:** One of the major issues of most AI-based models is that they are acting as a black box, where the input and output data can be seen and observed but the processes and operations working in between cannot be seen. Explainable Artificial Intelligence (XAI) enables the conversion of black-box models to glass-box models by generating explanations. XAI-based models outperform experts, and these models are more dependable and trustworthy. The goals of XAI systems are not only to improve a task's competence and accuracy but also to provide explanations for how a specific decision is made [60]. Integrating ML with XAI can provide more effective BEC phishing detection models where XAI can be used for global and local interpretation to empower the AI-based system with trust and reliability. Hence, more studies are required to investigate the development of ML with XAI tools that can provide effective BEC phishing detection models that can outperform existing models.
- **Real-Time BEC Phishing Detection:** One of the necessities in every corporation is having the ability to provide real-time BEC phishing detection. Before disclosing a user's personal information on a phishing website, the prediction of a phishing detection approach must be provided. The fraudulent email must be blocked by a trustworthy phishing detection tool without disclosing the user's credentials to the hackers [61]. There is a need for more studies to develop a technique that can be easily used by everyone to detect non-legitimate and BEC phishing emails accurately in real time.

7. Conclusions

Research efforts have mostly focused on finding ways to stop basic phishing emails that use text as their medium. In recent years, attacks have come up with new and creative tactics to utilise BEC phishing emails to attack organisations and businesses. BEC phishing email is a legitimate-looking email meant to trick the receiver. These emails may download harmful software if the receiver clicks on dangerous links in the body. Tricking a user involves telling them their business email user information have changed and asking them to check in to evaluate the changes. Once users click on an obfuscated link, they are led to a rogue site, which steals their information and redirects them to the corporate site. Although there are some efforts made to create effective methods to detect BEC phishing emails, there is still a need for more work to investigate this topic further and to provide better and more effective solutions. This paper presents a systematic literature review and analysis of the state of the art of BEC phishing attacks. This paper systematically analyses journal articles and conference proceedings published between 2012 and 2022. Based on the selected search strategy, 38 articles (out of 950 articles) were chosen for a closer examination in terms of recent BEC phishing detection models, ML-based algorithms used to build these models, common datasets used to develop these models, and common features utilised to detect BEC phishing emails. The results provide a summarised version of selected articles to give readers a basic view of the state of the art of BEC phishing attacks. The results indicate that several researchers are interested in utilising ML-based techniques for detecting BEC attacks, as the number of BEC attacks is increasing daily and the attacks' measures are changing and evolving daily, with DT, SVM, ANN, NB, and Logistic algorithms being the most common techniques used by various researchers. In addition, there is a large number of researchers who have utilised the body and header features to detect BEC phishing attacks, with 28 articles utilising the body features, 23 articles utilising the header features, and 14 articles using a combination of both header and body features. The paper also presents challenges and future research directions related to BEC phishing detection based on ML. There is a need for more research studies on dynamic feature selection, creating real-life datasets, integrating NLP with deep learning, and combining ML with XAI to develop an effective and optimised BEC phishing detection system.

Author Contributions: Conceptualization, H.F.A. and O.O.; methodology, H.F.A.; Implementation, H.F.A. and O.O.; validation, H.F.A.; investigation, H.F.A. and O.O.; resources, O.O.; writing—original draft preparation, H.F.A. and O.O.; writing—review and editing, H.F.A.; visualization, H.F.A. and O.O.; supervision, H.F.A.; project administration, H.F.A. All authors have read and agreed to the published version of the manuscript.

Funding: This research received no external funding.

Data Availability Statement: No new data were created or analyzed in this study. Data sharing is not applicable to this article.

Conflicts of Interest: The authors declare no conflict of interest.

References

1. Cidon, A.; Korshun, N.; Schweighauser, M.; Tsitkin, A.; Gavish, L.; Bleier, I. High Precision Detection of Business Email Compromise High Precision Detection of Business Email Compromise. In Proceedings of the 28th USENIX Security Symposium (USENIX Security 19), California, USA, 14–16 August 2019; Available online: https://www.usenix.org/system/files/sec19-cidon.pdf (accessed on 17 August 2020).
2. Cross, C.; Gillett, R. Exploiting trust for financial gain: An overview of business email compromise (BEC) fraud. *J. Financ. Crime* **2020**, *27*, 871–884. [CrossRef]
3. Jang-Jaccard, J.; Nepal, S. A survey of emerging threats in cybersecurity. *J. Comput. Syst. Sci.* **2014**, *80*, 973–993. [CrossRef]
4. Nisha, T.N.; Bakari, D.; Shukla, C. Business E-mail Compromise—Techniques and Countermeasures. In Proceedings of the 2021 International Conference on Advance Computing and Innovative Technologies in Engineering (ICACITE), Greater Noida, India, 4–5 March 2021. [CrossRef]
5. Teerakanok, S.; Yasuki, H.; UEHARA, T. A Practical Solution Against Business Email Compromise (BEC) Attack using Invoice Checksum. In Proceedings of the 2020 IEEE 20th Innternational Conference on Software Quality, Reliability and Security Companion (QRS-C), Macau, China, 11–14 December 2020. [CrossRef]
6. Compsysplus. Business Email Compromise Attacks-Computer Systems Plus. Available online: https://www.compsysplus.com/2021/07/the-10-stages-of-a-business-email-compromise-attack/ (accessed on 17 November 2022).
7. Cornish, D.B.; Clarke, R.V. Opportunities, precipitators and criminal decisions: A reply to Wortley's critique of situational crime prevention. *Crime Prev. Stud.* **2003**, *16*, 41–96.
8. Butt, U.A.; Amin, R.; Aldabbas, H.; Mohan, S.; Alouffi, B.; Ahmadian, A. *Cloud-Based Email Phishing Attack Using Machine and Deep Learning Algorithm*; Springer: New York, NY, USA, 2022. [CrossRef]
9. Karim, A.; Azam, S.; Shanmugam, B.; Kannoorpatti, K.; Alazab, M. A comprehensive survey for intelligent spam email detection. *IEEE Access* **2019**, *7*, 168261–168295. [CrossRef]
10. Dewis, M.; Viana, T. Phish Responder: A Hybrid Machine Learning Approach to Detect Phishing and Spam Emails. *Appl. Syst. Innov.* **2022**, *5*, 73. [CrossRef]
11. Chakraborty, S.; Mondal, B. Spam Mail Filtering Technique using Different Decision Tree Classifiers through Data Mining Approach-A Comparative Performance Analysis. *Int. J. Comput. Appl.* **2012**, *47*, 26–31. [CrossRef]
12. Qasem, S.N.; Shamsuddin, S.M.; Zain, A.M. Multi-objective hybrid evolutionary algorithms for radial basis function neural network design. *Knowl. Based Syst.* **2012**, *27*, 475–497. [CrossRef]
13. Dhanaraj, S.; Karthikeyani, V. A study on e-mail image spam filtering techniques. In Proceedings of the 2013 International Conference on Pattern Recognition, Informatics and Mobile Engineering, Salem, India, 21–22 February 2013. [CrossRef]
14. Shams, R.; Mercer, R.E. Classifying Spam Emails Using Text and Readability Features. In Proceedings of the 2013 IEEE 13th International Conference on Data Mining, Dallas, TX, USA, 7–10 December 2013. [CrossRef]
15. Laorden, C.; Ugarte-Pedrero, X.; Santos, I.; Sanz, B.; Nieves, J.; Bringas, P.G. Study on the effectiveness of anomaly detection for spam filtering. *Inf. Sci.* **2014**, *277*, 421–444. [CrossRef]
16. Rathod, S.B.; Pattewar, T.M. Content-based spam detection in email using Bayesian classifier. In Proceedings of the 2015 International Conference on Communications and Signal Processing (ICCSP), Melmaruvathur, India, 2–4 April 2015. [CrossRef]
17. Zhu, S.; Dong, W.; Liu, W. Hierarchical Reinforcement Learning Based on KNN Classification Algorithms. *Int. J. Hybrid Inf. Technol.* **2015**, *8*, 175–184. [CrossRef]
18. Daeef, A.; Ahmad, R.B.; Yacob, Y.; Yaakob, N.; Bin, M.N.; Warip, M. Phishing Email Classifiers Evaluation: Email Body and Header Approach. *J. Theor. Appl. Inf. Technol.* **2015**, *80*, 354–361.
19. Yasin, A.; Abuhasan, A. An Intelligent Classification Model for Phishing Email Detection. *Int. J. Netw. Secur. Its Appl.* **2016**, *8*, 55–72. [CrossRef]
20. Zweighaft, D. Business email compromise and executive impersonation: Are financial institutions exposed. *J. Invest. Compliance* **2017**, *18*, 1–7. [CrossRef]

21. Rawal, S.; Rawal, B.; Pilani, B.; Goa, I.; Shaheen; Malik, S. ISSN: 2249-0868 Foundation of Computer Science FCS. *Int. J. Appl. Inf. Syst. (IJAIS)* **2017**, *12*, 21–24.
22. Zeng, Y.G. Identifying email threats using predictive analysis. In Proceedings of the 2017 International Conference on Cyber Security and Protection of Digital Services (Cyber Security), London, UK, 19–20 June 2017. [CrossRef]
23. Moradpoor, N.; Clavie, B.; Buchanan, B. Employing machine learning techniques for detection and classification of phishing emails. In Proceedings of the 2017 Computing Conference, London, UK, 18–20 July 2017. [CrossRef]
24. Niu, W.; Zhang, X.; Yang, G.; Ma, Z.; Zhuo, Z. Phishing Emails Detection Using CS-SVM. In Proceedings of the 2017 IEEE International Symposium on Parallel and Distributed Processing with Applications and 2017 IEEE International Conference on Ubiquitous Computing and Communications (ISPA/IUCC), Guangzhou, China, 12–15 December 2017. [CrossRef]
25. Peng, I.T.; Harris, I.; Sawa, Y. Detecting Phishing Attacks Using Natural Language Processing and Machine Learning. In Proceedings of the 2018 IEEE 12th International Conference on Semantic Computing (ICSC), Laguna Hills, CA, USA, 31 January 2018–2 February 2018. [CrossRef]
26. Baykara, M.; Gurel, Z.Z. Detection of phishing attacks. In Proceedings of the 2018 6th International Symposium on Digital Forensic and Security (ISDFS), Antalya, Turkey, 22–25 March 2018. [CrossRef]
27. Sahoo, P.K. Data mining a way to solve Phishing Attacks. In Proceedings of the 2018 International Conference on Current Trends towards Converging Technologies (ICCTCT), Coimbatore, India, 1–3 March 2018. [CrossRef]
28. Hiransha, M.; Unnithan, N.A.; Vinayakumar, R.; Soman, K.P. Deep Learning Based Phishing E-mail Detection. In Proceedings of the 1st Antiphishing Shared Pilot 4th ACM International Workshop on Security and Privacy Analytics (IWSPA). Arizona, USA, 21 March 2018.
29. Singh, M.; Pamula, R.; shekhar, S.k. Email Spam Classification by Support Vector Machine. In Proceedings of the 2018 International Conference on Computing, Power and Communication Technologies (GUCON), Greater Noida, India, 28–29 September 2018. [CrossRef]
30. Aassal, A.E.; Moraes, L.; Baki, S.; Das, A.; Verma, R. Anti-Phishing Pilot at ACM IWSPA 2018. In Proceedings of the 1st Antiophishing Shared Pilor 4th ACM International Workshop on Security and Privacy Analytics (IWSPA), Tempe, AZ, USA, 21 March 2018; Available online: http://www2.cs.uh.edu/~{}shahryar/files/IWSPA-AP.pdf (accessed on 26 October 2022).
31. Unnithan, N.A.; Harikrishnan, N.B.; Vinayakumar, R.; Soman, K.P. Detecting Phishing E-mail using Machine learning techniques. In Proceedings of the 1st AntiPhishing Shared Pilot at 4th ACM International Workshop on Security and Privacy Analytics (IWSPA 2018), Tempe, AZ, USA, 21 March 2018.
32. Fomunyam, D.K.G. Machine Learning and the Business of Cyber Security. *Int. J. Civil Eng. Technol. (IJCIET)* **2019**, *10*, 353–359.
33. Oña, D.; Zapata, L.; Fuertes, W.; Rodríguez, G.; Benavides, E.; Toulkeridis, T. Phishing Attacks: Detecting and Preventing Infected E-mails Using Machine Learning Methods. In Proceedings of the 2019 3rd Cyber Security in Networking Conference (CSNet), Quito, Ecuador, 23–25 October 2019. [CrossRef]
34. Maleki, N. A Behavioral Based Detection Approach for Business Email Compromises. Master's Thesis, University of New Brunswick, Fredericton, NB, Canada, 2019.
35. Yang, Y.; Qiao, C.; Kan, W.; Qiu, J. Phishing Email Detection Based on Hybrid Features. *IOP Conf. Ser. Earth Environ. Sci.* **2019**, *252*, 042051. [CrossRef]
36. Garces, I.O.; Cazares, M.F.; Andrade, R.O. Detection of phishing attacks with machine learning techniques in cognitive security architecture. In Proceedings of the 2019 International Conference on Computational Science and Computational Intelligence (CSCI), Las Vegas, NV, USA, 5–7 December 2019. [CrossRef]
37. Rendall, K.; Nisioti, A.; Mylonas, A. Towards a Multi-Layered Phishing Detection. *Sensors* **2020**, *20*, 4540. [CrossRef] [PubMed]
38. Alam, M.N.; Sarma, D.; Lima, F.F.; Saha, I.; Ulfath, R.-E.; Hossain, S. Phishing Attacks Detection using Machine Learning Approach. In Proceedings of the 2020 Third International Conference on Smart Systems and Inventive Technology (ICSSIT), Tirunelveli, India, 20–22 August 2020. [CrossRef]
39. Alotaibi, R.; Al-Turaiki, I.; Alakeel, F. Mitigating Email Phishing Attacks using Convolutional Neural Networks. In Proceedings of the 2020 3rd International Conference on Computer Applications & Information Security (ICCAIS), Riyadh, Saudi Arabia, 19–21 March 2020. [CrossRef]
40. Salahdine, F.; El Mrabet, Z.; Kaabouch, N. Phishing Attacks Detection A Machine Learning-Based Approach. In Proceedings of the 021 IEEE 12th Annual Ubiquitous Computing, Electronics & Mobile Communication Conference (UEMCON), New York, NY, USA, 1–4 December 2021. [CrossRef]
41. Ripa, S.P.; Islam, F.; Arifuzzaman, M. The Emergence Threat of Phishing Attack and The Detection Techniques Using Machine Learning Models. In Proceedings of the 2021 International Conference on Automation, Control and Mechatronics for Industry 4.0 (ACMI), Rajshahi, Bangladesh, 8–9 July 2021. [CrossRef]
42. Dutta, A.K. Detecting phishing websites using machine learning technique. *PLoS ONE* **2021**, *16*, e0258361. [CrossRef]
43. Mughaid, A.; AlZu'bi, S.; Hnaif, A.; Taamneh, S.; Alnajjar, A.; Abu Elsoud, E. An intelligent cyber security phishing detection system using deep learning techniques. *Clust. Comput.* **2022**, *25*, 3819–3828. [CrossRef]
44. Mridha, K.; Hasan, J.; Saravanan, D.; Ghosh, A. Phishing URL Classification Analysis Using ANN Algorithm. In Proceedings of the 2021 IEEE 4th International Conference on Computing, Power and Communication Technologies (GUCON), Kuala Lumpur, Malaysia, 24–26 September 2021. [CrossRef]

45. Li, X.; Zhang, D.; Wu, B. Detection method of phishing email based on persuasion principle. In Proceedings of the 2020 IEEE 4th Information Technology, Networking, Electronic and Automation Control Conference (ITNEC), Chongqing, China, 12–14 June 2020. [CrossRef]
46. Magdy, S.; Abouelseoud, Y.; Mikhail, M. Efficient spam and phishing emails filtering based on deep learning. *Comput. Netw.* **2022**, *206*, 108826. [CrossRef]
47. Bagui, S.; Nandi, D.; Bagui, S.; White, R.J. Classifying Phishing Email Using Machine Learning and Deep Learning. In Proceedings of the 2019 International Conference on Cyber Security and Protection of Digital Services (Cyber Security), Oxford, UK, 3–4 June 2019. [CrossRef]
48. Mantas, C.J.; Castellano, J.G.; Moral-García, S.; Abellán, J. A comparison of random forest based algorithms: Random credal random forest versus oblique random forest. *Soft Comput.* **2018**, *23*, 10739–10754. [CrossRef]
49. Bagui, S.; Nandi, D.; Bagui, S.; White, R.J. Machine Learning and Deep Learning for Phishing Email Classification using One-Hot Encoding. *J. Comput. Sci.* **2021**, *17*, 610. Available online: https://www.academia.edu/75025334/Machine_Learning_and_Deep_Learning_for_Phishing_Email_Classification_using_One_Hot_Encoding (accessed on 30 August 2022). [CrossRef]
50. Posevkin, R.; Bessmertny, I. Translation of natural language queries to structured data sources. In Proceedings of the 2015 9th International Conference on Application of Information and Communication Technologies (AICT), Rostov on Don, Russia, 14–16 October 2015. [CrossRef]
51. Simpson, G.; Moore, T. Empirical Analysis of Losses from Business-Email Compromise. In Proceedings of the 2020 APWG Symposium on Electronic Crime Research (eCrime), Boston, MA, USA, 16–19 November 2020. [CrossRef]
52. Spamassassin, P.C.; Index of /old/publiccorpus. spamassassin.apache.org. Available online: https://spamassassin.apache.org/old/publiccorpus/ (accessed on 16 November 2022).
53. Dada, E.G.; Bassi, J.S.; Chiroma, H.; Abdulhamid, S.M.; Adetunmbi, A.O.; Ajibuwa, O.E. Machine learning for email spam filtering: Review, approaches and open research problems. *Heliyon* **2019**, *5*, e01802. [CrossRef] [PubMed]
54. Schäfer, C. Detection of compromised email accounts used for spamming in correlation with mail user agent access activities extracted from metadata. In Proceedings of the 2015 IEEE Symposium on Computational Intelligence for Security and Defense Applications (CISDA), Verona, NY, USA, 26–28 May 2015. [CrossRef]
55. Bountakas, P.; Xenakis, C. Helphed: Hybrid Ensemble Learning Phishing Email Detection. *J. Netw. Comput. Appl.* **2023**, *210*, 103545. [CrossRef]
56. Salloum, S.; Gaber, T.; Vadera, S.; Shaalan, K. A Systematic Literature Review on Phishing Email Detection Using Natural Language Processing Techniques. *IEEE Access* **2022**, *10*, 65703–65727. [CrossRef]
57. Al-Musib, N.S.; Al-Serhani, F.M.; Humayun, M.; Jhanjhi, N.Z. Business email compromise (BEC) attacks. Materials Today: Proceedings. *Mater. Today Proc.* **2021**. [CrossRef]
58. Ahmed, C.M.; MR, G.R.; Mathur, A.P. Challenges in Machine Learning based approaches for Real-Time Anomaly Detection in Industrial Control Systems. In Proceedings of the 6th ACM on Cyber-Physical System Security Workshop, Taipei, Taiwan, 6 October 2020. [CrossRef]
59. Catal, C.; Giray, G.; Tekinerdogan, B.; Kumar, S.; Shukla, S. Applications of deep learning for phishing detection: A systematic literature review. *Knowl. Inf. Syst.* **2022**, *64*, 1457–1500. [CrossRef]
60. Aslam, N.; Khan, I.U.; Mirza, S.; AlOwayed, A.; Anis, F.M.; Aljuaid, R.M.; Baageel, R. Interpretable Machine Learning Models for Malicious Domains Detection Using Explainable Artificial Intelligence (XAI). *Sustainability* **2022**, *14*, 7375. [CrossRef]
61. Aljofey, A.; Jiang, Q.; Rasool, A.; Chen, H.; Liu, W.; Qu, Q.; Wang, Y. An effective detection approach for phishing websites using URL and HTML features. *Sci. Rep.* **2022**, *12*, 8842. [CrossRef]

Disclaimer/Publisher's Note: The statements, opinions and data contained in all publications are solely those of the individual author(s) and contributor(s) and not of MDPI and/or the editor(s). MDPI and/or the editor(s) disclaim responsibility for any injury to people or property resulting from any ideas, methods, instructions or products referred to in the content.

MDPI
St. Alban-Anlage 66
4052 Basel
Switzerland
Tel. +41 61 683 77 34
Fax +41 61 302 89 18
www.mdpi.com

Electronics Editorial Office
E-mail: electronics@mdpi.com
www.mdpi.com/journal/electronics

www.ingramcontent.com/pod-product-compliance
Lightning Source LLC
LaVergne TN
LVHW070410100526
838202LV00014B/1429